T0271197

Latent Curve Models

WILEY SERIES IN PROBABILITY AND STATISTICS

Established by WALTER A. SHEWHART and SAMUEL S. WILKS

A complete list of the titles in this series appears at the end of this volume.

Latent Curve Models

A Structural Equation Perspective

KENNETH A. BOLLEN

University of North Carolina
Department of Sociology
Chapel Hill, North Carolina

PATRICK J. CURRAN

University of North Carolina
Department of Psychology
Chapel Hill, North Carolina

WILEY-INTERSCIENCE

A JOHN WILEY & SONS, INC., PUBLICATION

Published by John Wiley & Sons, Inc., Hoboken, New Jersey.
Published simultaneously in Canada.

Library of Congress Cataloging-in-Publication Data:

Bollen, Kenneth A.
Latent curve models: a structural equation perspective / Kenneth A. Bollen, Patrick J. Curran.
p. cm.—(Wiley series in probability and statistics)
Includes bibliographical references and index.
ISBN-13 978-0-471-45592-9 (cloth)
ISBN-10 0-471-45592-X (cloth)
1. Latent structure analysis. 2. Latent variables. I. Curran, Patrick J., 1965- II. Title.
QA278.6.B65 2005
519.5′35—dc22

2005047028

10 9 8 7 6

Contents

Preface

The last 20 years has witnessed a greater accumulation of longitudinal data in the social sciences than any other period. Accompanying this growth of data has been an interest in methods to analyze such data. Growth curve models or latent curve models have been among some of the most recently popular longitudinal techniques. Although new to some researchers, the historic roots of these techniques are deep. Indeed, these models have their origins in biostatistics, statistics, demography, and other disciplines. Their different disciplinary origins have given the same technique different orientations.

Our book is about growth curve models as they have developed in the latent variable and factor analysis tradition. Borrowing from Meredith and Tisak (1984, 1990), we refer to these as latent curve models (LCMs). They are models in which we have random intercepts and random slopes that permit each case in the sample to have a different trajectory over time. The random coefficients are incorporated into structural equation models (SEMs) by considering them as latent variables. By so doing, we are able to capitalize on all of the strengths of SEMs and to apply them to latent curve models. These include the ability to use maximum likelihood techniques for missing data, to estimate a variety of nonlinear forms of trajectories, to have measures of model fit and diagnostics to determine the source of ill-fit, the inclusion of latent covariates and latent repeated variables, and so on.

One of our goals is to provide a reference source on the LCM approach to longitudinal data. This is a field that is undergoing rapid development, but we do our best to provide many of the most recent developments. The book synthesizes a large amount of work on LCMs as well as providing original results. For instance, we develop new results on the identification of several types of LCMs. We include some of our recently proposed work on the autoregressive latent trajectory (ALT) model and suggest new models for method factors in multiple indicator, repeated latent variable models. Researchers with a background in structural equation models and an interest in longitudinal data are ideally suited for this book.

We have said what the book is about. It also is worthwhile to state what it is not. It is *not* a comprehensive guide to all approaches to growth curve models. There is a valuable literature in biostatistics on growth curve models and another tradition from multilevel models. There is a great deal of overlap in these approaches and

LCMs, and in parts of the book we point out the connections and sometimes the equivalency of models and results. However, a reader looking for comprehensive treatments of growth curve models from these other traditions will not find it here. Those seeking an extensive treatment of these models from a structural equation perspective are our intended audience.

Our book is a multiyear project that would not have happened without the help of numerous people. The most important group has been the Carolina Structural Equation Modeling (CSEM) group that we co-direct. This rotating group of intelligent graduate students and postdocs has been indispensible. They provided feedback on all of the chapters. They provided research assistance in developing empirical examples and a critical audience with whom to discuss results. We consider ourselves incredibly fortunate to have had such a talented group with which to work.

Personal support is at least as important as the professional support in making this book happen. We are lucky to have wives who provide both. We also have the good fortune to each have two daughters who helped us to maintain our perspective on the important things in life. We dedicate the book to our families.

KENNETH A. BOLLEN
PATRICK J. CURRAN

Chapel Hill, North Carolina

CHAPTER ONE

Introduction

Every discipline in the social sciences seeks to understand the sources of stability and change in variables. A psychologist, for instance, maps the development of cognitive abilities in children. An economist traces the economic growth of nations or the sales of companies. A sociologist tracks crime rates across communities. The time interval for change might be daily, weekly, monthly, or annually. The unit of analysis might be children, businesses, or countries. Regardless of the subject area or the time interval, social and behavioral scientists have a keen interest in describing and explaining the time trajectories of their variables.

Although the analysis of longitudinal data has a long and rich history in the social sciences, the past several decades have witnessed a sharp increase in interest in this topic. This resurgence of concern has a variety of sources. First, there is a growing appreciation of the limits of relying solely on cross-sectional data. Many theoretical hypotheses inherently focus on change, and longitudinal data are necessary to fully evaluate such hypotheses. Second, given the wider appreciation of the advantages of longitudinal research designs, there is much greater availability of panel data. Many longitudinal data sets exist that follow children, adults, communities, or countries over many time periods. The increased availability of large and complex longitudinal data sets demands the development of increasingly rigorous statistical analytic methods. Finally, there seems to be greater dissatisfaction with many traditional longitudinal analytic designs. For example, a developmental theory may predict that delinquent behavior develops naturally in children over time and a treatment intervention is designed to deflect this natural trajectory meaningfully. Using a statistical model that evaluates the treatment effect by comparing pre- and posttest group means may not provide an optimal test of the theory that hypothesized change in terms of individual developmental trajectories. Analytical methods are desired that allow for a better correspondence between the theoretical model that gave rise to the research hypotheses and the statistical model that is used to test these hypotheses empirically. For these and many other reasons, the analysis of longitudinal data has become an increasingly important topic in social science research.

Latent Curve Models: A Structural Equation Perspective, by Kenneth A. Bollen and Patrick J. Curran

There is a remarkable array of analytical techniques available for longitudinal empirical data. Examples of these methods include autoregressive models, repeated measures multivariate analysis of variance (MANOVA), raw difference scores, residualized change scores, and random and fixed effects panel data models. Rarely can we conclude that one of these models is inherently correct or incorrect. Rather, these models differ in how well they capture the theoretical orientation guiding the research and how well their assumptions correspond to the characteristics of the empirical data. From this perspective, a particular analytic approach may be well suited for one application but quite poorly suited to another. In this book, the focus is on methods to analyze the temporal trajectories of cases where each case can have a distinct trajectory.

The goal of this chapter is to orient the reader to this type of longitudinal model and to provide background on its origins. In the next section we introduce the conceptualization of these models by contrasting it with that of autoregressive methods and by introducing an empirical data set. In the third section we discuss three basic questions that characterize nearly all trajectory analyses. A brief review of the history of these methods follows, and a discussion of the organization of the rest of the book concludes the chapter.

1.1 CONCEPTUALIZATION AND ANALYSIS OF TRAJECTORIES

The majority of traditional analytical methods focus on change by examining the relation between two adjacent time points and predicting change using parameters that are common across cases. For example, the simple autoregressive model mentioned above is

$$y_{it} = \alpha_t + \rho_{t,t-1} y_{i,t-1} + \epsilon_{it} \tag{1.1}$$

where y_{it} and $y_{i,t-1}$ are the observed values of y for individual i at time t and time $t-1$, respectively; α_t is the time-specific intercept; and $\rho_{t,t-1}$ is the autoregressive coefficient pooled over all cases in the sample. The disturbance (ϵ_{it}) has a mean of zero, is not autocorrelated, and is uncorrelated with $y_{i,t-1}$. The i subscript indexes the cases in the sample, and the t subscript indicates time. Notice that the autoregressive parameter, $\rho_{t,t-1}$, has subscripts that permit it to change over time, but that for a given time point it takes the *same value* for all cases. We can modify Eq. (1.1) to include additional lagged values of y (e.g., the influence of $y_{i,t-2}$) or other explanatory variables with their coefficients (e.g., the influence of an exogenous measure $x_{i,t-1}$), but from this analytical perspective the forces governing change are its past values and the values of covariates.

This book represents a different perspective. The approach posits the existence of *continuous underlying* or *latent trajectories*. The pattern of change in the repeated measures provides information on the trajectories. *Latent* means a process that is not observed directly. The trajectory process is observed only indirectly using the repeated measures. Importantly, this trajectory can differ by individual case. For example, say that we had assessed reading ability on a sample of children every other year over a period of eight years. A theory of cognitive development might predict that there exists an unobserved underlying ability to read that develops as a continuous function of time. Our interest in the time-specific measures of reading ability is because these repeated measures enable us to estimate the underlying reading trajectory that gives rise to the measures over time. The focus is not on the dependence of current reading level on past reading level across all children, but instead, on the estimation of an underlying developmental trajectory across all time points within *each* child. Our goal, then, will be to estimate a line, or trajectory, that best fits the repeated observations over time for each case. That is, we would like to smooth over our repeated observations to get a more parsimonious estimate of the underlying trajectory that gave rise to the repeated measures. It is this trajectory estimate that is of primary interest in subsequent modeling. An empirical example helps to further illustrate these issues.

1.1.1 Trajectories of Crime Rates

Our first empirical example consists of crime rates among 359 communities in New York State taken from 1995 *Uniform Crime Reports* (UCR) data.[1] The *crime rate* is the sum of the number of offenses reported to police departments for murder, robbery, aggravated assault, rape, burglary, motor vehicle theft, larceny, and arson converted to rates per 100,000 population. Taking the natural logarithm (ln) of this variable reduces its levels of skew and kurtosis. Crime rates spike upward on weekends, and computing bimonthly figures smoothes out the fluctuations that occur due to differing numbers of weekends within months (e.g., Hipp et al., 2004). Therefore, we use bimonthly crime rates per 100,000 population for the first eight months of 1995 for police units in New York State. Two predictors of these crime trajectories measured in 1990 are *poverty*, as measured by the percentage of the city population whose income is at or below 125% of the poverty level, and *population density*, formed as the community population divided by the square kilometers in the community.[2]

Table 1.1 presents the means, standard deviations, and correlations for the four repeated measures of ln(crime rates) taken on the 359 communities. Several characteristics can be observed. First, the correlations among the repeated measures are moderately high (ranging from 0.71 to 0.83). Lagged crime measures are clearly associated. Second, the means of the repeated measures increase over time, whereas the standard deviations do not have such a consistent trend. This trend reflects a tendency for crime rates to increase from the winter to the summer months. It will be important to select a statistical model that allows for these patterns in the means and the standard deviations.

[1]The UCR consists of reports on crime activity from all police units in the United States to the Federal Bureau of Investigation. These police units include city police, university police, and county sheriffs. In general, the UCR has a very high response rate, as police units covering about 97% of the population respond (U.S. Department of Justice, 1995).

[2]Both of these variables are taken from the 1990 U.S. Census Summary Tape File (STF) 3-A (U.S. Department of Justice, 2000).

Table 1.1 Means, Standard Deviations, and Correlations for Four Repeated Measures of ln(Crime Rate) for 359 Communities in New York State

	Time 1 Crime	Time 2 Crime	Time 3 Crime	Time 4 Crime
Time 1 crime	1.0			
Time 2 crime	0.833	1.0		
Time 3 crime	0.790	0.805	1.0	
Time 4 crime	0.709	0.728	0.819	1.0
Mean	5.319	5.516	5.611	5.763
Standard deviation	0.791	0.754	0.753	0.781

Table 1.2 Poverty, Population Density, and Four Waves of ln(Crime Rates) for Five Communities from the *Uniform Crime Reports*

Community	Poverty	Population Density	Time 1 Crime	Time 2 Crime	Time 3 Crime	Time 4 Crime
1	16.04	0.48	4.01	4.01	5.79	4.69
2	23.05	1.82	6.55	6.68	6.77	7.03
3	18.52	0.77	5.48	5.83	5.82	5.81
4	16.68	0.62	4.43	5.81	5.74	6.62
5	19.34	1.49	6.12	6.35	5.80	4.79

It is very important to consider not only the summary statistics on the repeated measures pooling over all cases, but also to closely examine the individual-level data. Table 1.2 presents the observed data on poverty, population density, and the four repeated measures of crime rates for the first five communities in the sample. These data allow us to consider several interesting questions. First, is there any monotonic increase or decrease in the crime rates over this period? Examining the raw data for the first four communities suggests an upward tendency, but it is far from consistent. For example, for the first community, the four repeated crime measures were 4.01, 4.01, 5.79, and 4.69, indicating no change for the first two times and a larger crime rate followed by one not as large as the third time. The fifth community exhibits a tendency to decrease. Examining just these five communities reveals potentially interesting characteristics of these crime rates across the communities.

Although it is interesting to begin with an examination of the time-specific measures of ln(crime rate) for each community taken at each time point, recall the posited underlying or latent trajectory of crime for each community. That is, it is of less interest that the first community received scores of 4.01, 4.01, 5.79, and 4.69 at each of the four time points than it is to use these repeated observations to estimate an underlying trajectory that *gave rise* to these specific repeated measures over time. A logical starting point is to consider fitting a line

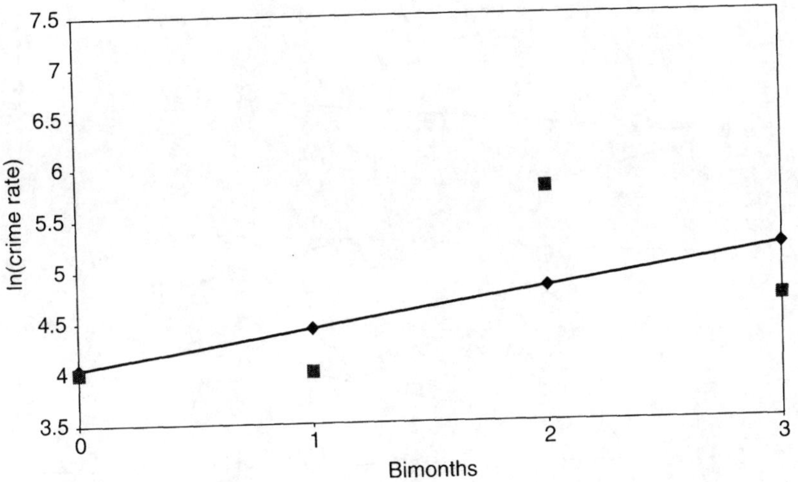

FIGURE 1.1 Observed and predicted crime rates for the first community.

that best fits the repeated measures observed within the first community. Chapter 2 treats this in much greater detail, but we can use ordinary least-squares (OLS) regression to estimate this line (or trajectory) for the first community. The OLS regression line relating ln(crime rate) to time within the first community is presented in Figure 1.1.

Although the repeated observations did not fall on a perfectly straight line, we are estimating the existence of the trajectory for the first community by finding a line that best fits the first community's four bimonthly data points. That is, the line in Figure 1.1 is nothing more than the regression of ln(crime rate) on time within the first community. As with any line, it is characterized completely by an intercept (say, $\widehat{\alpha}$) and a slope (say, $\widehat{\beta}$). Because the initial assessment (January–February) is set equal to zero, the intercept represents the model-implied value of ln(crime rate) at the initial time period.[3] Keep in mind that the $\widehat{\alpha}$ and $\widehat{\beta}$ values characterize the first community. This means that the first community has its *own* estimated intercept term and its *own* estimated slope term. This trajectory (sometimes called a *growth curve* or a *time path*) is an estimate of *intracommunity* (or *within-community*) change for the first community. We have thus *smoothed over* the four repeated measures and have a more parsimonious estimate of the process of change for the first community that is consistent with our theoretical model.

This provides an estimate of intracommunity change in crime rate for the first community, but also of interest are *intercommunity* differences in *intracommunity* change. That is, it is helpful to have a trajectory for *each* community (e.g., intracommunity change) and then to examine the characteristics of these trajectories

[3]We present a detailed discussion of alternative methods for coding time in Chapter 4.

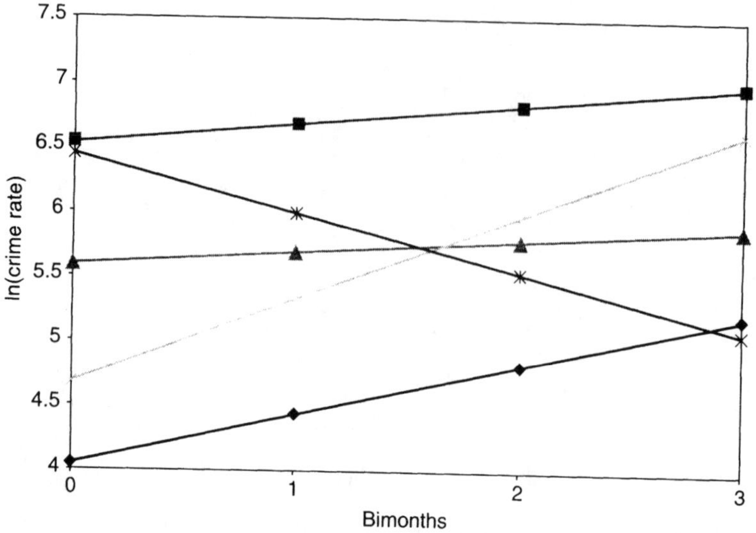

FIGURE 1.2 Model-implied linear trajectories of crime rates for the first five communities.

across all of the communities (e.g., intercommunity differences). To work toward this goal, a useful starting point is simply estimating a unique trajectory for each community using the same OLS regression method as above. So for 359 communities there would be 359 individual regression lines, one for each community. To demonstrate this, we fit individual regression lines to the first five communities in our data set, and these five trajectory estimates are displayed in Figure 1.2.

Figure 1.2 highlights several important concepts. First, the fitted trajectory lines are displayed, but the time-specific crime rates are not. The regression lines represent a *smoothed* version of the repeated observations over time, and a comparison of the observed points to the fitted line would reveal the quality of the fit of the linear trajectory to the observed data. Second, the estimated underlying trajectories are different for each of the five communities. So each line in Figure 1.2 represents the OLS regression of crime rate on time estimated separately for each community, and this necessitates a separate subscript to distinguish the intercept and slope for each community. These parameter estimates now become $\widehat{\alpha}_i$ and $\widehat{\beta}_i$, respectively, where $i = 1, 2, 3, 4, 5$ indicate which $\widehat{\alpha}$ and $\widehat{\beta}$ are associated with which specific community i.

Finally, given the existence of five estimated $\widehat{\alpha}$ values and five estimated $\widehat{\beta}$ values, it is possible to find the mean of the $\widehat{\alpha}$'s and the mean of the $\widehat{\beta}$'s. More specifically, combining these individual trajectory estimates allows estimation of the overall trajectory for the entire set of communities. Further, given the estimation of the mean trajectory, we can examine variability of each community's trajectory around these group means. This variability reflects that some communities are reporting higher levels of initial crime rates, while others are reporting lower initial levels; further, some communities are reporting steeper increases in crime over time, while others are reporting less steep increases and even decreases. These estimates of variability in community trajectories around the group mean trajectory capture the presence of intercommunity differences in intracommunity change. The larger the magnitude of these variance estimates, the greater evidence there is for individual community variability in trajectories. In Chapter 2 we present these results in detail.

Thus far, the repeated observations allow estimation of a trajectory for each community and for estimating a mean trajectory for the group. The variability of the individual trajectories around this group mean is also available. This model is commonly referred to as an *unconditional trajectory model* because we estimate the means and variances of the trajectories but are not modeling the trajectories as a function of other predictor variables.

However, a key analytic goal is to determine if the individual variability of the trajectories is predictable using other community characteristics. That is, given the presence of significant variability in our set of $\widehat{\alpha}$ estimates, does the magnitude of these $\widehat{\alpha}$'s vary as a function of, say, community poverty? Are higher levels of poverty associated with higher initial levels of crime? Similarly, are higher levels of poverty associated with steeper increases in ln(crime rate) over time? Given that we are interested in the distributions of the $\widehat{\alpha}$ and $\widehat{\beta}$ estimates *conditioned on* one or more predictor variables (e.g., using other community measures to predict our individual trajectory estimates), this model is commonly referred to as a *conditional trajectory model*. We discuss methods by which covariates can predict these trajectory parameters in conditional growth models in Chapter 5.

1.1.2 Data Requirements

As we explicate more fully in Chapter 2, a minimum of three repeated observations are necessary to identify and to estimate underlying linear trajectories and the other parameters in the model. In general, to estimate a polynomial trajectory of degree d, it is necessary to have obtained $d + 2$ repeated observations for identification purposes. So three repeated measures permit a straight line (e.g., a polynomial of degree $d = 1$); four observations fit a quadratic trajectory ($d = 2$); and so on. Conceptually, this is easy to understand. With only two observations, the line of best fit simply connects the two observations and thus reflects a perfect fit of the trajectory to the data. In contrast, the line of best fit for three observations in general reflects some imperfect relation between time and the repeated variables and thus allows a test of the adequacy of model fit. Because of this, our emphasis is on data where there are at least three repeated observations for a large proportion of observed units.[4]

1.1.3 Summary

Conceptualizing stability and change in terms of individual trajectories allows for the articulation and assessment of a wide array of research questions that are not

[4]As we discuss in Chapter 3, we can have missing data for some observations. So it is possible to estimate these models even if fewer than three repeated measures are available for some cases. See Chapter 3 for details.

easily accommodated with other techniques. For example, each case can have a separate trajectory, and a mean group trajectory and variability around the mean are available. We can then incorporate predictor variables and model individual differences in trajectories across cases.

As we explore throughout the book, many interesting and complex questions can be posed about stability and change in a variable. As a starting point, it is helpful to develop a general organizing framework to begin thinking about these many questions of change over time.

1.2 THREE INITIAL QUESTIONS ABOUT TRAJECTORIES

Our focus is on the trajectories of variables across cases. Although many questions about the trajectories are possible, three questions are common in many latent trajectory analyses. These questions relate to (1) the characteristics of the mean trajectory of the entire group, (2) the evaluation of individual differences in trajectories, and (3) the potential incorporation of predictors of individual differences in trajectories. To give substance to these issues, we return to our ln(crime rate) example. Recall that we have four repeated measures of crime rates over eight months for 359 communities in New York State. Further, we have time 1 measures of the percent poverty and the population density for each community. We propose considering three initial questions about the trajectories underlying crime rates.

1.2.1 Question 1: What Is the Trajectory for the Entire Group?

This initial question simply asks: When pooling over all 359 communities in the New York sample, what is the mean trajectory of crime over the four bimonthly measurements? That is, what is the average initial level of the crime rate trajectory? On average, does crime change as a function of time? If so, what is the functional form of change over time? Is this rate of change strictly linear, or is there a curvilinear component in which the rate of change accelerates or decelerates with the passage of time? Evaluation of this first question thus provides an initial insight into the characteristics of the mean trajectory of the crime rates for the *entire sample* of New York communities under study. The mean values of the trajectory parameter estimates are sometimes referred to as the *fixed-effects components* of the trajectory model.

1.2.2 Question 2: Do We Need Distinct Trajectories for Each Case?

Understanding the mean trajectory of crime rates is important, but there is additional information yet to be gained. Of particular interest is whether there is meaningful variability of community trajectories around the mean trajectory. That is, is the mean trajectory reflective of every community in the sample, or are there cases that depart from the group-level trajectory? Depending on our sampling design, we may find that although the mean initial level of crime rate for the group is a particular value, some communities are reporting initial values well above this, whereas others are well below. Further, although the mean rate of change in crime rates for the group is a specific number, some communities may be increasing at a more rapid rate and some are not increasing at all. The degree to which it is necessary to allow each community to take on a distinct trajectory is a question of *intercommunity* differences in *intracommunity* change. If all communities are within sampling fluctuations of the mean trajectory, there is no evidence of community differences in trajectories; the mean trajectory is thus representative of the trajectory of every community. In contrast, the more the individual community trajectories diverge from the overall group, the greater evidence there is for individual differences in trajectories of change. These variances are sometimes referred to as the *random-effects components* of the trajectory model.

1.2.3 Question 3: If Distinct Trajectories Are Needed, Can We Identify Variables to Predict These Individual Trajectories?

Say that the response to Question 1 indicates that there is a mean trajectory of crime rates of some particular form (e.g., an overall linear trajectory). Further, say that the response to Question 2 indicates that there is meaningful variability of the individual trajectories around the mean intercept and mean slope. The third question addresses whether we can incorporate other explanatory variables to better understand the variability observed in individual trajectories. Do some variables predict which communities start with a higher crime rate relative to those communities that start at lower levels? Similarly, can we use a set of explanatory measures to understand what type of community will increase more steeply in crime relative to those that increase less steeply? This third and final question thus allows us to explore our ability to model individual variability in trajectories and to make an attempt at predicting which cases will start high versus low and which will increase steeply and which will not. The incorporation of predictors leads to the *conditional* trajectory model to which we referred previously.

1.2.4 Summary

There are many questions that we could pose when evaluating trajectories of change. However, three initial questions are a helpful starting point to understanding trajectory analysis. The first question allows us insight into group-level change, the second question explores the presence of individual variability in change, and the third question attempts to model individual differences in change as a function of other explanatory variables of interest. Taken together, this allows us to begin working toward the development of a comprehensive understanding of the individual and group processes of change.

1.3 BRIEF HISTORY OF LATENT CURVE MODELS

The methods that we discuss in this book are not all new. In fact, parts of the methodology that we present have roots more than a century old. In this section we sketch some of the major developments that have led to the contemporary latent curve models (LCMs) that are the focus of our book. It is not our intention to present an exhaustive literature review or to present the background to all approaches to analyzing growth curve models. Rather, our goal is to focus on LCMs and to show that they derive from over a century of tradition of studying individual and group change. Even the idea of using latent variables or factor analysis to study such change has roots that are nearly a half century old. See Bollen (2005) for a more detailed discussion of the relation between growth curve models and factor analysis. Thus, the LCM is newly popular, but it is not a new idea. Below we identify several time periods that correspond to developments that contributed to the contemporary form of LCM.

1.3.1 Early Developments: The Nineteenth Century

Although philosophical discussions of change have been traced back as far as Aristotle (Zeger and Harlow, 1987), the earliest development of statistical analysis of data over time appears to be early in the nineteenth century. The studies were oriented toward the change in aggregates or groups rather than in individuals. Gompertz (1820) presents one of the first discussions of modeling stability and change from a trajectory perspective. The key development of Gompertz (1820, 1825) was the method of estimating a single trajectory that would characterize continuous mortality rates over time for a group of individuals. The general goal of this work was to use mathematical modeling to estimate the "nature of the function of mortality" (p. 230), a topic that today we would classify as a topic of demography.

In the early nineteenth century, predictions of mortality were based on extrapolating values from actuarial tables which were prone to significant error. Gompertz (1820) proposed using polynomial mortality curves to provide better estimates of predicted mortality rates. He stated that "... besides the usual mode of ... working from calculated tables ... subject to the errors of numerous interpolations ... it may be easier to work them directly without such reduction, or by reducing the terms to geometrical progressions for short periods ..." (p. 235). Further, he considered his proposed methods to be rather general, such that his object was to "...propose a plan of analysis and notation, which I conceive may be applied with utility to most problems likely to occur, and capable of suggesting a variety of new speculations in the pursuit of this science" (p. 216). This is the earliest application that we have identified that addresses what we might consider to be "trajectory analysis" with the focus on finding "laws" of development for a group of individuals.

During this same time period, Verhulst (1845, 1847) was working on a similar problem. Specifically, Verhulst was attempting to develop mathematical models of population growth over time, and he proposed that population increases may best be described using a logarithmic function of time. As with Gompertz, Verhulst was primarily interested in discovering "laws" of population growth over time, and he, too, focused on a demographic topic. That is, his focus was on the trajectory of population growth for the entire group, and he did not consider variability between individual trajectories.

Finally, Quetelet (1835) focused on mathematical models of growth of the human body. He opens his 1835 paper by stating, "Man is born, grows up, and dies, according to certain laws which have never been properly investigated" (p. 317). A key difference between Quetelet and other early theorists was that in addition to seeking lawful developmental processes, he also expressed an interest in difference in human development as a function of other mitigating influences. For example, he describes the relation between physical development and factors such as socioeconomic status and gender. However, he still did not explicitly move from the desire to find lawful relations over time into the estimation of interindividual differences in intraindividual change.

In sum, there are clear lines of development of trajectory analysis that can be traced back to the early nineteenth century. The work of Gompertz, Verhulst, and Quetelet combine to provide the theoretical foundation upon which later developments in trajectory modeling rest.

1.3.2 Fitting Group Trajectories: 1900–1937

Studies from the early twentieth century continued to focus on aggregate change. The majority of these early applications utilized rather complex functional forms of growth (e.g., nonlinear polynomials and logistic curves) to examine change for an entire group of individuals. That is, these early works attempted to address the *group-level trajectory* described in Question 1. For example, Robertson (1908) used logistic curves to model the growth of white rats and then attempted to apply these same models to the growth of humans. He concluded that human development required multiple logistic curves superimposed upon one another to adequately characterize the patterns of change. As with earlier theorists, Robertson was working under the assumption that there is an underlying law of growth that is to be discovered and modeled.

Reed, Pearl, and colleagues drew upon Robertson's early work as they evaluated developmental trajectories in problems as diverse as the weight of the human brain, national food consumption, influenza, and the population problem (e.g., Reed, 1921; Pearl, 1924, 1925; Reed and Pearl, 1927; Reed and Love, 1932). Their goal was to develop mathematical equations to describe growth processes with greater accuracy than was possible with other methods available at that time. Although they believed that the fitting of an equation to summarize a growth process increased the accuracy of prediction, they also believed that a given mathematical model did not necessarily aid in an understanding of the causes and regulatory factors of growth.

In sum, the first third of the twentieth century gave rise to many alternative formulations of mathematical forms of trajectories applied to a wide array of biological and social science problems. Despite the advances made in the variety of functional forms of trajectories, the goal was still to estimate a single underlying trajectory that best characterized the process for the entire group under consideration.

1.3.3 Fitting Individual and Group Trajectories: 1938–1950s

Whereas the analytical developments and empirical applications of the early twentieth century focused primarily on the trajectory of change for the entire group, the 1930s gave rise to the significant development of estimating unique trajectories for each individual within the group and modeling these trajectories as a function of other predictor variables. Of key importance is the work of Wishart (1938), who applied growth curves to weight gain in bacon pigs. His focus was on an experimental design with pigs randomly assigned to diets of low, medium, or high protein content, and the trajectory of the weight of the pigs was the outcome of interest. Wishart fit by hand a parabolic equation to the weight of each pig over a 16-week period. He then used these pig-specific growth trajectory estimates as the dependent variables in an ANOVA model to predict the rate of growth from the experimental conditions and gender. To our knowledge, this is the first attempt to consider not only the developmental trajectory for the entire group of individual cases, but to treat the individual trajectory estimates as varying randomly over individual cases.

Some years later, Griliches (1957) used a similar approach to modeling the rate of growth of hybrid corn in various regions of the United States. Griliches fit S-curves to the percentage of hybrid corn acreage of each of the regions and computed the origin, slope, and ceiling of these trajectories. These trajectory estimates for each region were then regressed on various measures to predict growth and yield over time.

The work of Wishart (1938) and Griliches (1957) highlights the major shift in moving from focusing on the estimation of the trajectory for the entire group to estimating the specific trajectory for each individual in the group. This allowed both the examination of individual differences in trajectories and the prediction of trajectories by other explanatory variables.

1.3.4 Trajectory Modeling with Latent Variables: 1950s–1984

Until the early 1950s, all trajectory models were estimated using ANOVA, ANCOVA, and MANOVA types of analytic approaches. That is, the individual trajectory estimates were typically computed by hand, and differences in the means of the trajectory estimates as a function of group membership were then examined. A significant shift in this general analytic strategy occurred in the middle of the twentieth century, and several key developments moved the estimation of trajectory models into the latent variable framework. In this framework, the growth process is governed by processes that are not observed directly, only indirectly through the repeated measures.

To our knowledge, the first person to propose using latent variables within the factor analytic framework to do trajectory modeling was Baker (1954). Baker reasoned that a factor analytic model could "indicate separate influences operating on the relative growth of individuals. These separate influences may be identified as environmental, physiological, or genetical effects. In any case the structure of the growth phenomenon is clearly exhibited with respect to time" (p. 137).

To illustrate this approach, he factor-analyzed data from 20 repeated measures of growth of 75 peaches taken at seven-day intervals. He extracted four underlying latent factors and concluded that each factor represented different stages of growth throughout the growing season. Baker (1954) concluded that factor analysis may be a useful method for reducing complex repeated measures to relatively fewer latent factors that could help us to better understand patterns of change.

Whereas Baker (1954) used an unrestricted factor analytic model, it was Tucker (1958) who first used latent variable factor analysis to estimate the individual parts of a formal function of change with respect to time. That is, Baker (1954) simply extracted four factors underlying the 20 repeated measures and interpreted the loadings from the factor pattern matrix in terms of differential stages of growth. In contrast, Tucker (1958) proposed a method for using latent factors to estimate known functional forms relating time to the repeated measures. His general goal was to develop a factorial approach to estimating nonlinear functional relations such that "If a factorial approach could be developed, it would have considerable application to experimental problems such as learning curves, work decrement curves, dark adaptation curves, etc. This note gives a theoretical basis for determination of parameters by factor analysis for many nonlinear functions" (Tucker, 1958, p. 19). He presented factor analytic models to estimate a variety of nonlinear functions and concluded that this approach may be quite beneficial for modeling various types of trajectories, although he noted that several issues, such as communality estimation and rotation, remained to be resolved.

Similarly, Rao (1958, pp. 9–12) noted the relationships between traditional growth curve models, factor analysis, and principal component analysis. He suggested that procedures used in factor analysis and principal components analysis could be useful in the growth curve modeling context. His emphasis was on group differences in growth curves, but he clearly notes the similarities between growth curve models and factor analysis.

1.3.5 Current Latent Curve Modeling: 1984–present

Baker (1954) introduced the notion of extracting latent factors from a set of repeated measures, and Tucker (1958) and Rao (1958) proposed parameterizing these factors to allow for the estimation of specific functional forms of growth. Drawing on this early work in exploratory factor analysis, Meredith and Tisak (1984, 1990) proposed embedding trajectory modeling within the confirmatory latent variable framework. This work drew on the power of structural equation models (SEMs) to estimate and test a wide variety of latent curve models. One of the basic tenets of the general latent variable modeling strategy is that there exists some underlying unobserved latent factor that gave rise to the covariances and means of the observed indicators. The analytic interest is not specifically in the indicators but on the *unobserved* latent factors that gave rise to the relationship among observed indicators. Meredith and Tisak demonstrated how this general latent variable approach applies directly to the longitudinal trajectory setting. That is, the analytic interest is not specifically in the repeated measures observed; instead, the interest is in the *unobserved latent trajectory factors* that lead to the repeated measures observed. From this perspective, trajectory modeling fits quite nicely into the SEM latent variable framework.

Meredith and Tisak's (1984, 1990) placement of growth curve models in the context of SEMs laid the foundation for further synthesis of these two methodologies, which take advantage of the many features of SEMs. Recent developments allow for various modeling strategies, including the use of multiple-indicator latent factors within each time period, the estimation of multiple-group models to evaluate interactions in development over time, the inclusion of mediating influences on growth processes, detecting latent group membership in LCMs, and the ability to model trajectories simultaneously in two or more variables. These topics are covered later in the book.

1.3.6 Summary

There is a long and rich history of developments that eventually gave rise to our modern approaches to latent curve models. Again, our brief review is not intended to be exhaustive, and several other influential lines of work are not presented here. However, it is important to realize that attempts to overcome the challenge of modeling change in a construct dates back nearly 200 years. The earliest work focused exclusively on the discovery of ordered "laws" that governed growth and development in humans and in animals. The goal was to compute the best estimates of these laws that were assumed to hold for all individuals. The next key development was the conceptualization that in addition to a trajectory that characterized the entire group, an individual trajectory could be estimated for each case within the group. These individual trajectories could then be used to assess potential individual differences in change, or could be predicted by one or more explanatory variables in later analyses. The final key development was the placement of the entire set of trajectory analytic techniques within a latent variable modeling framework. Our goal is to present an integration of much of the growing literature on latent curve models while filling in some of the gaps in the literature (e.g., discussions of identification).

1.4 ORGANIZATION OF THE REMAINDER OF THE BOOK

The goal of this first chapter was to present a conceptual introduction to latent trajectory modeling and to explore briefly the history of developments that gave rise to the analytic methods of today. In Chapter 2 we present several methods available for estimating trajectory models and a variety of component and omnibus tests of model fit. In Chapter 3 we discuss how to handle missing data in LCMs and the choice of the time metric to measure change. Modeling nonlinear trajectories and transformations of the metric of time are the topics of Chapter 4. Chapter 5 covers the conditional growth model and the incorporation of one or more exogenous predictors of the random trajectory parameters. In Chapter 6 we examine the multiple-group latent trajectory model, in which we estimate models for two or more discrete groups simultaneously. In Chapter 7 we extend the standard LCM to consider modeling trajectories in two or more constructs, and we present both unconditional and conditional models. Finally, in Chapter 8 we discuss two advanced issues, incorporating multiple indicators of repeated latent variables and using dichotomous and ordinal indicators.

CHAPTER TWO

Unconditional Latent Curve Model

Latent curve models come in a variety of forms. They can be complicated ones that include the trajectories of multiple variables, each measured with error and with exogenous variables driving the process. Or they can be as simple as a single repeated measure. In this chapter we emphasize the latter and examine *unconditional latent curve models*. These models are *unconditional* in that covariates (explanatory variables) that affect the trajectory are not included. The term *latent curve model* comes from the structural equation modeling (SEM) perspective, where the case-specific parameters that determine the trajectories are treated as latent variables. Growth curve models or growth curves are two other terms that refer to essentially the same model. The chapter notes the equivalency of the model that underlies these terms, although the emphasis is on the SEM approach to latent curve models. The chapter begins with an overview of repeated measures and how to model them. A general model follows that holds for repeated measures for any number of cases. Next is a discussion of model identification. For largely didactic purposes, we then discuss a case-by-case approach to estimating the unconditional model. Included is a discussion of the assumptions of this approach, estimation, assessing fit, and the limitations of case-by-case methods. This section is followed by the SEM approach to the unconditional latent curve model and a brief section comparing the SEM approach to a multilevel approach. We conclude by summarizing the major points raised in the chapter. The range of topics in this chapter is broad, but the ideas are fundamental to much of the rest of the book.

2.1 REPEATED MEASURES

Consider the 1995 New York crime data introduced in Chapter 1. Figure 2.1 plots the four repeated observations of the logged crime rate for the first case in our sample, and superimposes the ordinary least-squares (OLS) regression line fitted to the logged crime rate observed. (Recall that the logged transformation reduced the skewness and kurtosis that were present in the original data.)

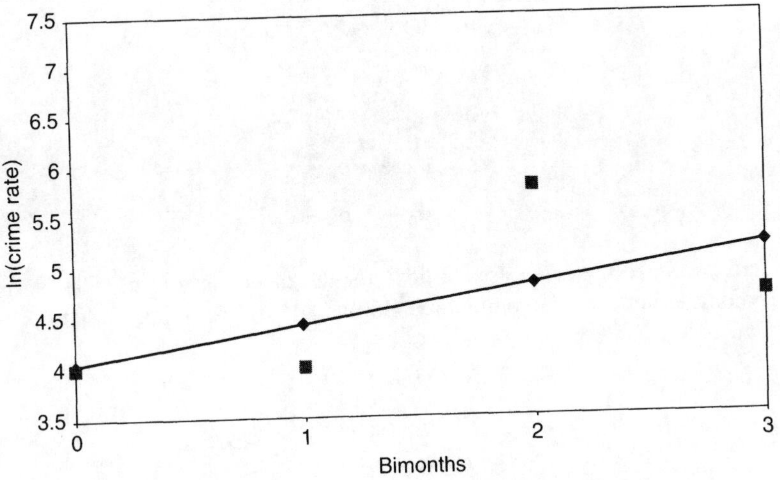

FIGURE 2.1 OLS regression line superimposed on ln(crime rate) for the first case.

If we call the intercept term α, the time trend variable λ_t, the slope β, the logged crime rate at each time y_t, and the disturbance at each time ϵ_t, we have

$$y_t = \alpha + \lambda_t\beta + \epsilon_t \tag{2.1}$$

as the model describing the trajectory of the logged crime rate [henceforth, ln(crime rate)] for this one community over the four time periods of observation. We make the usual OLS assumptions about the disturbance that its mean is zero [$E(\epsilon_t) = 0$ for all t], it is uncorrelated with the trend variable, λ_t [$\text{COV}(\lambda_t, \epsilon_t) = 0$], and it is homoscedastic [$E(\epsilon_t^2) = \sigma_{\epsilon\epsilon}$ for all t] and nonautocorrelated [$\text{COV}(\epsilon_t, \epsilon_{t+j}) = 0$ for $j \neq 0$]. For this single case, the OLS sample estimates of the intercept and slope are $\widehat{\alpha} = 4.050$ and $\widehat{\beta} = 0.383$. These values reflect that in January–February 1995 ($t = 0$) the model implied ln(crime rate) was 4.05 and this ln(crime rate) increased by 0.383 for each two months afterward over the four repeated assessments (recall that these are estimated as bimonthly crime totals).

The example corresponds to an extremely simple time series with a single unit traced over four time points. If these were all the data, a researcher might be more satisfied with just having the four ln(crime rate) values plotted against time rather than going through the trouble of estimating a regression line. However, the problem becomes more interesting when we add more communities.

The model-predicted trajectories, including four more communities, are shown in Figure 2.2. For each community in Figure 2.2 there is a separately estimated regression line, as with the first community. The graphs suggest that each community's trajectory requires a separate regression line.

If we index the five communities by the numbers $i = 1, 2, 3, 4, 5$, the regression model equations describing the community trajectories are

$$\begin{aligned} y_{1t} &= \alpha_1 + \lambda_t \beta_1 + \epsilon_{1t} \\ y_{2t} &= \alpha_2 + \lambda_t \beta_2 + \epsilon_{2t} \\ y_{3t} &= \alpha_3 + \lambda_t \beta_3 + \epsilon_{3t} \\ y_{4t} &= \alpha_4 + \lambda_t \beta_4 + \epsilon_{4t} \\ y_{5t} &= \alpha_5 + \lambda_t \beta_5 + \epsilon_{5t} \end{aligned} \tag{2.2}$$

Here we make the same assumptions for the disturbance terms that we made when considering a single community. But we add the assumptions that

$$\begin{aligned} \mathrm{COV}(\epsilon_{it}, \alpha_j) &= 0 \\ \mathrm{COV}(\epsilon_{it}, \beta_j) &= 0 \\ \mathrm{COV}(\alpha_i, \alpha_j) &= 0 \\ \mathrm{COV}(\beta_i, \beta_j) &= 0 \\ \mathrm{COV}(\alpha_i, \beta_j) &= 0 \\ \mathrm{COV}(\epsilon_{it}, \epsilon_{jt}) &= 0 \end{aligned} \tag{2.3}$$

for $i, j = 1, 2, 3, 4, 5$, $i \neq j$. In essence, we must now add subscripts i, j to keep track of the distinct intercepts and slopes that govern the trajectories for each community.

The regression lines in Figure 2.2 suggest different trajectories of ln(crime rate) over time. The regression line that we estimated for the first community would

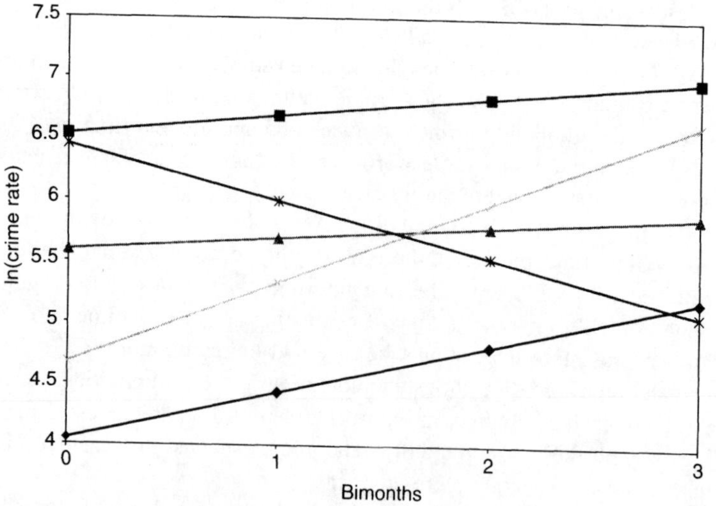

FIGURE 2.2 Regression-predicted trajectories of ln(crime rate) for the first five cases.

not necessarily be reflective of the trajectories for the other four communities. The same would be true for any of the five communities. Thus, the intercepts of the model-implied trajectories for some communities are higher than for others, and the rate of change is steeper for some communities than for others.

2.2 GENERAL MODEL AND ASSUMPTIONS

The figures above present the model-implied trajectory for a single community (Figure 2.1) and the trajectories for five communities (Figure 2.2). However, recall that we have complete data on a sample of 359 communities. Although it is relatively easy to understand the nature of change over time for a small number of cases, this task gets exceedingly difficult when considering a larger number of cases. For example, Figure 2.3 presents the model-implied trajectories based on the OLS estimates for the first 50 communities in the sample.

In this more realistic situation, it is useful to have a general notation to keep track of the separate trajectories of each case and to have summaries of these trajectory parameters, such as the mean and standard deviation. To facilitate this task, consider the *unconditional latent curve model*. The *trajectory equation* for this model is

$$y_{it} = \alpha_i + \lambda_t \beta_i + \epsilon_{it} \tag{2.4}$$

where y_{it} is the value of the trajectory variable y for the ith case at time t, α_i is the random intercept for case i, and β_i is the random slope for case i. As was true for

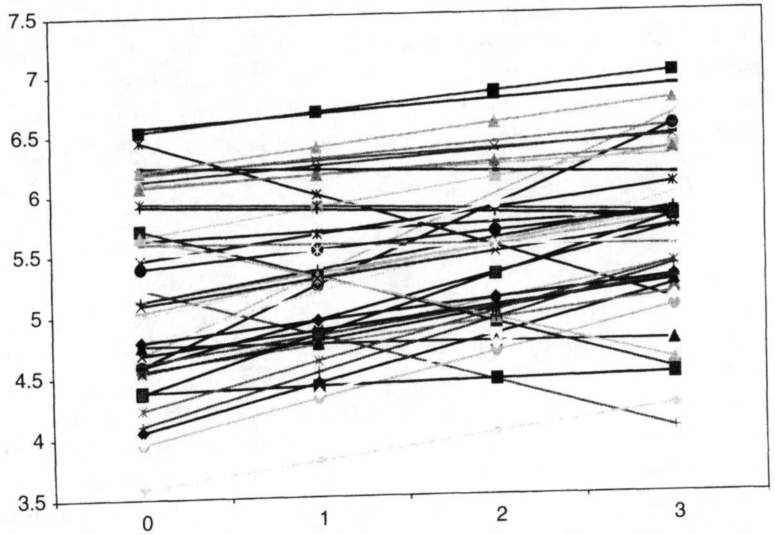

FIGURE 2.3 OLS-predicted trajectories for the first 50 communities for logged crime rates.

our models in Section 2.1, the latent curve model (LCM) allows each case (i) to have a distinct intercept and slope to describe the path of a variable over time (t). This is captured by indexing the intercepts (α_i) and slopes (β_i) by i to show that they can differ across cases in the sample. The λ_t is a constant where a common coding convention is to have $\lambda_1 = 0$ and $\lambda_2 = 1$. The remaining values of λ_t allow the incorporation of linear or nonlinear trajectories. There are a variety of ways in which to code time via λ_t, and we explore these options in detail in Chapter 4. For now, we set $\lambda_1 = 0$ so that $E(\alpha_i)$ represents the mean of the trajectory at the initial time point. In the case of a linear latent curve model, λ_t equals $t - 1$ for all t. This is equivalent to the coding that we used when tracking the simple time trend for the New York crime data, and it assumes that a linear trend is a good approximation.

We assume that $E(\epsilon_{it}) = 0$ for all i and t, an assumption that is familiar from regression analysis, where we assume that the disturbance has a mean of zero. Also in a fashion analogous to regression, another assumption is that the right-hand-side random "variables" in Eq. (2.4) (i.e., α_i, β_i) are uncorrelated with the equation disturbance. More formally, this assumption is that $\text{COV}(\epsilon_{it}, \beta_i) = 0$ and $\text{COV}(\epsilon_{it}, \alpha_i) = 0$ for all i and $t = 1, 2, 3, \ldots, T$, and $E(\epsilon_{it}\epsilon_{jt}) = 0$ for all t and $i \neq j$. The variance of the disturbance is $E(\epsilon_{it}^2) = \text{VAR}(\epsilon_{it})$ for each t. Most researchers assume that all cases have the same variance for each time period [i.e., $E(\epsilon_{it}\epsilon_{jt}) = \text{VAR}(\epsilon_t)$ or $\sigma_{\epsilon_t\epsilon_t}$], although there are situations where this assumption is not made. Sometimes an even more restrictive assumption is made that the error variances are constant over time and over cases [i.e., $E(\epsilon_{it}^2) = \text{VAR}(\epsilon)$ or $\sigma_{\epsilon\epsilon}^2$]. However, for now we keep the assumptions general, noting when the assumptions require modification to permit estimation. Another assumption is that $\text{COV}(\epsilon_{it}, \epsilon_{i,t+s}) = 0$ for $s \neq 0$, so that the errors are not correlated over time. In addition, the random intercepts and slopes for one case are assumed to be uncorrelated with those of another [$\text{COV}(\alpha_i, \alpha_j) = 0$, $\text{COV}(\beta_i, \beta_j) = 0$, $\text{COV}(\alpha_i, \beta_j) = 0$ for $i \neq j$]. The final assumption is that the disturbances for different individuals are uncorrelated, so that $\text{COV}(\epsilon_{it}, \epsilon_{j,t+s}) = 0$ for $i \neq j$ and all s.

The mean intercept and mean slope are of interest and this leads us to the *intercept* and *slope equations* for the unconditional latent curve model:

$$\alpha_i = \mu_\alpha + \zeta_{\alpha i} \tag{2.5}$$

$$\beta_i = \mu_\beta + \zeta_{\beta i} \tag{2.6}$$

where μ_α and μ_β are the mean intercept and mean slope across all cases. The *intercept equation* (2.5) represents the individual intercept α_i as a function of the mean of the intercepts for all cases (μ_α) and a disturbance $\zeta_{\alpha i}$. Similarly, the *slope equation* [Eq.(2.6)] treats the individual slope β_i as a function of the mean of the slopes for all cases (μ_β) and a disturbance $\zeta_{\beta i}$. So we have introduced means of the intercepts and slopes across cases as measures of central tendencies in the trajectories. The $\zeta_{\alpha i}$ and $\zeta_{\beta i}$ are disturbances with means of zero and variances of $\psi_{\alpha\alpha}$ and $\psi_{\beta\beta}$ and covariance of intercepts and slopes of $\psi_{\alpha\beta}$. Furthermore, $\zeta_{\alpha i}$ and $\zeta_{\beta i}$ are assumed to be uncorrelated with ϵ_{it}. Note that in this unconditional model, the variance of α is equivalent to the variance of ζ_α or $\psi_{\alpha\alpha}$ and the variance of β is equivalent to the variance of ζ_β or $\psi_{\beta\beta}$. This will not be true in the conditional model of Chapter 5.

If most cases in a sample have very similar intercepts and very similar slopes, we expect $\psi_{\alpha\alpha}$ and $\psi_{\beta\beta}$ to be small. In the most extreme case where the variances are zero, all cases have the same intercept and slope for their trajectories. In contrast, more diversity in case-specific intercepts and slopes will lead to higher values for these variances. This is an *unconditional* trajectory model in that the intercept and slope equations have only the mean intercept and mean slope as determinants. In other words, the individual intercepts and slopes are not a function of other variables in the model.

We can combine the trajectory, intercept, and slope equations into a single equation by substituting the right-hand sides of the intercept equation (2.5) and slope equation (2.6) for α_i and β_i, respectively, in the trajectory equation (2.4). This results in

$$y_{it} = (\mu_\alpha + \lambda_t \mu_\beta) + (\zeta_{\alpha_i} + \lambda_t \zeta_{\beta i} + \epsilon_{it}) \tag{2.7}$$

This is the *combined model*, in that it combines the three equations into a single equation. Another term for this expression is the *reduced- form equation* for the trajectory model. It is a reduced form in that the endogenous random coefficients, α_i and β_i, are replaced by their exogenous determinants and disturbances. In this form of the equation we see that the trajectory of y_{it} is a function of the mean intercept, the trend variable times the mean slope, and a complex composite disturbance term. It is noteworthy that this composite disturbance is heteroscedastic over time, due to the presence of $\lambda_t \zeta_{\beta i}$, which has a variance that depends on λ_t. Often, the first term in parentheses in Eq.(2.7) is referred to as the *fixed component*, and the second term, the *random component*. The fixed component represents the mean structure, and the random component represents various sources of individual variability.

Table 2.1 provides a convenient summary of the definitions and assumptions for the unconditional latent curve model. It repeats what is in the text but in a more concise form. Having an explicit form for the model does not guarantee that we will be able to estimate it. One requirement is that all of the parameters in the model be identified. It is to this topic that we now turn.

2.3 IDENTIFICATION

Establishing *model identification* comes prior to estimation of the trajectory, intercept, and slope equations. *Identification* refers to whether there are unique values of the model parameters that are determined by the model structure and the information that we have about the variables. For instance, suppose that we knew that $\mu_x = \theta_1 + \theta_2$, where μ_x is the mean of x and is known. If no further information on θ_1 and θ_2 is available, these parameters are not identified or are *underidentified*. But if there is another equation, say, $\mu_z = \theta_1$, where μ_z is known, the two equations together will identify θ_1 and θ_2 ($\theta_1 = \mu_z$ and $\theta_2 = \mu_x - \mu_z$).

Table 2.1 Definitions and Assumptions.

Model

$y_{it} = \alpha_i + \lambda_t \beta_i + \epsilon_{it}$	*Trajectory Equation*
$\alpha_i = \mu_\alpha + \zeta_{\alpha i}$	*Intercept Equation*
$\beta_i = \mu_\beta + \zeta_{\beta i}$	*Slope Equation*
$y_{it} = (\mu_\alpha + \lambda_t \mu_\beta) + (\zeta_{\alpha_i} + \lambda_t \zeta_{\beta i} + \epsilon_{it})$	*Combined Equation*

where

$i = 1, 2, \ldots, N$, where N is the total number of cases
$t = 1, 2, \ldots, T$, where T is the total number of time points

Definitions

y_{it} = trajectory variable for case i, time t
α_i = intercept of trajectory for case i
β_i = slope of trajectory for case i
λ_t = value of trend variable for time t
ϵ_{it} = disturbance of y_{it}
$\mu_\alpha = E(\alpha_i)$ = mean of the intercepts
$\mu_\beta = E(\beta_i)$ = mean of the slopes
$\zeta_{\alpha i}$ = disturbance deviation from μ_α for case i
$\zeta_{\beta i}$ = disturbance deviation from μ_β for case i
$\psi_{\alpha\alpha} = \text{VAR}(\zeta_{\alpha i})$ = variance of intercepts
$\psi_{\beta\beta} = \text{VAR}(\zeta_{\beta i})$ = variance of slopes
$\psi_{\alpha\beta} = \text{COV}(\zeta_{\alpha i}, \zeta_{\beta i})$ = covariance of intercepts and slopes
$\theta_{\epsilon_{it}} = \text{VAR}(\epsilon_{it})$ = variance of disturbance at time t for case i

Assumptions

$E(\epsilon_{it}) = 0$ for $i = 1, 2, \ldots, N$; $t = 1, 2, \ldots, T$
$E(\zeta_{\alpha i}) = 0$ for $i = 1, 2, \ldots, N$
$E(\zeta_{\beta i}) = 0$ for $i = 1, 2, \ldots, N$
$\text{COV}(\epsilon_{it}, \zeta_{\alpha i}) = 0$
$\text{COV}(\epsilon_{it}, \zeta_{\beta i}) = 0$
$\text{COV}(\zeta_{\alpha i}, \zeta_{\alpha j}) = 0$ for $i \neq j$
$\text{COV}(\zeta_{\beta i}, \zeta_{\beta j}) = 0$ for $i \neq j$
$\text{COV}(\zeta_{\beta i}, \zeta_{\alpha j}) = 0$ for $i \neq j$
$\text{COV}(\epsilon_{it}, \epsilon_{jt}) = 0$ for $i \neq j$
$\text{COV}(\epsilon_{it}, \epsilon_{i,t+s}) = 0$ for $i \neq j$
[$\text{COV}(\epsilon_{it}, \epsilon_{i,t+s}) = 0$ for $s \neq 0$ (Optional assumption)]

Following Bollen (1989a, pp. 88–89), we can distinguish between parameters that are known to be identified and those whose identification statuses are unknown. For short we can refer to these as the *known* and *unknown parameters* with respect to identification. The key identified moments of the observed variables with which to work are the means, variances, and covariances, assuming that these moments are defined for the random variables.[1] Consider these moments as known to be identified. If the unknown parameters are unique functions of those known to be identified moments, this establishes their identification as well. If all parameters in a model are so identified, the model is identified and model estimation is the next step.

A useful starting point is to first make a list of the parameters known to be identified and the unknown parameters in the model. The parameters known to be identified for the unconditional latent curve model are the means, variances, and covariances of the y's: $E(y_{it})$, $\text{VAR}(y_{it})$, and $\text{COV}(y_{it}, y_{i,t-s})$ for $s > 0$ and $t - s \geq 1$. For the linear trajectory model, the unknown parameters are μ_α, μ_β, $\text{VAR}(\epsilon_{it})$, $\psi_{\alpha\alpha}$, $\psi_{\beta\beta}$, $\psi_{\alpha\beta}$, and λ_t, where the symbols are as defined in Table 2.1. One necessary condition for model identification is that there are at least as many parameters known to be identified as there are unknown parameters.[2]

In general, we have a population mean (μ_{y_t}) for each unit of time and population variances and covariances of the y_{it} leading to $\frac{1}{2}T(T+3)$ identified parameters with which to work. The $\frac{1}{2}T(T+3)$ formula is general and provides the number of means, variances, and covariances available no matter the number of waves of data. There will be NT parameters for $\text{VAR}(\epsilon_{it})$, T parameters for λ_t, and $\frac{1}{2}K(K+3)$ parameters for $\mu_\alpha, \mu_\beta, \psi_\alpha, \psi_{\beta\beta}$, and $\psi_{\alpha\beta}$, where K equals the total number of latent variables to describe the trajectory. For instance, $K = 1$ if there only is a random intercept, $K = 2$ if there is a random intercept and random slope, $K = 3$ when a random quadratic term is added, and so on.

Without further restrictions, we will not be able to identify all of the model parameters. A common set of assumptions alleviates this problem. One is to assume that the trend values captured by λ_t are known, such as in the linear trend case, where $\lambda_t = t - 1$. Thus, λ_t would be known, not estimated. Another assumption is that each case in the sample has the same error variance in the same time period, although the variances can differ over time; that is, $\text{VAR}(\epsilon_{it}) = \text{VAR}(\epsilon_t)$. Under these two assumptions, there are now $T + 5$ unknowns for the linear model. The necessary condition for identification will not be satisfied with only two waves of data because there will be seven unknown parameters and only five known parameters with which to work. Three waves of data satisfy this *necessary condition* (eight unknowns and nine knowns), but this alone is not sufficient to establish model identification.[3]

Instead, we use an algebraic approach to investigate identification. Throughout the section the notation is simplified by not using the i subscript [e.g., $E(y_1)$ instead

[1]Moments are characteristics of the distribution of random variables. Depending on the distribution from which the observed variable is drawn and the model structure, it is sometimes theoretically possible to use higher-order moments of the observed variables to aid in parameter identification. However, in virtually all practical applications it is the mean, variances, and covariances of the observed variables that are considered. We follow this convention here.

[2]Note that this assumes no inequality constraints that could obtain identification in some special cases.

[3]The condition of having more knowns than unknowns is often referred to as the *t-rule* (e.g., Bollen, 1989b, pp. 93, 242–244, 328), but note that in this usage t does not refer to time. To avoid confusion, we do not refer to this as the t-rule in the text.

of $E(y_{i1})$]. Not showing i implies that the moment or parameter is the same across cases. The means of the variables observed in terms of the model parameters come from taking the expected value of both sides of the trajectory equation (2.4):

$$E(y_t) = \mu_\alpha + \lambda_t \mu_\beta \tag{2.8}$$

Equation (2.8) gives the *model-implied equation* for the means of the observed variables. Consider the case of three waves of data where we have a linear trajectory model ($\lambda_t = t - 1$) and we substitute $\lambda_t = 0, 1, 2$. This leads to

$$E(y_1) = \mu_\alpha \tag{2.9}$$

$$E(y_2) = \mu_\alpha + \mu_\beta \tag{2.10}$$

$$E(y_3) = \mu_\alpha + 2\mu_\beta \tag{2.11}$$

From equations (2.9) and (2.10), we can establish the identification of the mean of the intercepts and mean of the slopes as

$$\mu_\alpha = E(y_1) \tag{2.12}$$

$$\mu_\beta = E(y_2) - E(y_1) \tag{2.13}$$

We must turn to the model-implied variances and covariances to identify the other model parameters. The variances of the observed variables are

$$\text{VAR}(y_t) = \text{VAR}(\alpha) + \lambda_t^2 \text{VAR}(\beta) + 2\lambda_t \text{COV}(\alpha, \beta) + \text{VAR}(\epsilon_t) \tag{2.14}$$

and the lagged covariances are

$$\text{COV}(y_t, y_{t-s}) = \text{VAR}(\alpha) + \lambda_t \lambda_{t-s} \text{VAR}(\beta) + (\lambda_t + \lambda_{t-s})\text{COV}(\alpha, \beta) \tag{2.15}$$

for $s \neq 0$. The variances and covariances of the random intercepts and slopes are

$$\text{COV}(\alpha, \beta) = \text{COV}(y_1, y_3) - \text{COV}(y_1, y_2) \tag{2.16}$$

$$\text{VAR}(\alpha) = 2\text{COV}(y_1, y_2) - \text{COV}(y_1, y_3) \tag{2.17}$$

$$\text{VAR}(\beta) = [\text{COV}(y_2, y_3) - \text{COV}(y_1, y_4)]/2 \tag{2.18}$$

The error variances are identified by

$$\text{VAR}(\epsilon_t) = \text{VAR}(y_t) - \text{VAR}(\alpha) - \lambda_t^2 \text{VAR}(\beta) - 2\lambda_t \text{COV}(\alpha, \beta) \tag{2.19}$$

where all of the right-hand-side elements are identified by the preceding equations, and hence the $\text{VAR}(\epsilon_t)$'s are identified. In the three-wave model we have six variances and covariances of the observed variables and three means, a total of nine elements. The parameters that require identification are the mean intercept and mean slope, the variances and covariance of the intercepts and slopes, and three error variances, leading to a total of eight. In cases such as this where there is a unique solution for each model parameter and where there are more means, variances, and covariances than there are model parameters, the model is *overidentified.* Thus, the three-wave model is overidentified. If the number of waves of data increases, so does the degree of overidentification. This example illustrates that having three or more waves of data on y_{it} for an unconditional latent curve model leads to an overidentified model.

The New York ln(crime rate) data illustrate the preceding ideas. The complete data set has 359 communities that are tracked over $T = 4$ time points. From the sample data we have observed 14 values known to be identified: $(\frac{1}{2})(T)(T + 3) = (\frac{1}{2})(4)(7) = 14$, reflecting four means and 10 variances and covariances. Given the assumptions of $\lambda_t = t - 1$ and $\text{VAR}(\epsilon_{it}) = \text{VAR}(\epsilon_t)$, a linear growth model for $T = 4$ contains $T + 5$, or nine, unknowns. Given 14 known values and nine unknown values, the standard unconditional latent curve model for the four-occasion crime example is overidentified. The difference between the known and unknown values gives the degrees of freedom for the model. In the four-wave example, there are 5 degrees of freedom (df).

In sum, this section shows that with a minimum of three waves of data, the unconditional linear LCM is overidentified. There being fewer waves creates an underidentified model, whereas more waves increase the degree of overidentification.

2.4 CASE-BY-CASE APPROACH

Once the researcher establishes the identity of the model, the next step is to estimate the parameters.[4] The trajectory equation and equations for the mean intercept and mean slope are

$$y_{it} = \alpha_i + \lambda_t \beta_i + \epsilon_{it} \tag{2.20}$$

with

$$\begin{aligned} \alpha_i &= \mu_\alpha + \zeta_{\alpha i} \\ \beta_i &= \mu_\beta + \zeta_{\beta i} \end{aligned} \tag{2.21}$$

Ideally, we would like to estimate the following parameters: μ_α, μ_β, $\text{VAR}(\epsilon_{it})$, $\psi_{\alpha\alpha}$, $\psi_{\beta\beta}$, and $\psi_{\alpha\beta}$. For now, assume a linear trajectory model such that $\lambda_t =$

[4]The identification process is slightly different from that emphasized in Section 2.3 since, as we note below, we make different assumptions about the error variances for the repeated measures in the case-by-case approach. Here we assume that the error variance is constant over time for each case but that the error variance can differ across cases. We could address identification here by noting that each simple regression for each case is identified. The mean and standard deviation of these intercepts and slopes also are identified.

$t - 1$. There are at least two approaches to estimation. The case-by-case approach to estimation of the latent curve model is the subject of this section. In Chapter 1 we introduced case-by-case regressions of the repeated measures on the time trend variable as a method to summarize individual trajectories. This estimation procedure has much intuitive appeal since for each case there is a separate regression fit to the same trend variable. The linear regression line provides an estimate of the slope and intercept for each member of the sample. The assumptions underlying estimation match those listed in Table 2.1 with the addition that $\text{VAR}(\epsilon_{it}) = \text{VAR}(\epsilon_i)$. In other words, we assume that the error variances for each case are equal over time, but the error variances can differ over individuals.[5] This assumption differs from the error variance assumption of Section 2.3, where the identification of the unconditional latent curve model was established. Specifically, here we are assuming that each case has equal error variances over time, but each individual's error variance can differ from others, whereas previously we assumed that the error variances differ over time, but the error variances were equal for each individual within a given time. Fortunately, this alternative assumption of error variances differing over individual, but the same over time, still leads to an identified model.

Furthermore, these assumptions satisfy the conditions of ordinary least squares (OLS), leading to the best linear unbiased estimator (BLUE) of the intercept and slope for each case considered one at a time. The availability of complete data (i.e., no missing data) across all time points is a simplifying assumption used throughout this chapter. In Chapter 3 we describe methods to handle missing data.

For each case, the OLS estimator of the slope is

$$\widehat{\beta}_i = \frac{\sum_{t=1}^{T}(\lambda_t - \bar{\lambda})(y_{it} - \bar{y}_i)}{\sum_{t=1}^{T}(\lambda_t - \bar{\lambda})^2} \tag{2.22}$$

and the intercept is

$$\widehat{\alpha}_i = \bar{y}_i - \widehat{\beta}_i \bar{\lambda} \tag{2.23}$$

where $\bar{\lambda}$ is the mean of the time trend variable and $\bar{y}_i$ is the mean of y_{it} for the ith case over the T time points. Estimating these simple regressions for all cases in the sample provides separate intercept and slope estimates for every case. Forming the sample means of the intercepts and slopes gives a simple way to estimate the mean intercept (μ_α) and mean slope (μ_β):

$$\widehat{\mu}_\alpha = \frac{\sum_{i=1}^{N}\widehat{\alpha}_i}{N} \tag{2.24}$$

[5]If the error variances differed over time for a case, we could use weighted least squares to estimate the model provided that we had a consistent estimator of the changing variance of the error (e.g., Johnson, 1984). Or we could use heteroscedastic-consistent standard errors with OLS (e.g., White, 1980). Given the common situation of a small number of waves of data, we should use these options only when we know that the heteroscedasticity is severe and we have confidence in the error variance estimates.

and

$$\widehat{\mu}_\beta = \frac{\sum_{i=1}^{N} \widehat{\beta}_i}{N} \tag{2.25}$$

respectively. Under the assumptions above, these are unbiased estimators so that $E(\widehat{\mu}_\alpha) = \mu_\alpha$ and $E(\widehat{\mu}_\beta) = \mu_\beta$. Given the computation of point estimates for $\widehat{\mu}_\alpha$ and $\widehat{\mu}_\beta$, it would be useful to compute a corresponding standard error for hypothesis testing or the construction of confidence intervals around these point estimates. These standard errors (s.e.'s) are

$$\text{s.e.}(\widehat{\mu}_\alpha) = \sqrt{\frac{\sum(\widehat{\alpha}_i - \widehat{\mu}_\alpha)^2/(N-1)}{N}} \tag{2.26}$$

$$\text{s.e.}(\widehat{\mu}_\beta) = \sqrt{\frac{\sum(\widehat{\beta}_i - \widehat{\mu}_\beta)^2/(N-1)}{N}} \tag{2.27}$$

The standard errors of the means together with the assumption that the means are normally distributed enable the construction of confidence intervals around the point estimates of the mean. For example, for a given mean slope value $\widehat{\mu}_\beta$ with standard error s.e.$(\widehat{\mu}_\beta)$, the 95% confidence interval is

$$\text{CI} = \widehat{\mu}_\beta \pm 1.96[\text{s.e.}(\widehat{\mu}_\beta)] \tag{2.28}$$

Estimated individual intercepts and linear slopes for each of the 359 communities in our sample are estimable. Using Eqs. (2.24) and (2.25), the means of the OLS intercepts and slopes were estimated to be $\widehat{\mu}_\alpha = 5.34$ and $\widehat{\mu}_\beta = 0.143$, respectively. These values imply that pooling over all 359 communities, the mean ln(crime rate) was 5.34 at the initial assessment, and the mean increase in the ln(crime rate) was 0.143 per bimonthly period. Further, using Eqs. (2.26) to (2.28), the 95% confidence interval for the mean intercepts was (5.26, 5.42), and the 95% confidence interval for the mean slopes was (0.122, 0.163).

At this point we have estimates of the means of the intercepts and slopes and their standard errors. This information provides answers to the first question about the group intercept and slope described in Chapter 1. The second question concerns whether there is a need to have separate intercepts and separate slopes for individuals or whether the mean intercept and mean slope are an adequate summary to apply to all individuals. More formally, this is a question of whether there is zero variance around the mean of the intercepts (or slopes). If there is variance, what is the covariance of the intercepts with the slopes? Estimates of the variances of α_i and β_i and their covariance (i.e., $\psi_{\alpha\alpha}$, $\psi_{\beta\beta}$, and $\psi_{\alpha\beta}$, respectively) are needed. Having estimates of the case-specific intercepts ($\widehat{\alpha}_i$) and slopes ($\widehat{\beta}_i$), it is tempting to estimate these as the sample variances of the $\widehat{\alpha}_i$'s and $\widehat{\beta}_i$'s, employing the usual formula for an unbiased estimate of variance:[6]

[6]Here and throughout the book, var(·) refers to a sample estimator of variance, whereas VAR(·) refers to population variance.

$$\text{var}(\widehat{\alpha}) = \frac{\sum_{i=1}^{N}(\widehat{\alpha}_i - \widehat{\mu}_\alpha)^2}{N-1} \tag{2.29}$$

$$\text{var}(\widehat{\beta}) = \frac{\sum_{i=1}^{N}(\widehat{\beta}_i - \widehat{\mu}_\beta)^2}{N-1} \tag{2.30}$$

For example, recall that earlier we computed the mean intercept ($\widehat{\mu}_\alpha = 5.34$) and slope ($\widehat{\mu}_\beta = 0.143$) values pooling over the 359 communities. Using Eqs. (2.29) and (2.30), sample variances around these mean estimates are $\text{var}(\widehat{\alpha}_i) = 0.598$ and $\text{var}(\widehat{\beta}_i) = 0.039$. Although these variances appear to be sizable relative to the corresponding mean estimates, this may or may not be the case.

Unfortunately, simply computing the variance of the OLS estimates of intercept and slope involves making an important and untenable assumption that these are estimated *without error*. That is, this method considers the sample estimates of the individual OLS $\widehat{\alpha}_i$'s and $\widehat{\beta}_i$'s without also considering the standard errors of these estimates. It is thus probably misleading to examine the OLS estimates of $\widehat{\alpha}_i$ and $\widehat{\beta}_i$ without incorporating some estimate of the *precision* of these estimates. Fortunately, use of the mean-squared-error estimate from each individual regression permits an unbiased estimate of the error variances, denoted $\text{var}(\epsilon_i)$. This is given as

$$\text{var}(\epsilon_i) = \frac{\sum_{t=1}^{T} e_{it}^2}{T-2} \tag{2.31}$$

where e_{it}^2 is the squared residual for individual i at time t. There is thus an estimate, $\text{var}(\epsilon_i)$, for each individual i, so there are a total of N individual estimates of variances.

Up to this point, we have allowed $\text{var}(\epsilon_i)$ to vary randomly over individuals. However, by imposing the assumption that the error variances are *equal* over individuals [i.e., $\text{VAR}(\epsilon_i) = \text{VAR}(\epsilon)$ for all i], these individual estimates can be pooled into a single estimate of error variance denoted

$$\text{var}(\epsilon) = \frac{\sum_{i=1}^{N} \text{var}(\epsilon_i)}{N} \tag{2.32}$$

This represents the average error in estimating the individual OLS trajectories pooled over all N cases in the sample, and this will allow us to estimate the variances of the true intercept and slope components of the estimated trajectories (e.g., Rogosa and Saner, 1995). Using this pooled estimate, we can adjust the sample variances of the intercepts and slopes to compute unbiased estimates of variability around the trajectories. This adjustment is given as

$$\widehat{\psi}_{\alpha\alpha} = \text{var}(\widehat{\alpha}) - \frac{\text{var}(\epsilon)\sum_{t=1}^{T}\lambda_t^2}{\left[\sum_{t=1}^{T}(\lambda_t - \bar{\lambda})^2\right]T} \tag{2.33}$$

$$\widehat{\psi}_{\beta\beta} = \text{var}(\widehat{\beta}) - \frac{\text{var}(\epsilon)}{\sum_{t=1}^{T}(\lambda_t - \bar{\lambda})^2} \tag{2.34}$$

where $\widehat{\psi}_{\alpha\alpha}$ and $\widehat{\psi}_{\beta\beta}$ represent unbiased estimators of the variance of the intercept and slope, respectively. We can rearrange Eq. (2.34) such that

$$\text{var}(\widehat{\beta}) = \widehat{\psi}_{\beta\beta} + \frac{\text{var}(\epsilon)}{\sum_{t=1}^{T}(\lambda_t - \overline{\lambda})^2} \tag{2.35}$$

to simply highlight that the sample variance of the individual $\widehat{\beta}_i$ estimates is an additive combination of true variance plus error. In the unlikely case that each trajectory is estimated with perfect precision, $\text{var}(\epsilon) = 0$ and $\text{var}(\widehat{\alpha}) = \widehat{\psi}_{\alpha\alpha}$ and $\text{var}(\widehat{\beta}) = \widehat{\psi}_{\beta\beta}$. However, in the typical situation in which $\text{var}(\epsilon) > 0$, $\text{var}(\widehat{\alpha})$ and $\text{var}(\widehat{\beta})$ must be adjusted accordingly.

To remove this error, we compute $\text{var}(\epsilon)$ via Eq. (2.32), and enter these values into Eqs. (2.33) and (2.34). These results indicate that the estimate of $\text{var}(\epsilon)$ over all N is 0.0982. For our sample data, $\Sigma\lambda_t^2 = 14$, $T = 4$, $\overline{\lambda} = 1.5$, and $\Sigma\left(\lambda_t - \overline{\lambda}\right)^2 = 5.0$; and entering these values into Eqs. (2.33) and (2.34) results in $\widehat{\psi}_{\alpha\alpha} = 0.529$ and $\widehat{\psi}_{\beta\beta} = 0.020$. Note that these values are lower than the simple unadjusted variances of $\text{var}(\widehat{\alpha}) = 0.598$ and $\text{var}(\widehat{\beta}) = 0.039$ that were obtained using Eqs. (2.29) and (2.30). The smaller values computed for our estimates of the true variability of the intercepts and slopes compared to the unadjusted sample variability of the OLS estimates reflects the degree of imprecision in estimation of the OLS trajectory estimates. These are accurate estimates of the variability that will inform us about Question 2 (i.e., the variance of the intercepts and slopes) raised in Chapter 1.[7]

2.4.1 Assessing Model Fit

Earlier in the chapter we reported the case-by-case regression results for the crime index data for some cases. In this section we report summary results for all cases as a way to introduce some of the methods of assessing the fit of a model. OLS regression techniques form estimates of the intercept and slope for each case using a linear trajectory model. Recall that the crime index was available for 359 crime reporting units for four points in time. This leads to 359 intercepts and slopes. Table 2.2 reports the estimates of the means of the intercepts and slopes, the error variance, and the corrected variances of the intercepts and slopes. The table shows that the mean initial value for the crime index is 5.338 and the mean of the slopes is a positive 0.143. In addition, the variance around the mean initial level of the crime index is 0.529 with a variance of the slope of 0.020. These summary statistics are helpful in giving information on the mean and variance of the intercepts and slopes, but there are additional ways to use the results of the case-by-case approach.

The case-by-case regression estimates provide predicted values of the individual intercepts ($\widehat{\alpha}_i$'s) and slopes ($\widehat{\beta}_i$'s) for all cases in the sample. These parameter estimates are useful in evaluating the model. A simple but useful approach is to

[7]Similarly, the covariance of $\widehat{\alpha}$ and $\widehat{\beta}$ calculated with the usual sample covariance formula will not be accurate. The SEM approach section presents ways to estimate the covariance and variances of the intercepts and slopes accurately without requiring further adjustment.

Table 2.2 Point Estimates and (Standard Errors) from OLS Regressions for Case-by-Case Approach for Crime Index Data ($N = 359$)

Parameter	Estimate
var(ϵ)	0.0982 (0.0103)
μ_α	5.338 (0.0408)
μ_β	0.143 (0.0105)
$\psi_{\alpha\alpha}$	0.529 (–)
$\psi_{\beta\beta}$	0.020 (–)

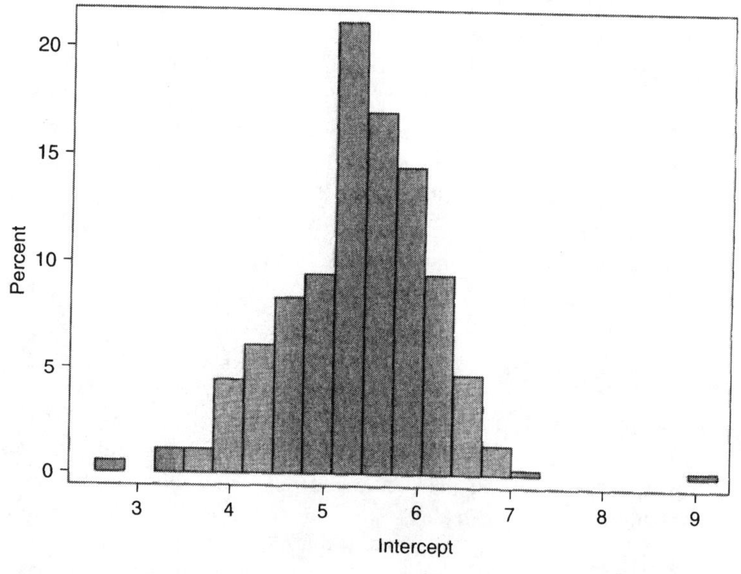

FIGURE 2.4 Histogram of OLS estimates of intercepts of ln(crime rate) ($N = 359$).

begin with univariate plots of the $\widehat{\alpha}_i$'s and the $\widehat{\beta}_i$'s. Figures 2.4 and 2.5 present the histograms of these parameter estimates for the crime example data that were analyzed in Section 2.3.

Several interesting characteristics about the OLS estimates of ln(crime rate) trajectories are evident in these plots. First, the histogram shows all positive intercepts, meaning that at the earliest time point all predicted regression lines have positive values. Since all communities have some crime, this positive value is anticipated. Also noteworthy is the intercept outlier of about 9, a value far larger than the other communities. The histogram of the slopes reveals mostly positive values, which signify an upward trend over this eight-month period. However, there are some communities with negative slopes that indicate a decline in ln(crime rate) over this

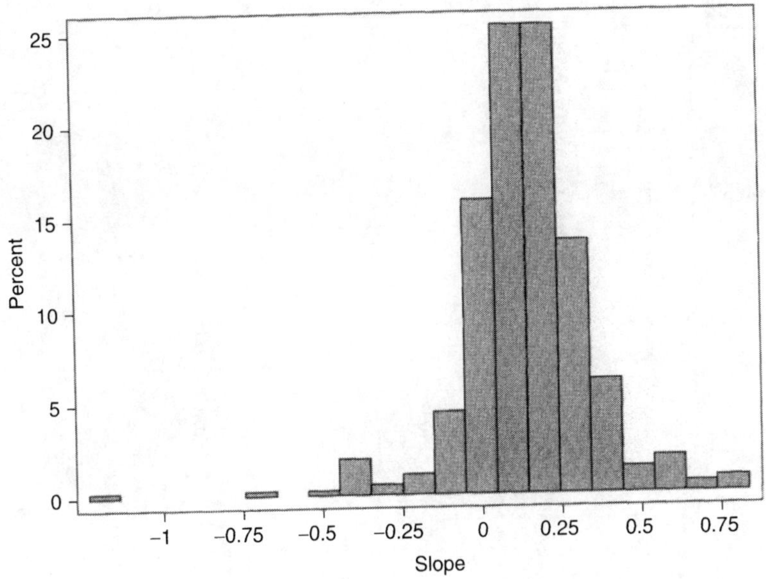

FIGURE 2.5 Histogram of OLS estimates of slopes for ln(crime rate) ($N = 359$).

period for those communities. One slope is less than -1, a value quite different from those for the other communities. Outlying values for the intercepts and slopes have the potential to unduly influence summary statistics such as means. Recalculating the mean of the intercepts and the mean of the slopes removing these two communities with extreme intercept and slope resulted in only negligible differences in the mean intercept (5.338 vs. 5.328) and mean slope (0.143 vs. 0.146) with and without the outliers included.

The R_i^2 values for each regression provide information on the closeness of the linear trajectory to the data points observed. A distribution of R_i^2's clustered at high values indicate that the variance in the points observed is well described by the linear trajectory. However, we need to be cautious, in that the R_i^2 values are sometimes affected by outliers and only take account of the linear association. Figure 2.6 presents the histogram of the R_i^2 values for each of the $N = 359$ individual regressions.

Here the mean R^2 value for all 359 regressions is 0.58, indicating that the estimation of an underlying linear time trend on average accounts for a fair amount of the observed variance in ln(crime rate) over time and within community. However, this distribution has considerable spread where many community regression lines have R^2 values of 0.7 or more, reflecting strong linear trends. Others are more moderate. Still others are quite low (< 0.1), indicating that in some communities the linear trend is weak. Additional analysis is needed to better understand why the trajectory model is not a good fit for these individual cases. Visual examination of the plots of those cases with small R^2 values reveals that typically, these communities had one repeated measure with a value that departs from the trend defined

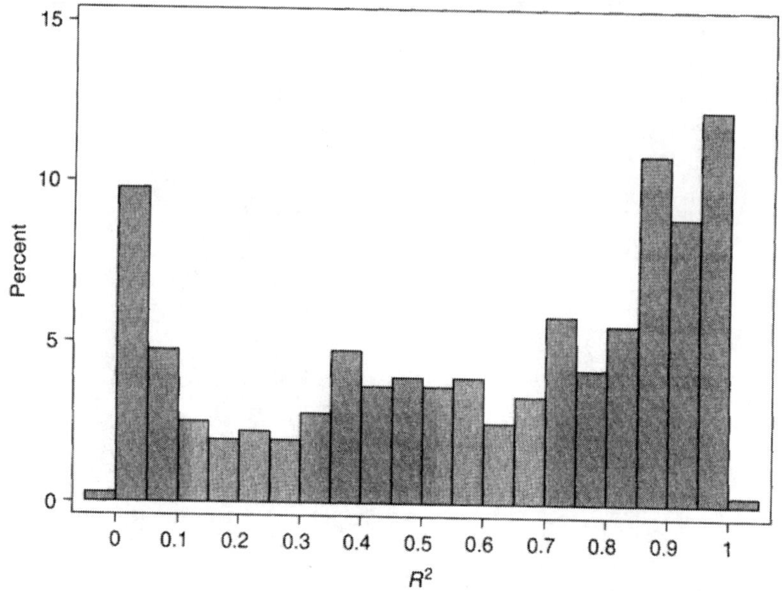

FIGURE 2.6 Histogram of R^2 values for OLS regressions of ln(crime rate) on time ($N = 359$).

by the other three waves of the repeated measure. With so few waves to define the trajectories there is insufficient information to justify modifying the linear trend analysis, but we do note the existence of these outlying ln(crime rate) values for some communities.

Another tool for assessing fit is to plot the trajectory line traced by the mean of the intercepts and the mean of the slopes against the mean value of the y values for each point in time. This reveals whether the mean trajectory line does an adequate job of tracing the means of the observed variables. If it does not, the researcher should consider alternative forms for the trajectory. The previous analysis of the distributions of the intercepts ($\widehat{\alpha}_i$'s) and slopes ($\widehat{\beta}_i$'s) could prove useful if outliers or other odd aspects of the distributions (e.g., excessive skewness) distort the mean values of the parameter estimates. Figure 2.7 superimposes the mean trajectory line on the mean values of each wave of data.

It can be seen that the OLS trajectory fitted to the time-specific means of ln(crime rate) over all 359 cases appears to reflect a good fit to the data means observed. To supplement the mean trajectory plots, the researcher can take a small subsample of cases and examine the fitted regression trajectory and the data points observed for each case. If the sample is small (e.g., $N < 30$), it might be feasible to do this for all cases. But in large samples, a random sample of the full set of cases might be the only feasible alternative. We presented several of these plots above (e.g., the trajectories for $N = 5$ cases and for $N = 50$ cases), so we will not present these again here. However, it is important to realize that visual examination of the fit of the trajectory to the observed repeated measures within each case is a powerful

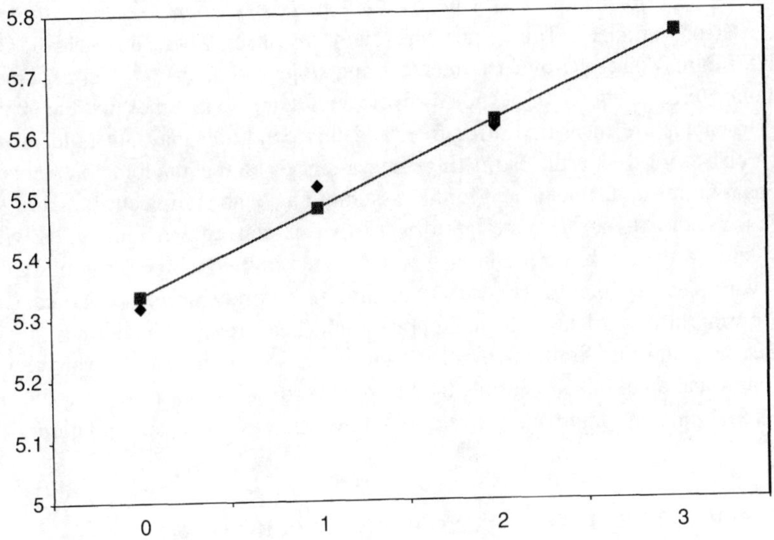

FIGURE 2.7 OLS-fitted trajectory line superimposed on means of ln(crime rate).

method for informally evaluating the adequacy of model fit on a case-by-case basis (Carrig et al., 2004).

2.4.2 Limitations of Case-by-Case Approach

There are a variety of advantages in using the case-by-case approach to estimate trajectories over time. First, it is intuitively appealing and explicates many of the key conditions and assumptions made within the latent curve modeling framework. Predictions of the parameters for individual trajectory estimates are calculated for each case in the sample. Summary statistics can be computed on these estimates and they can be graphically plotted or used in a variety of other analytic frameworks (e.g., as criterion measures in regression analysis, clustering trajectories).

However, there are several limitations of this approach. First, *overall* tests of fit are not readily available for the case-by-case approach. Recall that this method involves estimating a separate regression for each individual in the sample. It is difficult to make overall assessments of model fit based on the combined case-by-case results. Second, when a case-by-case approach uses OLS estimation, it imposes a rather restrictive structure on the error variances. Recall that the OLS estimation for an individual trajectory assumed equal error variances over time within each individual, but that these were free to vary over individual. However, to allow examination of the variances and covariances of the estimates across individuals, we needed to impose the additional restriction that the error variances from each regression were equal over individuals. This assumption allowed for the computation of a single error variance estimate across time and over individuals. This assumption restricts a variety of more complex error structures that might be present. Adjustments would be required to take account of such differences in the error variances. Third, there are many instances where it would be desirable to regress our case-by-case intercept and slope estimates on other explanatory variables (e.g., to address Question 3). However, doing so assumes that the explanatory variables are measured error free, and this condition may not hold in many research settings. Fourth, estimating the variances of the random intercepts and slopes requires adjustments and further assumptions; and significance tests of these variance estimates are less straightforward than is convenient. Finally, as will be seen later in the book, we might also be interested in the incorporation of explanatory variables that themselves vary over time (e.g., time-varying covariates). It is not a straightforward task to incorporate such measures within the case-by-case framework, and this limits many of the questions we might like to evaluate using our empirical data. In sum, despite the many strengths of the case-by-case approach, alternative analytic approaches overcome all of the case-by-case limitations.

2.5 STRUCTURAL EQUATION MODEL APPROACH

In Section 2.4 we pointed out some of the limitations of the case-by-case approach to estimating trajectories over time. In this section we describe how a latent curve model approach can overcome many of these limitations. One of the goals of trajectory modeling is to utilize the set of repeated measures observed to estimate an unobserved trajectory that gave rise to the repeated measures. In this sense, the trajectory is latent in that it was not observed directly, but we infer its existence from the repeated measures observed (see, e.g., Bollen, 2002). As noted in our historical summary of these models, Baker (1954) was first to suggest the use of factor analysis, a latent variable technique, to analyze panel data. Tucker (1958) and Rao (1958) gave a more technical expression of this idea for exploratory factor analysis. Meredith and Tisak (1984, 1990) took this to the confirmatory factor analysis framework and demonstrated that trajectory modeling fit naturally into these types of models.

One of the core concepts of latent variable analysis is that some underlying and unobservable latent variable exists, which is evidenced by the interrelations among observed variables. For example, consider a 10-item scale of depressive feelings in a sample of elderly men and women, and the goal is to examine the factors that predict feelings of depression. Although there are 10 items assessing depressive symptoms, inherent interest is not in these 10 specific items. Rather, interest lies in the ability to infer the existence of an unobserved latent construct of depression from the associations of these 10 items. It is this unobserved latent variable of depression that then becomes the focus of analysis.

Note the close links between the standard latent variable SEM approach and the goals of the trajectory models. With the depression example, the observed variables are secondary to the interest in estimating the unobserved variable that gave rise to these measures. Analogously, in trajectory analysis the set of observed measures taken within an individual over time are secondary to the interest in using these to

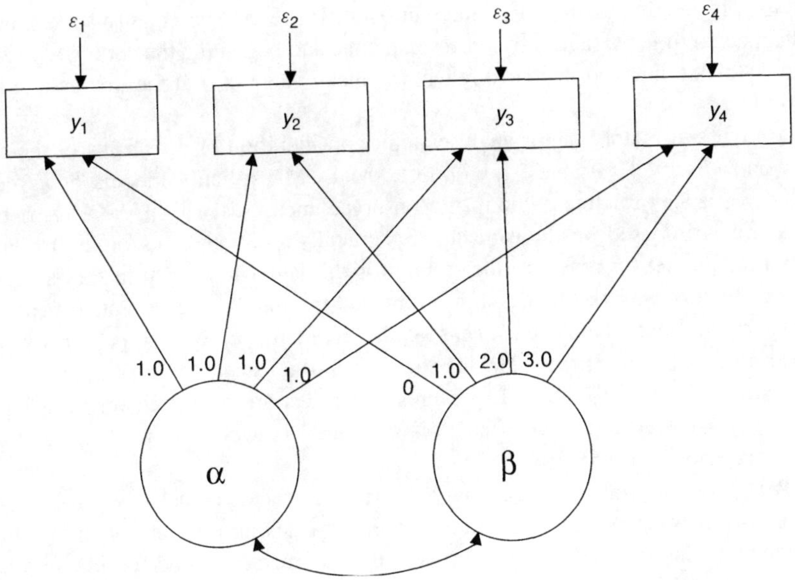

FIGURE 2.8 Unconditional linear latent trajectory model for four time points.

estimate an unobserved trajectory that gives rise to these repeated measures. It is this estimated latent trajectory that becomes the focus of analysis.

The starting point is constructing a two-factor multiple-indicator confirmatory factor analysis (CFA) model (Jöreskog, 1969) within the usual SEM framework (see Figure 2.8). In the figure, latent variables are enclosed in circles or ellipses, observed variables appear in rectangular boxes, and error terms are not enclosed in any circle or box. The straight single-headed arrow signifies the impact of the variable at the base of the arrow on the variable at the head of the arrow. The curved two-headed arrow represents the linear covariance of the two variables connected by the arrow. The path diagram is a *pictorial representation* of the equations of the latent curve model with four repeated measures.

The multiple indicators will be the repeated measures of y for every individual i over all time points t. Two underlying latent factors (sometimes referred to as "growth factors") form the basic linear trajectory model. The first latent factor represents the intercept component of our trajectory, and the second latent factor will represent the slope component of our trajectory. The observed repeated measures are related to the underlying latent factors through the factor loading matrix, $\mathbf{\Lambda}$. Whereas in the usual CFA model, factor loadings, other than the loading for the scaling indicator, are generally estimates, the latent curve model (LCM) fixes these loadings to specific a priori values. For the intercept factor, the loadings from the factor to each of the repeated measures are fixed to values of 1.0. That is, the intercept factor equally influences all repeated measures across all the waves of assessment. For the slope factor, there are a variety of codings for time and we explore some of these options later in the chapter. However, a logical starting point is to fix the loadings to values of $\lambda_t = 0, 1, 2, 3$. The equally spaced units reflect equal time passage between assessments, and beginning the coding with zero allows for the intercept factor to reflect the mean value of y at the first assessment period.

Many confirmatory factor analyses omit consideration of the means of the factors and observed variables. In contrast, the LCM explicitly models *both* mean and covariance structures among the observed measures. However, a restrictive structure is imposed on these means. Specifically, the intercepts of the repeated measures are set to zero, and the means for the latent trajectory factors are estimated. In this way the model-implied mean structure of the repeated measures is determined entirely by the means of the latent trajectory factors. Finally, the covariance structure of the LCM is estimated in the usual way; residual variances are estimated for the observed measures, variances are estimated for each latent trajectory factor, and the covariance is estimated between the intercept and the slope trajectory factors.

This parameterization of the usual confirmatory factor model will provide estimates of all trajectory model parameters. It provides an estimate of the residual variance of the repeated measures; this is the variance in a repeated measure that is not explained by the underlying growth process. It leads to estimates of means of both the intercept and slope factors; these represent the mean starting point and mean rate of change in the repeated measures across all individuals in the sample. Also available are estimates of the variance of the intercept and slope factors, reflecting the degree of variability of the random intercepts and random slopes around their means. Finally, an estimate of the covariance between the initial point (intercept) and the slope of the trajectory is available.

In sum, the traditional confirmatory factor analytic framework, which is a special case of SEM, is exceptionally well suited for the estimation and testing of all of the key components of the latent curve model. Indeed, as demonstrated below, the LCM approach exactly replicates the OLS regression results for the case-by-case approach described above, but the latent curve model can then draw on all of the strengths associated with the general SEM framework (e.g., overall fit measures, covariates with measurement error, alternative estimators). Prior to discussing parameter estimation in greater detail, we next express the standard trajectory model in terms of the general structural equation modeling matrix notation.

2.5.1 Matrix Expression of the Latent Curve Model

Meredith and Tisak (1984, 1990) demonstrated that the random effects trajectory model could be estimated using a traditional structural equation modeling (SEM) framework. A number of authors have expanded on this framework [e.g., McArdle (1988); Browne and du Toit (1991); Muthén (1996)]. We illustrate this matrix representation here.

We will begin by considering a $T \times 1$ vector $\mathbf{y}$ that contains the set of T repeated measures of y for individual i. We can express $\mathbf{y}$ in terms of an underlying confirmatory factor analytic model in which the latent factors represent the latent curve components. In matrix terms, the general expression is

$$\mathbf{y} = \mathbf{\Lambda}\boldsymbol{\eta} + \boldsymbol{\epsilon} \tag{2.36}$$

where $\mathbf{y}$ is a $T \times 1$ vector of repeated measures, $\mathbf{\Lambda}$ is a $T \times m$ matrix of factor loadings, $\boldsymbol{\eta}$ is an $m \times 1$ vector of m latent factors, and $\boldsymbol{\epsilon}$ is a $T \times 1$ vector of residuals. For a simple linear trajectory model fit to T repeated measures, the elements of Eq.(2.36) are

$$\begin{pmatrix} y_{i1} \\ y_{i2} \\ \vdots \\ y_{iT} \end{pmatrix} = \begin{pmatrix} 1 & 0 \\ 1 & 1 \\ \vdots & \vdots \\ 1 & T-1 \end{pmatrix} \begin{pmatrix} \alpha_i \\ \beta_i \end{pmatrix} + \begin{pmatrix} \epsilon_{i1} \\ \epsilon_{i2} \\ \vdots \\ \epsilon_{iT} \end{pmatrix} \tag{2.37}$$

Thus, each repeated observation of y for individual i at time t is a weighted combination of a random intercept and linear slope term plus an individual and time-specific residual. $\boldsymbol{\eta}$ can be expressed in terms of a mean and deviation

$$\boldsymbol{\eta} = \boldsymbol{\mu}_\eta + \boldsymbol{\zeta} \tag{2.38}$$

where $\boldsymbol{\mu}_\eta$ is an $m \times 1$ vector of factor means, and $\boldsymbol{\zeta}$ is an $m \times 1$ vector of residuals. For the linear trajectory model presented in Eq.(2.37), the matrix elements of Eq.(2.38) are

$$\begin{pmatrix} \alpha_i \\ \beta_i \end{pmatrix} = \begin{pmatrix} \mu_\alpha \\ \mu_\beta \end{pmatrix} + \begin{pmatrix} \zeta_{\alpha i} \\ \zeta_{\beta i} \end{pmatrix} \tag{2.39}$$

Equation (2.38) can be substituted into Eq.(2.36) to result in the reduced-form expression of $\mathbf{y}$,

$$\mathbf{y} = \mathbf{\Lambda}\left(\boldsymbol{\mu}_\eta + \boldsymbol{\zeta}\right) + \boldsymbol{\epsilon} \tag{2.40}$$

The model-implied variance of the reduced form is

$$\mathbf{\Sigma} = \mathbf{\Lambda}\mathbf{\Psi}\mathbf{\Lambda}' + \Theta_\epsilon \tag{2.41}$$

where $\mathbf{\Sigma}$ is the covariance matrix of the y's and $\mathbf{\Theta}_\epsilon$ represents the covariance structure of the disturbances for the T repeated measures of y such that

$$\mathbf{\Theta}_\epsilon = \begin{pmatrix} \mathrm{VAR}(\epsilon_1) & 0 & 0 & \cdots & 0 \\ 0 & \mathrm{VAR}(\epsilon_2) & 0 & 0 & 0 \\ 0 & 0 & \ddots & 0 & 0 \\ \vdots & 0 & 0 & \ddots & \vdots \\ 0 & 0 & 0 & \cdots & \mathrm{VAR}(\epsilon_T) \end{pmatrix}$$

Although we assume that the disturbances are uncorrelated, it is possible to permit some correlated errors provided that the model is identified. Further, $\mathbf{\Psi}$ represents the covariance matrix of the equation errors, $\boldsymbol{\zeta}$, among the latent trajectory factors with elements

$$\mathbf{\Psi} = \begin{pmatrix} \psi_{\alpha\alpha} & \psi_{\alpha\beta} \\ \psi_{\beta\alpha} & \psi_{\beta\beta} \end{pmatrix}$$

In the case where there are no predictors of the random intercepts and random slopes, the variance of $\boldsymbol{\eta}$ is equal to the variance of $\boldsymbol{\zeta}$ (i.e., $\mathbf{\Sigma}_{\eta\eta} = \mathbf{\Psi}$). This will not be true when we introduce predictors of the random intercepts and random slopes in Chapter 5.

Finally, the expected value of the reduced-form trajectory model is

$$\mathbf{E}(\mathbf{y}) = \mathbf{\Lambda}\boldsymbol{\mu}_{\eta} \tag{2.42}$$

where $\boldsymbol{\mu}_{\eta}$ represents an $m \times 1$ vector of factor means and $\mathbf{\Lambda}$ is defined as before.

For example, consider the matrix expressions for the hypothetical trajectory model with $T = 4$ repeated observations presented in Figure 2.8. There are a total of four matrices needed to define this unconditional trajectory model in terms of the SEM framework: $\mathbf{\Lambda}$, $\mathbf{\Theta}_{\epsilon}$, $\mathbf{\Psi}$, and $\boldsymbol{\mu}_{\eta}$. For an equally spaced set of $T = 4$ repeated assessments and a linear trajectory, the 4×2 factor loading matrix is

$$\mathbf{\Lambda} = \begin{pmatrix} 1 & 0 \\ 1 & 1 \\ 1 & 2 \\ 1 & 3 \end{pmatrix} \tag{2.43}$$

with a 4×4 diagonal residual matrix

$$\mathbf{\Theta}_{\epsilon} = \begin{pmatrix} \text{VAR}(\epsilon_1) & 0 & 0 & 0 \\ 0 & \text{VAR}(\epsilon_2) & 0 & 0 \\ 0 & 0 & \text{VAR}(\epsilon_3) & 0 \\ 0 & 0 & 0 & \text{VAR}(\epsilon_4) \end{pmatrix} \tag{2.44}$$

and a 2×2 symmetric covariance matrix among the random intercepts and slopes,

$$\mathbf{\Psi} = \begin{pmatrix} \psi_{\alpha\alpha} & \psi_{\beta\alpha} \\ \psi_{\alpha\beta} & \psi_{\beta\beta} \end{pmatrix} \tag{2.45}$$

and finally, a 2×1 vector of factor means,

$$\boldsymbol{\mu}_{\eta} = \begin{pmatrix} \mu_{\alpha} \\ \mu_{\beta} \end{pmatrix} \tag{2.46}$$

Together, these four matrices fully define the linear unconditional trajectory model within the standard SEM framework. We now describe the maximum likelihood estimator, the most common estimator, of the parameters of the LCM.

2.5.2 Maximum Likelihood Estimation

Maximum likelihood (ML) estimation provides a method with which to estimate the parameters of the unconditional LCM in the SEM approach that we have just described. ML is the most popular estimator to use with SEMs and it has a number of useful properties that we describe below. But before we discuss estimation, we emphasize one assumption about the error variances. In the general list of definitions and assumptions in Table 2.1, we have a term $\text{VAR}(\epsilon_{it})$, which is the error variance for the tth time point and ith individual. For the ML estimator, we assume that $\text{VAR}(\epsilon_{it}) = \text{VAR}(\epsilon_t)$ for all i. Note the contrast with the OLS estimator for the case-by-case approach, which assumed that the error variances could differ by individual but for a given individual they were the same over time $[\text{VAR}(\epsilon_{it}) = \text{VAR}(\epsilon_i)]$. In contrast, here the error variances can differ over time, but for each time period the error variance is the same for all individuals.

A helpful starting point to explain the ML estimator is to return to the *implied mean structure* and the *implied covariance structure* from Section 2.3. Taking the expected value of both sides of the level 1 equation, given as

$$y_{it} = \alpha_i + \lambda_t \beta_i + \epsilon_{it} \tag{2.47}$$

implies that the means of the y values (μ_{y_t}) will be

$$\mu_{y_t} = \mu_\alpha + \lambda_t \mu_\beta \tag{2.48}$$

The model-implied variances of the observed variables are

$$\text{VAR}(y_t) = \psi_{\alpha\alpha} + \lambda_t^2 \psi_{\beta\beta} + 2\lambda_t \psi_{\alpha\beta} + \text{VAR}(\epsilon_t) \tag{2.49}$$

where $\psi_{\alpha\alpha} = \text{VAR}(\alpha)$, $\psi_{\beta\beta} = \text{VAR}(\beta)$, and $\psi_{\alpha\beta} = \text{COV}(\alpha, \beta)$. The model implied covariances are

$$\text{COV}(y_t, y_{t+j}) = \psi_{\alpha\alpha} + \lambda_t \lambda_{t+j} \psi_{\beta\beta} + (\lambda_t + \lambda_{t+j}) \psi_{\alpha\beta} \tag{2.50}$$

for $j \neq 0$. The parameters that correspond to the LCM are μ_α, μ_β, λ_t, $\text{VAR}(\epsilon_t)$, $\psi_{\alpha\alpha}$, $\psi_{\beta\beta}$, and $\psi_{\alpha\beta}$. Define $\boldsymbol{\theta}$ to be a vector that contains these model parameters. We can write the mean structure equation for the model in vector notation as

$$\boldsymbol{\mu} = \boldsymbol{\mu}(\boldsymbol{\theta})$$

$$\begin{bmatrix} \mu_{y_1} \\ \mu_{y_2} \\ \vdots \\ \mu_{y_T} \end{bmatrix} = \begin{bmatrix} \mu_\alpha + \lambda_1 \mu_\beta \\ \mu_\alpha + \lambda_2 \mu_\beta \\ \vdots \\ \mu_\alpha + \lambda_T \mu_\beta \end{bmatrix} \tag{2.51}$$

This equation is the *mean structure equation*, where $\boldsymbol{\mu}$ is the vector of means of the y values, and $\boldsymbol{\mu}(\boldsymbol{\theta})$ is the implied means that are a function of the model parameters in $\boldsymbol{\theta}$.

Analogously, the *covariance structure equation* represents the covariances and variances of the observed variables as functions of the parameters in $\boldsymbol{\theta}$:

$$\boldsymbol{\Sigma} = \boldsymbol{\Sigma}(\boldsymbol{\theta}) \tag{2.52}$$

where

$$\boldsymbol{\Sigma} = \begin{bmatrix} \text{VAR}(y_1) & \text{COV}(y_1, y_2) & \cdots & \text{COV}(y_1, y_T) \\ \text{COV}(y_2, y_1) & \text{VAR}(y_2) & \cdots & \text{COV}(y_2, y_T) \\ \vdots & \vdots & \ddots & \vdots \\ \text{COV}(y_T, y_1) & \text{COV}(y_T, y_2) & \cdots & \text{VAR}(y_T) \end{bmatrix} \tag{2.53}$$

and

$$\boldsymbol{\Sigma}(\boldsymbol{\theta}) =$$

$$\begin{bmatrix} \psi_{\alpha\alpha} + \lambda_1^2\psi_{\beta\beta} + 2\lambda_1\psi_{\alpha\beta} + \text{VAR}(\epsilon_1) & \cdots & \psi_{\alpha\alpha} + \lambda_1\lambda_T\psi_{\beta\beta} + (\lambda_1 + \lambda_T)\psi_{\alpha\beta} \\ \psi_{\alpha\alpha} + \lambda_2\lambda_1\psi_{\beta\beta} + (\lambda_2 + \lambda_1)\psi_{\alpha\beta} & \cdots & \psi_{\alpha\alpha} + \lambda_2\lambda_T\psi_{\beta\beta} + (\lambda_2 + \lambda_T)\psi_{\alpha\beta} \\ \vdots & \ddots & \vdots \\ \psi_{\alpha\alpha} + \lambda_T\lambda_1\psi_{\beta\beta} + (\lambda_T + \lambda_1)\psi_{\alpha\beta} & \cdots & \psi_{\alpha\alpha} + \lambda_T^2\psi_{\beta\beta} + 2\lambda_T\psi_{\alpha\beta} + \text{VAR}(\epsilon_T) \end{bmatrix} \tag{2.54}$$

Thus, $\boldsymbol{\Sigma}$ is the population covariance matrix of the observed variables and $\boldsymbol{\Sigma}(\boldsymbol{\theta})$ is the model-implied covariance matrix that is a function of the parameters of the model. The model-implied mean vector $[\boldsymbol{\mu}(\boldsymbol{\theta})]$ and model implied covariance matrix $[\boldsymbol{\Sigma}(\boldsymbol{\theta})]$ give general expressions that demonstrate that the LCM implies that the means, variances, and covariances of the observed variables are functions of the parameters of the model.

To illustrate these implied moment matrices, consider the case of three time points for a linear trajectory model (i.e., $\lambda_t = t - 1; t = 1, 2, 3$). Here the implied mean structure equation $[\boldsymbol{\mu} = \boldsymbol{\mu}(\boldsymbol{\theta})]$ is

$$\begin{bmatrix} \mu_{y_1} \\ \mu_{y_2} \\ \mu_{y_3} \end{bmatrix} = \begin{bmatrix} \mu_\alpha \\ \mu_\alpha + \mu_\beta \\ \mu_\alpha + 2\mu_\beta \end{bmatrix} \tag{2.55}$$

and the implied covariance structure equation $[\boldsymbol{\Sigma} = \boldsymbol{\Sigma}(\boldsymbol{\theta})]$ is

$$\begin{bmatrix} \text{VAR}(y_1) & \text{COV}(y_1, y_2) & \text{COV}(y_1, y_3) \\ \text{COV}(y_2, y_1) & \text{VAR}(y_2) & \text{COV}(y_2, y_3) \\ \text{COV}(y_3, y_1) & \text{COV}(y_3, y_2) & \text{VAR}(y_3) \end{bmatrix}$$

$$= \begin{bmatrix} \psi_{\alpha\alpha} + \text{VAR}(\epsilon_1) & \psi_{\alpha\alpha} + \psi_{\alpha\beta} & \psi_{\alpha\alpha} + 2\psi_{\alpha\beta} \\ \psi_{\alpha\alpha} + \psi_{\alpha\beta} & \begin{matrix} \psi_{\alpha\alpha} + \psi_{\beta\beta} + 2\psi_{\alpha\beta} \\ + \text{VAR}(\epsilon_2) \end{matrix} & \psi_{\alpha\alpha} + 2\psi_{\beta\beta} + 3\psi_{\alpha\beta} \\ \psi_{\alpha\alpha} + 2\psi_{\alpha\beta} & \psi_{\alpha\alpha} + 2\psi_{\beta\beta} + 3\psi_{\alpha\beta} & \begin{matrix} \psi_{\alpha\alpha} + 4\psi_{\beta\beta} + 4\psi_{\alpha\beta} \\ + \text{VAR}(\epsilon_3) \end{matrix} \end{bmatrix} \tag{2.56}$$

Equations (2.55) and (2.56) reveal that the latent trajectory parameter values will determine the means, variances, and covariances of the observed variables. For instance, the (1, 2) element of the implied covariance structure in Eq. (2.56) shows that $\text{COV}(y_1, y_2)$ equals $\psi_{\alpha\alpha} + \psi_{\alpha\beta}$, so the latter parameters will determine the covariance of y_1 and y_2. In a similar fashion we could match up each element of the implied moment matrices to the corresponding element in the mean vector or covariance matrix of the observed variables.

The implied moment matrices, $\boldsymbol{\mu} = \boldsymbol{\mu}(\boldsymbol{\theta})$ and $\boldsymbol{\Sigma} = \boldsymbol{\Sigma}(\boldsymbol{\theta})$, also form the foundation of an approach to estimating the values of the model parameters. The idea that underlies the estimation is to select values for the LCM parameters that will reproduce as closely as possible the means and covariance matrix of the observed variables. The sample means ($\bar{\mathbf{y}}$) and sample covariance matrix ($\mathbf{S}$) serve in place of the unavailable population mean vector ($\boldsymbol{\mu}$) and population covariance matrix ($\boldsymbol{\Sigma}$). Representing the estimated model parameters [i.e., $\widehat{\mu}_\alpha, \widehat{\mu}_\beta, \widehat{\text{VAR}(\epsilon_t)}, \widehat{\psi}_{\alpha\alpha}, \widehat{\psi}_{\beta\beta}$, and $\widehat{\psi}_{\alpha\beta}$] as $\widehat{\boldsymbol{\theta}}$, the goal is to choose values of $\widehat{\boldsymbol{\theta}}$ to make $\boldsymbol{\mu}(\widehat{\boldsymbol{\theta}})$ close to $\bar{\mathbf{y}}$ and $\boldsymbol{\Sigma}(\widehat{\boldsymbol{\theta}})$ close to $\mathbf{S}$.

What is the definition of close? This is where the ML principle comes into play. The classical derivation of the ML estimator assumes that the observed variables come from a multivariate normal distribution (Lawley, 1940; Bollen, 1989b, pp. 131–135). For now, retain this assumption, although alternative assumptions are considered later. The ML fitting function (F_{ML}) is[8]

$$
\begin{aligned}
F_{\text{ML}} &= \ln|\boldsymbol{\Sigma}(\boldsymbol{\theta})| - \ln|\mathbf{S}| + \text{tr}[\boldsymbol{\Sigma}^{-1}(\boldsymbol{\theta})\mathbf{S}] - \mathbf{p} \\
&\quad -[\bar{\mathbf{y}} - \boldsymbol{\mu}(\boldsymbol{\theta})]'\boldsymbol{\Sigma}^{-1}(\boldsymbol{\theta})[\bar{\mathbf{y}} - \boldsymbol{\mu}(\boldsymbol{\theta})]
\end{aligned}
\tag{2.57}
$$

All of the symbols in the fitting function have been defined earlier. F_{ML} has a minimum of zero and no fixed maximum. It approaches its minimum as $\boldsymbol{\Sigma}(\widehat{\boldsymbol{\theta}}) \to \mathbf{S}$ and $\boldsymbol{\mu}(\widehat{\boldsymbol{\theta}}) \to \bar{\mathbf{y}}$, and it is zero when these relations match exactly. One case where $F_{\text{ML}} = 0$ is when the LCM is identified exactly. For these exactly identified models, we are not testing overidentification constraints and we are not testing overall model fit. In overidentified models, it is highly unusual for the fitting function to be exactly zero. An exactly identified LCM would be highly unusual since, as we noted earlier, a two-wave model is generally underidentified, whereas a three-wave model is overidentified.

In general, there are no explicit analytic solutions for $\widehat{\boldsymbol{\theta}}$ written in terms of $\bar{\mathbf{y}}$ and $\mathbf{S}$. Therefore, numerical minimization procedures find those values of $\widehat{\boldsymbol{\theta}}$ that minimize F_{ML}. There are a variety of minimization procedures, and interested readers can refer to SEM software manuals for the specific details on the numerical minimization options available for a specific package and to Bollen (1989b, pp. 136–144) for a simple hypothetical illustration.

As a ML estimator, $\widehat{\boldsymbol{\theta}}$ has a number of desirable properties. The ML $\widehat{\boldsymbol{\theta}}$ is consistent, asymptotically unbiased, asymptotically normal, and asymptotically efficient.

[8]The *maximum* likelihood function here that is typical of SEM is formulated so that this function is *minimized*.

Furthermore, the asymptotic covariance (ACOV) matrix of the parameter estimates $(\widehat{\boldsymbol{\theta}})$ is

$$\text{ACOV}(\widehat{\boldsymbol{\theta}}) = \frac{2}{N-1}\left[E\left(\frac{\partial^2 F_{\text{ML}}}{\partial\boldsymbol{\theta}\partial\boldsymbol{\theta}'}\right)\right]^{-1} \tag{2.58}$$

Thus, tests of statistical significance are available when we substitute the estimated values of $\widehat{\boldsymbol{\theta}}$ into the asymptotic covariance matrix.

Robustness to Nonnormality The ML estimator maintains its desirable asymptotic properties when the observed variables have the same *multivariate kurtosis* as a multivariate normal distribution (Browne, 1984), a condition referred to as *no excess multivariate kurtosis*. Mardia (1970, 1974) provides measures and tests of multivariate kurtosis, and these are implemented in most major SEM software programs so that researchers can check the kurtosis of their data. Fortunately, there are additional conditions under which the ML estimator will be robust even with observed variables from multivariate distributions with excess kurtosis and there are procedures that take account of excess kurtosis when present. Suffice it to say here that the main potential consequence is that asymptotic standard errors and significance test statistics might not be accurate, but that corrected asymptotic standard errors, corrected test statistics, and bootstrap procedures are available that will enable the researcher to overcome these issues. See the appendix to this chapter for more details and references.

2.5.3 Empirical Example

In this section we illustrate the ML estimator with the crime index data and the OLS estimates from the case-by-case approach. Recall that the crime index was measured at four time points on 359 communities. OLS estimation utilized the case-by-case estimation method described above with corrections for the variances of the intercepts and slopes. ML estimation utilized the four-indicator two-factor trajectory model presented in Figure 2.8. Table 2.3 reports the estimates using both of these methods.

The parameter estimates are essentially identical for OLS and ML. The results provide interesting insight into the three questions posed in Chapter 1. Our first question related to the characteristics of the growth trajectory for the overall group. The ML (and OLS) estimate of the mean intercept of the trajectory is $\widehat{\mu}_\alpha = 5.338$, reflecting that, on average, communities have a ln(crime rate) of 5.338 at the first time point in January–February. The estimate of the mean slope of the trajectory is $\widehat{\mu}_\beta = 0.143$, reflecting that, on average, the ln(crime rate) increases 0.143 between each assessment period (where each period is two months).

Recall that Question 2 asks whether there is variability in trajectory estimates across individuals (i.e., is there evidence for interindividual differences in intraindividual change?). The ML provides estimates of the variance of the intercepts $(\widehat{\psi}_{\alpha\alpha} = 0.529)$ and the variance of the slopes $(\widehat{\psi}_{\beta\beta} = 0.020)$. Confidence intervals for each fail to include zero, implying that there is significant variability of

Table 2.3 Point Estimates and (Standard Errors) from Case-by-Case OLS and SEM ML Estimators of Trajectory Models

	Estimator	
Parameter	OLS	ML
$VAR(\epsilon_t)$	0.098 (–)	0.099 (0.005)
μ_α	5.338 (0.041)	5.338 (0.041)
μ_β	0.143 (0.011)	0.143 (0.011)
$\psi_{\alpha\alpha}$	0.529 (–)	0.529 (0.045)
$\psi_{\beta\beta}$	0.020 (–)	0.020 (0.003)
$\psi_{\alpha\beta}$	– (–)	−0.034 (0.009)

both intercept and slopes around their mean values. The significant covariance between the intercept and slope factors ($\widehat{\psi}_{\alpha\beta} = -0.034$) implies that there is a negative association between the ln(crime rate) at the initial time period and the rate of change in crime over subsequent time periods. Finally, the confidence interval for the residual variance estimate ($\widehat{VAR(\epsilon_t)} = 0.099$) does not come close to zero, indicating that there remains unexplained variability in the repeated measures above and beyond that explained by the growth factors.

In sum, these results provide insight into our first two questions. We now have an estimate of the initial starting point and rate of change in ln(crime rate) across all 359 communities. Furthermore, we have evidence to suggest that there is substantial individual variability of trajectories around these group mean values. That is, communities vary in their initial crime index levels and in their rates of change. This finding allows us to move naturally to the third question of interest: Can we incorporate explanatory variables to help understand what type of community starts higher versus lower, and increases more steeply versus less steeply? This is a question to which we return in Chapter 5.

However, prior to moving toward the incorporation of one or more explanatory variables, an important distinction must be drawn between the case-by-case and SEM approaches to estimation. Although there are estimated means, variances, and covariances among the individual trajectories, there is no determination as to the adequacy with which the model successfully reproduces the observed data. It is thus premature to interpret the resulting model estimates substantively without first establishing adequate overall model fit.

As was presented earlier, one of the limitations of OLS estimation is that because the trajectory estimates are gathered together after the estimation of 359 individual regression equations, it is quite difficult to compute a test of overall fit. In contrast, because the LCM approach is embedded within the more general SEM framework, a variety of both global and local tests of fit are available. The ability to assess overall model fit is a feature available with the SEM approach that is not well developed when using the case-by-case OLS trajectory method.

2.5.4 Assessing Model Fit

In the case-by-case approach to analyzing the trajectories, we were able to examine several components of the model fit, such as the distribution of the intercept and slope estimates or the R^2 values. The latent variable approach has component fit measures but also has overall fit measures that assess how well the model corresponds to the data as a whole. The latter are summary measures of fit that are based primarily on quantifying the success of the hypothesized model in reproducing the variances, covariances, and the means of the observed variables in terms of the model parameters. As with any summary measure, these measures lack information on more specific components of the model. The component fit measures take over where the summary measures leave off. They give more specific assessments of the fit for parts of the model rather than for the model as a whole. In this section we discuss both the overall fit and component fit measures for the SEM approach.

Overall Fit Measures There are numerous overall fit indices for these models (see, e.g., Bollen and Long, 1993). Here we present a few of these measures that should prove helpful in LCM fit assessment. More detail and references are provided in the chapter appendix. The most popular is a test statistic defined as

$$T_{\mathrm{ML}} = (N-1)F_{\mathrm{ML}} \tag{2.59}$$

for the ML fitting function estimator. The fitting function is evaluated at the final parameter estimates and then multiplied by the sample size minus one. The simultaneous null hypotheses being tested are the mean and covariance structure hypotheses of

$$H_0 : \boldsymbol{\mu} = \boldsymbol{\mu}(\boldsymbol{\theta}) \text{ and } \boldsymbol{\Sigma} = \boldsymbol{\Sigma}(\boldsymbol{\theta}) \tag{2.60}$$

If the null hypotheses are true and the distributional assumptions justifying the ML estimator holds, the asymptotic distribution of the test statistics are central chi-square variates with degrees of freedom (df) of $\frac{1}{2}T(T+3) - u$, where u is the number of unconstrained parameters estimated in the model. In an exactly identified model where there are as many parameters to estimate as there are means, variances, and covariances of the observed variables, the fit functions and the test statistic will be zero, so the test statistic is of little use. We can test the fit of the model only when the model is overidentified with positive degrees of freedom. In these situations, a statistically significant test statistic suggests that the model specification does not exactly reproduce the means or the covariance matrix of the observed variables.

There are two properties of these test statistics that lead researchers not to rely solely on them to assess overall model fit. One issue is that the observed variables might derive from multivariate distributions that exhibit excess kurtosis, which can lead the test statistic to be too high or too low. Another concerns the power of the chi-square test, especially in large samples. Considering the excess kurtosis issue first, fortunately there are adjustments to the test statistic and bootstrap procedures widely available that will take account of excess kurtosis. If the sample is not too small, these corrective procedures should properly adjust the test statistic so that the impact of kurtosis on the results is minimized (see the chapter appendix).

Statistical power refers to the probability of rejecting a false null hypothesis. In large samples, the chi-square test statistic often leads to rejection of the null hypothesis not due to a low number of cases or excess kurtosis, but rather, due to the *excessive* statistical power that the researcher has to detect even trivial departures from the mean structure or covariance structure hypotheses. For instance, suppose that the covariance structure hypothesis holds but that the mean structure hypothesis is incorrect for one of the variables. That is, the implied structure from the model does not exactly reproduce one of the means of the observed variables and fails to do so by a small amount. In a strict sense the null hypothesis is not true. Although the model is not perfect, it might be quite adequate for many purposes and an improvement over the models proposed in an area. Yet in a sufficiently large sample, the test statistic could be statistically significant despite the substantively small deviation. Large samples provide the statistical power to detect even small errors in specification. The chapter appendix gives a brief description and references on techniques to estimate statistical power in SEMs so that researchers can examine whether statistical power is an issue in a particular application.

Other fit indices do not measure statistical power, but they do provide another means by which to calibrate model fit. The means of the sampling distribution for several of these are relatively free of sample-size effects.[9] There are dozens of fit indices from which to choose. Here we sample several of these, some of which have quite different rationales. The first few we refer to as *baseline fit indices*.

Baseline Fit Indices The fit indices from this family use a baseline model with which to compare the hypothesized model. The baseline model is one that is more restricted than is the hypothesized structure. The general idea is to place the fit of the hypothesized model on a continuum that ranges from the low point of the chi-square test statistic of the baseline model to that of a saturated model that has a zero chi-square test statistic. The most common baseline model is one where the variances of the observed variables are free parameters and the covariances of the variables are zero. There is less consensus on what the mean structure should be for the baseline model, but we use a baseline where the means are free parameters. Other baselines could be more relevant in substantive areas where there is an acceptable alternative, more restrictive model that is relevant to current debates. However, in many fields the model with zero covariances and free variances and means is a suitable baseline.

Other than the chi-square test statistic, the *Tucker–Lewis index* (Tucker and Lewis, 1973) is probably the oldest fit index in structural equation modeling. Using

[9]Sample-size effects are sometimes used in at least two different ways in the SEM literature (Bollen, 1990). Here we refer to a relative lack of association between the mean of the sampling distribution of a fit index and N.

T_b to represent the test statistic for the baseline model, T_h for the hypothesized model, df_b for the degrees of freedom for the baseline model, and df_h for the degrees of freedom of the hypothesized model, the Tucker–Lewis index (TLI) is

$$\mathrm{TLI} = \frac{T_b/\mathrm{df}_b - T_h/\mathrm{df}_h}{T_b/\mathrm{df}_b - 1} \tag{2.61}$$

The TLI generally ranges between zero and 1, although it is possible to fall outside these bounds. A value of 1 is an ideal fit. Values much lower than 1 (e.g., < 0.9) raise concerns about the adequacy of a model. Values much greater than 1 (e.g., 1.2) suggest the possibility of overfitting the data or having a model with too many parameters, with some of the parameters capitalizing on chance variations in the data. The TLI takes account of the df values for the models by dividing each test statistic by the model degrees of freedom before subtracting. So the comparison is of the fit per degree of freedom for the baseline and hypothesized models. Among the desirable properties of the TLI is that the means of its sampling distributions tend to be the same for the same model across different sample sizes. A less desirable characteristic is that it tends to have a larger variance than do other fit indices with a similar range (Anderson and Gerbing, 1984; Bentler, 1990).

The *incremental fit index* (IFI) is another baseline fit index (Bollen, 1989a). Its formula is

$$\mathrm{IFI} = \frac{T_b - T_h}{T_b - \mathrm{df}_h} \tag{2.62}$$

where no new symbols are introduced. Like the TLI, the IFI generally ranges between zero and 1, where 1 is an ideal fit. It is possible to exceed 1, but values much larger than 1 suggest overfitting the data. Generally, values less than 0.9 are unacceptable. Other things being equal, a hypothesized model that uses up more degrees of freedom will lead to a bigger denominator and a smaller IFI than will another model that has the same test statistic but more degrees of freedom. In this sense there is a penalty for using up degrees of freedom. Monte Carlo evidence suggests that the means of the sampling distributions of IFI tends to be relatively stable across different sample sizes (Bentler, 1990; Gerbing and Anderson, 1993).

The *relative noncentrality index* (RNI) is a third baseline fit index that appears to behave very similar to the IFI in practice (Bentler, 1990; McDonald and Marsh, 1990). Its formula is

$$\mathrm{RNI} = \frac{(T_b - \mathrm{df}_b) - (T_h - \mathrm{df}_h)}{T_b - \mathrm{df}_b} \tag{2.63}$$

Bentler's (1990) *comparative fit index* (CFI) is the same as the RNI except that it caps the maximum to 1 by resetting to 1 any values greater than 1. Bentler (1990) shows that the IFI, RNI, and CFI have the same probability limits and hence converge to the same value as the sample size grows larger. Our experience has been that the values of these indices are often quite close in practice.

Stand-Alone Indices The *root-mean-square error of approximation* (RMSEA) is a stand-alone index of fit; that is, there is no baseline model comparison. Steiger and Lind (1980) proposed the RMSEA as a measure of fit that builds on the noncentral chi-square distribution. The noncentral chi-square distribution is the asymptotic distribution of the test statistic when H_0 is invalid and the degree of misspecification is not too severe (Browne and Cudeck, 1993). The RMSEA formula is

$$\text{RMSEA} = \sqrt{\frac{T_h - \text{df}_h}{(N-1)\text{df}_h}} \tag{2.64}$$

The numerator of the RMSEA under the square-root sign, $T_h - \text{df}_h$, is an asymptotically unbiased estimator of the noncentrality parameter for the noncentral chi-square distribution underlying T_h. The division by the number of cases minus 1 $(N - 1)$ is an adjustment to take account of the sample-size effect on the noncentrality parameter. The additional degrees of freedom term (df_h) provides a penalty for using up model degrees of freedom. The RMSEA has a minimum of zero (and is fixed to zero if the numerator is negative) and no upper limit. The closer to zero it is, the better is the model fit. Another advantage of the RMSEA is that confidence intervals are available, so that upper and lower values of the interval are available.

Steiger (1989) and Browne and Cudeck (1993) suggest guidelines such that values of less than 0.05 indicate a very good fit, those greater than 0.10 represent a poor fit, and those values in between, a moderate fit. The fact that lower values indicate better fit runs counter to the baseline fit indices that we described earlier.[10] We would caution the reader that any such guidelines are somewhat arbitrary and that a full assessment of fit requires more than calculating the values of the fit indices (see, e.g., Bollen and Long, 1993a).

Two information theory–based measures of model fit are the Akaike information criterion (Akaike, 1973, 1987),

$$\text{AIC} = T_h + 2u \tag{2.66}$$

and the Bayesian information criterion (Schwarz, 1978; Raftery, 1993),

$$\text{BIC} = T_h - u\ln(N) \tag{2.67}$$

where u is the number of "free" parameters estimated in the model and T is the number of waves of data.[11] For both of these indices, smaller values signify better fit than that of higher values.

[10] A simple recoding of the RMSEA puts it on a very similar scale (Bollen, 1999),

$$1 - \text{RMSEA} \tag{2.65}$$

Staying with the Steiger–Browne–Cudeck guidelines, values of 1 – RMSEA of less than 0.9 suggest poor fit, and values greater than 0.95 signify very good fit. Coincidentally, these guidelines are similar to those sometimes used for the baseline fit indices.

[11] These information-based measures are sometimes defined differently in different sources. The different formulas do not alter the ordering of the fit of the models.

Overall Fit of Crime Model Consider the overall fit of the four-indicator two-factor LCM presented in Figure 2.8 using maximum likelihood estimation. Recall that six parameters are estimated in this model: two means (of the intercept and the slope factors), three variances (of the intercept factor, the slope factor, and a single estimate of the residual variance of the four repeated measures over time), and one covariance (between the intercept and the slope factors). The overall fit of the model using the likelihood ratio statistic T_{ML} (or "chi-square" test) is $T_{ML} = 16.20$ with df $= 8$ and $p = 0.0396$. This overall fit test slightly misses the usual p-value cutoff of 0.05. Recall that the null hypothesis is that the population observed moment structure is equal to the implied moment structure. The chi-square test is significant at the 0.05 level, so that, strictly speaking, we would reject the null hypothesis. However, given the several hundred cases in the sample and the other factors that might influence the chi-square test, we should examine the other fit indices as well.

As described earlier, there are numerous alternative measures of overall fit. The TLI, RNI, and IFI are all 0.999. All three of these baseline fit indices are excellent. Further, RMSEA is 0.054 with a 90% confidence interval of 0.011 to 0.091. This value straddles the recommended 0.05 value for the RMSEA. The AIC and BIC measures are less helpful here since we are not comparing multiple models.

Taken together, although the chi-square likelihood ratio test statistic is marginally significant, the overall fit indices strongly suggest that a linear trajectory model fits the observed mean and covariance structure of the repeated crime index data well. Most researchers would judge the model to be a reasonable approximation to the data as far as the overall fit is of concern. It is important to note that these overall measures of fit are not readily available for the OLS case-by-case regression models, and for some measures are not available at all. The availability of these measures is an advantage of the SEM approach to trajectory modeling compared to the case-by-case approach.

2.5.5 Components of Fit

It is possible to have a model that has good overall fit on several or all of the overall fit measures, but that has poor fit in terms of its components (or vice versa). As such, we would not want to keep the model despite its favorable performance on the overall fit indices. Component fit measures refer to a broad class of diagnostics that examine the fit of pieces of the model in place of a single summary statistic for the entire model. These tend to be more diverse than the overall fit measures, and unlike the overall fit measures, they are available for the case-by-case regression estimator as well as for the ML estimator.

A starting point for the component fit for the ML estimator is to examine the parameter estimates and their asymptotic standard errors. In the linear trajectory models we have considered so far, the only estimated parameters are the variances, the covariance, the means of the random intercepts and slopes, and the error variances. The researcher should check all variance estimates for "improper solutions," that is, negative error variances. Although such values are impossible in the population, it is possible to estimate variances to be negative. Similarly, it is possible that converting the covariance matrix of the random intercept and random slope to a correlation matrix would reveal a correlation of absolute magnitude greater than one, a value impossible in the population. These improper solutions bear close examination. They have several possible causes, including sampling fluctuations, nonconvergence, outliers, model misspecification, and empirical underidentification.

Chen et al. (2001) suggest several approximate significance tests to determine whether the negative estimates are within sampling fluctuations of zero. The simplest approximate test for a single error variance is to use the usual z-test of the null hypothesis that the error variance is zero. If there are two or more error variances to test, Chen et al. (2001) suggest that a likelihood ratio, Lagrangian multiplier, or Wald are approximate tests. We will have more to say about the latter three tests shortly, but suffice it to say here that they permit the simultaneous testing of more than one hypothesis. Strictly speaking, the negative error variances do not lead to proper ML estimates, and the significance tests must be regarded with caution and treated as heuristic tools.[12] Fortunately, the crime index data example had no negative error variance estimates.

Beyond screening for improper solutions, it is important to examine the magnitude and statistical significance of the variances and covariance. Two key variance estimates are those for the random intercept ($\widehat{\psi}_{\alpha\alpha}$) and random slope ($\widehat{\psi}_{\beta\beta}$). The variances around these trajectory parameters give the variability in intercepts and slopes for the individuals in the sample. Confidence intervals around these estimates reveal whether they include zero. If a confidence interval includes zero, it suggests that the variance around the mean is not significant and that the group has the same intercept or slope of the trajectory. Alternatively, confidence intervals that exclude zero indicate significant variability in the intercepts or slopes. As was presented in Table 2.2, the ML estimates for the intercept and slope variances were $\widehat{\psi}_{\alpha\alpha} = 0.529$ and $\widehat{\psi}_{\beta\beta} = 0.020$, respectively. Given that the confidence intervals of both estimates do not include zero, we conclude that there was significant variability in both the intercept and slope components of the trajectory.

The estimate of the error variances for each wave of data [e.g., $\text{var}(\epsilon_1)$, $\text{var}(\epsilon_2)$, $\text{var}(\epsilon_3)$, ..., $\text{var}(\epsilon_T)$] also have asymptotic standard errors that permit the formation of confidence intervals. Beyond checking whether any improper estimates occur, most interest in the error variances lies in what they imply about the $R^2_{y_t}$ for each wave of data. These $R^2_{y_t}$ values indicate the variance in the observed variable explained by the random intercept and slopes. Low values suggest that the trajectories have a weak impact on the panel data, whereas large values suggest that the trajectories are a good prediction of the values of y_t over time. The $R^2_{y_t}$ values for the crime data within each time point range from 83% to 84%, indicating that the underlying latent trajectory factors account for a substantial proportion of the observed variability in ln(crime rate). One difference from the case-by-case approach is that the R^2 values we get with the latent variable approach are the same for each case in a given time period, although the R^2 values can differ over

[12]The problem is that a zero error variance is a boundary solution where the usual significance tests are often not exact, even asymptotically.

time. In the case-by-case approach we had different R^2 values for different cases, but R^2 value stayed the same for the same case over time.

Residual moments provide another statistic with which to assess the model's component fit. The moments that the ML fitting function tries to fit are the means, variances, and covariances of the observed variables. The chi-square test statistic described above gives an overall assessment of the degree to which the model-implied moments fit the moments of the observed variables. However the residual moments can highlight elements of these means, variances, and covariances that are poorly predicted by a model structure. If $\bar{\mathbf{y}}$ is the vector of the sample means of the observed variables and $\boldsymbol{\mu}(\widehat{\boldsymbol{\theta}})$ is the corresponding vector of model-implied means at the final estimates, the residual mean vector is simply

$$\bar{\mathbf{y}} - \boldsymbol{\mu}(\widehat{\boldsymbol{\theta}}) \tag{2.68}$$

Positive values in the residual mean vector correspond to means that the LCM underpredicts; negative residuals occur when the model-implied means are overpredicted. A common diagnostic that we used for the case-by-case regression method is to plot the observed means against the model-implied means over time. The residual difference between the observed mean and the mean reproduced at each of the four time periods is -0.019, 0.035, -0.012, and -0.003, respectively. It can be seen that the linear trajectory is overestimating the means at the first, third, and last time points but is underestimating these means at the second time point.

In an analogous fashion, the residual covariance matrix is

$$\mathbf{S} - \boldsymbol{\Sigma}(\widehat{\boldsymbol{\theta}}) \tag{2.69}$$

Although these are variance and covariance residuals, positive and negative values have interpretations similar to the mean values where the observed was under- or overpredicted by the model. The panel nature of these data permits the creation of some plots that might be useful in the diagnostics of model fit. For instance, order the sample variances from the earliest to the latest in time and plot both the sample variance and the model-implied variances at $\widehat{\boldsymbol{\theta}}$. This plot will then reveal how well the model predicts the variances. In this case, and in others that use the ML estimator, it is typical that these variances are predicted perfectly. The main situation where this is not true are those in which parameter constraints on, for example, the error variances prevent the model from exactly reproducing the sample variance. More informative are covariance plots where the sample and model-implied covariances are plotted. A straight 45° angle line indicates a perfect match of the covariances observed and those implied. Figure 2.9 is the plot for the crime example.

Figure 2.9 reflects the fact that although the covariances observed and those reproduced among the repeated measures of ln(crime rate) are similar in value, they are clearly not exact. The graph reveals that the model-implied values for some covariances underestimate the observed counterparts [e.g., $s_{14} = 0.436$, $\sigma_{14}(\widehat{\theta}) = 0.424$]; other model-implied values overestimate the sample counterparts [e.g., $s_{24} = 0.427$, $\sigma_{24}(\widehat{\theta}) = 0.449$]. Deviations for two of the covariances are notable, but there is no obvious modification to the model that would minimize these.

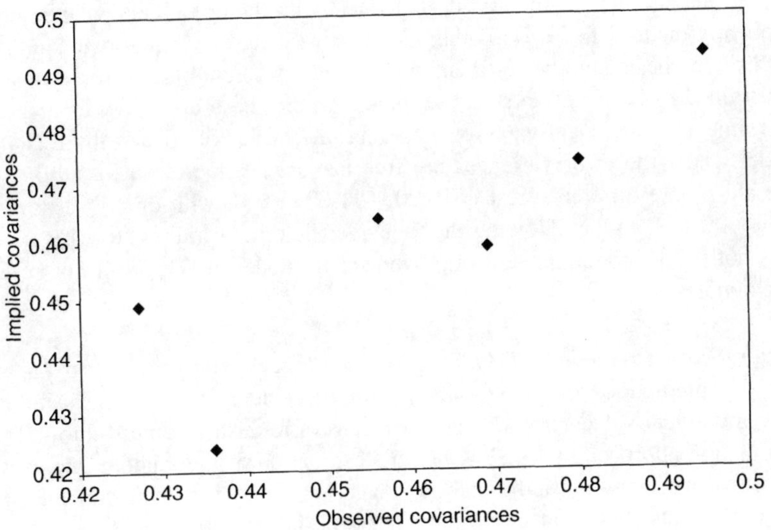

FIGURE 2.9 Scatter plot of observed and model-implied covariances.

Significance Tests for Component Fit Testing the statistical significance of individual parameters is straightforward with the ML estimator. It is also possible to test hypotheses about two or more parameters with these estimators. For instance, a researcher may want to test whether the error variances for a variable are equal over four waves of data. In this and other situations, we can use a *likelihood ratio* (LR) *test*, also referred to as the *chi-square difference test*.[13] The LR test requires that the parameters of one model be *nested* in the parameters of a second model. *Nested* means that the parameters of the nested model are a restrictive form of the parameters of the second model (see, e.g., Bollen, 1989a, pp. 289–296). An example is two models that are identical except that in the nested model, several parameters are set to zero, whereas they are free parameters in the other model. Or, two models might only differ in that equality constraints on two or more parameters hold in one model but not the other.

Suppose that T_r is the usual chi-square test statistic with degrees of freedom df_r from the nested, more restricted model. T_u is the chi-square test statistic of the less restrictive model with degrees of freedom df_u. The more restricted model will generally have a larger test statistic and it can be no smaller than the test statistic for the less restrictive model. A new chi-square test statistic is formed by taking the difference, $T_r - T_u$, and forming degrees of freedom as $\text{df}_r - \text{df}_u$. The null hypothesis for the new chi-square test is that the restrictions that distinguish the restrictive model from the less restrictive one are true. Rejection of the null hypothesis implies that at least one of the restrictions is false.

[13]Closely related to the likelihood ratio test are the Wald and Lagrangian multiplier tests for hypotheses that involve one or more parameters. We do not discuss them here, but see Bollen (1989a, pp. 298–303) for an overview of these tests in SEMs.

We illustrate the chi-square difference test with the crime data. One assumption frequently made with LCMs is that the error variances are equal over time. A test of this hypothesis involves estimating one model where all four error variances are constrained to be equal and another model where these are freely estimated. For the crime data, the equal-variance model has a T_r value of 16.20 with $\mathrm{df}_r = 8$. The model where the error variances are free has a T_u value of 11.02 with $\mathrm{df}_u = 5$. The chi-square difference test is $16.20 - 11.02 = 5.18$ with $\mathrm{df} = 3$. This has a p value of 0.159, so that allowing the time-specific error variances to vary over time does not result in a significant improvement in model fit. This will not be true in all examples.

Factor Score Prediction of Intercepts and Slopes The case-by-case approach provides predictions of the α_i and β_i for individual cases in the sample, and these individual values provide another means of assessing component fit. The full information approach using SEM permits estimation of the random intercepts and random slopes with an additional step. The key idea is that the full information approach treats the latent curve parameters as latent variables or factors from a factor analysis. There are established techniques for predicting the values of the latent variables (or factors) from such an analysis. Under the rubric of factor score prediction are several different methods for predicting the individual values of the latent variables for each case. Below we represent three such methods:

$$\widehat{\eta}_{\mathrm{FR}i} = \boldsymbol{\Psi}\boldsymbol{\Lambda}'\boldsymbol{\Sigma}_{yy}^{-1}(\boldsymbol{\theta})[\mathbf{y}_i - \boldsymbol{\mu}_y(\boldsymbol{\theta})] + \boldsymbol{\mu}_\eta \tag{2.70}$$

$$\widehat{\eta}_{\mathrm{GLS}i} = (\boldsymbol{\Lambda}'\boldsymbol{\Theta}_\epsilon^{-1}\boldsymbol{\Lambda})^{-1}\boldsymbol{\Lambda}'\boldsymbol{\Theta}_\epsilon^{-1}[\mathbf{y}_i - \boldsymbol{\mu}_y(\boldsymbol{\theta})] + \boldsymbol{\mu}_\eta \tag{2.71}$$

$$\widehat{\eta}_{\mathrm{CC}i} = [\boldsymbol{\Psi}^{1/2}\boldsymbol{\Lambda}'\boldsymbol{\Theta}_\epsilon^{-1}\boldsymbol{\Sigma}_{yy}^{-1}(\boldsymbol{\theta})\boldsymbol{\Theta}_\epsilon^{-1}\boldsymbol{\Lambda}\boldsymbol{\Psi}^{1/2}]^{-1/2}\boldsymbol{\Psi}^{1/2}\boldsymbol{\Lambda}'\boldsymbol{\Theta}_\epsilon^{-1}[\mathbf{y}_i - \boldsymbol{\mu}_y(\boldsymbol{\theta})] + \boldsymbol{\mu}_\eta \tag{2.72}$$

Equation (2.70) is the factor regression method (Thomson, 1939), perhaps the most common method to predict factor scores. Equation (2.71) provides the generalized least-squares method or Bartlett (1937) scores, and Eq. (2.72) is the constrained covariance method (Anderson and Rubin, 1956). The usual formulas for these factor score predictions assume that all variables are deviated from the means. We use formulas from Biesanz and Bollen (2003) that permit means and raw scores, since we use the means in LCMs.

We can apply any of these formulas to our data in an unconditional LCM to create predictions of α_i and β_i for all cases in the sample. These values will not be identical across methods, but they will generally be highly correlated. To illustrate these further, consider the most common factor score prediction method, the factor regression method. In the case of an unconditional model with four repeated measures such as our ln(crime) data, we can write Eq.(2.70) as

$$\widehat{\eta}_{\mathrm{FR}i} = \begin{bmatrix} \mathrm{VAR}(\alpha) & \mathrm{COV}(\alpha,\beta) \\ \mathrm{COV}(\beta,\alpha) & \mathrm{VAR}(\beta) \end{bmatrix} \begin{bmatrix} 1 & 1 & 1 & 1 \\ 0 & 1 & 2 & 3 \end{bmatrix}$$

$$\times \begin{bmatrix} \mathrm{VAR}(y_1) & \mathrm{COV}(y_1, y_2) & \mathrm{COV}(y_1, y_3) & \mathrm{COV}(y_1, y_4) \\ \mathrm{COV}(y_2, y_1) & \mathrm{VAR}(y_2) & \mathrm{COV}(y_2, y_3) & \mathrm{COV}(y_2, y_4) \\ \mathrm{COV}(y_3, y_1) & \mathrm{COV}(y_3, y_2) & \mathrm{VAR}(y_3) & \mathrm{COV}(y_3, y_4) \\ \mathrm{COV}(y_4, y_1) & \mathrm{COV}(y_4, y_2) & \mathrm{COV}(y_4, y_3) & \mathrm{VAR}(y_4) \end{bmatrix}^{-1}$$

$$\times \begin{bmatrix} y_{i1} - \mu_{y_1}(\boldsymbol{\theta}) \\ y_{i2} - \mu_{y_2}(\boldsymbol{\theta}) \\ y_{i3} - \mu_{y_3}(\boldsymbol{\theta}) \\ y_{i4} - \mu_{y_4}(\boldsymbol{\theta}) \end{bmatrix} + \begin{bmatrix} \mu_\alpha \\ \mu_\beta \end{bmatrix} \tag{2.73}$$

Using this expression provides predicted values of α_i and β_i for each case. To implement it, we would need to use the sample estimates of each of the quantities in it where the variances and covariances of the y values are replaced by the sample estimates of the elements of $\boldsymbol{\Sigma}(\boldsymbol{\theta})$ that correspond to appropriate variances or covariances, and the other terms have already been defined.

In the case of our ln(crime rate) data empirical example, the sample estimate of the factor regression score for a case is

$$\widehat{\eta}_{\mathrm{FR}i} = \begin{bmatrix} 0.5274 & -0.034 \\ -0.034 & 0.0196 \end{bmatrix} \begin{bmatrix} 1 & 1 & 1 & 1 \\ 0 & 1 & 2 & 3 \end{bmatrix} \tag{2.74}$$

$$\times \begin{bmatrix} 0.626 & 0.493 & 0.459 & 0.424 \\ 0.493 & 0.577 & 0.464 & 0.449 \\ 0.459 & 0.464 & 0.567 & 0.474 \\ 0.424 & 0.449 & 0.474 & 0.596 \end{bmatrix}^{-1} \begin{bmatrix} y_{i1} - 5.3187 \\ y_{i2} - 5.5157 \\ y_{i3} - 5.6109 \\ y_{i4} - 5.7626 \end{bmatrix}$$

$$+ \begin{bmatrix} 5.3379 \\ 0.1427 \end{bmatrix} \tag{2.75}$$

where the researcher would insert the values for y_{i1} to y_{i4} for the ith case, and would do the same for all cases in the sample. For instance, the first case's repeated measures are 4.009, 4.009, 5.786, and 4.693. This leads to $\widehat{\alpha}_1$ and $\widehat{\beta}_1$ of 4.267 and 0.266. Using the values of the repeated measures for the other cases, the remaining random intercepts and random slopes could be predicted. The use of these predictions of intercepts and slopes is the same as that for the case-by-case estimates of these values. Keep in mind that these are *predictions* of the random intercepts and random slopes and are not the same as the actual values. In fact, different factor score methods will lead to correlated but different estimates of these latent variables.

Factor indeterminacy is an issue, widely discussed in the factor analysis literature, that points to the inability to obtain the exact values of the latent factors (e.g., McDonald and Mulaik, 1979). Furthermore, the OLS estimates in the case-by-case analysis provide different predictions of the random intercepts and slopes. For example, the OLS sample estimates of the intercept and slope are $\widehat{\alpha}_1 = 4.050$ and $\widehat{\beta}_1 = 0.383$ for the first case, values that contrast with the factor regression predictions of 4.267 and 0.266, respectively. Although 95% confidence intervals around the OLS estimates would include the factor regression predicted values, it is clear that different methods of predicting a case's intercept and slope can lead to different values. Predicted random intercepts and random slopes for cases are not identical to the latent variable values. Despite this limitation, the factor score predictions can provide useful diagnostics in the same way that the OLS estimates do.

2.6 ALTERNATIVE APPROACHES TO THE SEM

So far we have described the case-by-case and the structural equation model approaches to estimating LCMs. This does not exhaust the approaches to these models, nor is LCMs the only term used to refer to these models. For instance, consider Raudenbush's (2001, pp. 38–39) model:

$$y_{ti} = \pi_{0i} + \pi_{1i}a_{ti} + e_{ti} \qquad e_{ti} \sim N(0, \sigma^2) \tag{2.76}$$

$$\begin{aligned} \pi_{0i} &= \beta_{00} + u_{0i} \\ \pi_{1i} &= \beta_{01} + u_{1i} \end{aligned} \tag{2.77}$$

$$\begin{pmatrix} u_{0i} \\ u_{1i} \end{pmatrix} \sim N\left[\begin{pmatrix} 0 \\ 0 \end{pmatrix}, \begin{pmatrix} \tau_{00} & \tau_{01} \\ \tau_{10} & \tau_{11} \end{pmatrix}\right] \tag{2.78}$$

With the exception of having the same symbol for the repeated measure, the notation here departs from ours. However, if we set π_{0i} to α_i; π_{1i} to β_i; a_{ti} to λ_{ti}; e_{ti} to ϵ_{it}; β_{00} to μ_α; β_{01} to μ_β; u_{0i} to $\zeta_{\alpha i}$; u_{1i} to $\zeta_{\beta i}$; and τ_{00}, τ_{11}, and τ_{10} to $\psi_{\alpha\alpha}$, $\psi_{\beta\beta}$, and $\psi_{\alpha\beta}$, respectively, we will see the match of this model to the model of this chapter.[14] One small difference is that we list the case index first and the time index second on our variables with double subscripts. Another difference is that we do not generally assume that the disturbances from the level 1 and level 2 models are always normal and homoscedastic (the same variance) across all waves for the level 1 disturbance. With regard to the normality assumption, we rely on the SEM literature that shows that less restrictive assumptions justify the ML estimator and that we have corrections for the significance tests when excess kurtosis is present (see the text and the chapter appendix).

Leaving these points aside, the LCM is essentially the same as what is referred to as the growth curve model, multilevel model, or hierarchical linear model for longitudinal data. There are some differences in the ease with which researchers can estimate this model with a SEM approach versus a multilevel model. For example, at this point in time the SEM approach is more adept at handling measurement error in a variety of forms, providing alternative estimators and fit indices for continuous, dichotomous, or ordinal repeated measures. A multilevel approach is easier to use when each case is observed at different time points, if the repeated measure is a count variable, or if there are three or more levels of analysis. However, the basic model is the same across these approaches and we anticipate that the differences in features will diminish as the SEM and multilevel approaches are further merged and

[14]Thus far we have only considered LCMs with λ_t. In Chapter 3 we extend these models to include λ_{ti}.

developed (see, e.g., Willett and Sayer, 1994; MacCallum et al., 1997; Bauer, 2003; Curran, 2003). As we noted earlier, our emphasis will be on the SEM approach to LCMs, although we recognize the valuable contributions of alternative approaches.

2.7 CONCLUSIONS

This chapter has covered a great deal of ground. We began by presenting the fitting of a regression line, or trajectory, to a set of repeated measures on a single variable for a single case over time. Using the case-by-case approach and OLS regression, it was extended to the estimation of trajectories for each individual in the sample. Combining these enabled computation of a mean of the individual intercepts and a mean of the individual slopes, and these were unbiased estimates of the corresponding population means. However, although the variances of the individual intercepts and slopes were biased estimates of the true intercept and slope variances, this bias could be corrected with additional calculations. The SEM framework enabled the parameterization of the trajectory model as a confirmatory factor analysis model. We estimated this model using ML estimation methods, and discussed a variety of issues in omnibus and component model fit.

One aspect that was constant throughout this chapter was the sole focus on a *linear* trajectory model. That is, each trajectory was defined by an intercept and a linear slope, and thus the model-implied change was constant between any two equally spaced assessments. However, there are many situations in which we may want to consider more complex nonlinear trajectory functions over time. We turn to this topic in Chapter 4. Two other issues were choosing the metric of time for the repeated measures and handling missing data. Both of these issues are addressed in Chapter 3.

APPENDIX 2A: Test Statistics, Nonnormality, and Statistical Power

The classic derivation of the ML estimator assumed that the observed variables came from a multivariate normal distribution. However, having variables that come from multivariate normal distributions is unlikely. Fortunately, there are many situations where the ML estimator is appropriate even when the observed variables derive from nonnormal distributions. First, Browne (1984) proved that as long as the observed variables come from a distribution that has the *same multivariate kurtosis* as a normal distribution, the ML estimator will retain many of its desirable properties. This means that it is possible to have multivariate distributions with skewness as long as the multivariate kurtosis is close to that of a multivariate normal distribution. In many cases of multivariate skewness there is excess kurtosis, so we cannot routinely make use of this desirable situation.

Amemiya and Anderson (1990), Browne (1987), Shapiro (1987), Satorra (1990), and others have discovered conditions under which the ML estimator's significance tests are asymptotically valid for at least some of the parameter estimates. In the context of the unconditional LCM, the test statistic would be asymptotically robust when the disturbances (ϵ_{it}'s) are distributed independent of the latent α and β random coefficients. The independence assumption is stronger than the uncorrelated assumption that was presented in Table 2.1. For example, the ϵ_{it} values could be uncorrelated with α and β even if the variances of ϵ_{it} were not constant across cases but were related to the magnitudes of α and β. But this same condition would violate the independence condition that we described as necessary for a robust test statistic. Under the independence assumption, the asymptotic standard errors and significance tests are asymptotically correct for any parameter that corresponds to a normally distributed variable. For instance, if the latent random α and β are normally distributed, the asymptotic standard errors for their variances will be accurate in large samples and suitable for significance testing even if the disturbances (ϵ_{it}'s) are from nonnormal distributions. However, we could not depend on the asymptotic standard errors and usual significance tests for the variances and covariances of nonnormally distributed disturbances.

Although these asymptotic robustness conditions for the ML estimator are impressive, their main drawback is that we cannot easily determine whether disturbances are independent of the random coefficients. Fortunately, there are at least two alternatives. One is that several researchers have shown how to adjust the asymptotic standard errors from the ML estimator to take account of nonnormality (e.g., Browne, 1984, p. 67; Arminger and Schoenberg, 1989). These corrected standard errors permit large sample tests of significance that are less influenced by excessive kurtosis than are the uncorrected ones. Another alternative is to use resampling bootstrap techniques to form empirically simulated standard errors or to form confidence intervals for ML parameter estimates that are less influenced by nonnormality (Bollen and Stine, 1990). Though these procedures also rely on asymptotic properties, the bootstrapping procedures often attain their asymptotic properties at more modest sample sizes than other procedures. Similarly, there are corrected versions of the chi-square test statistic (Satorra and Bentler, 1994) and bootstrap procedures (Bollen and Stine, 1993b) that take account of the possible effects of excess kurtosis on the chi-square test statistic.

The chi-square test statistic is often supplemented with other measures of overall model fit. One reason is that the test statistic follows a chi-square distribution *asymptotically*. This means that the sample must be sufficiently large for the test statistic to follow a chi-square distribution. Monte Carlo simulation research has repeatedly found that in a model that has a small to moderate number of free parameters, the test statistic tends to be too big for a valid model when the sample size is small (say $N < 100$) (Boomsma, 1982; Anderson and Gerbing, 1984). This implies that rejection of the model would be too frequent in smaller samples.

In large samples the chi-square test statistic often leads to the rejection of the null hypothesis not due to a low number of cases or excess kurtosis of the variables, but rather, due to the *excess statistical power* that the researcher has to detect even trivial departures from the mean structure or covariance structure hypotheses. *Statistical power* refers to the probability of rejecting a false null hypothesis. In a sufficiently large sample, the test statistic could be statistically significant despite a substantively small deviation. Large samples provide the statistical power to detect even small errors in specification.

The most direct way to deal with this issue is to estimate the statistical power of the chi-square test. To investigate statistical power, it is useful to consider the distribution of the test statistic when the hypothesized model is not valid. Assuming that the error of specification is not too large relative to the sampling error in the data (Steiger et al., 1985), the test statistic (T) for an invalid model asymptotically follows a *noncentral* chi-square distribution. Unlike the central chi-square distribution, whose shape and location are governed just by its degrees of freedom, the shape and location of the noncentral chi-square is determined by the model degrees of freedom and by a noncentrality parameter.

We can consider the usual central chi-square distribution to be a special case of the noncentral chi-square where the noncentrality parameter is zero. If the model is valid and the distributional assumptions of the observed variables are satisfied, the test statistic asymptotically follows a central chi-square distribution regardless of the sample size. The test statistic is a function of the model degrees of freedom, which does not depend on the number of observations. In contrast, if the model is misspecified such that H_0 is false, the test statistic asymptotically follows a noncentral chi-square distribution and the mean value of the test statistic is positively related to sample size. This is because the noncentrality parameter grows as the number of cases (N) grows. The increase in the magnitude of the test statistic as sample size grows implies that it will be easier to detect a given model misspecification in larger samples than in smaller ones. So statistical power is greater in large samples.

Satorra and Saris (1985) and Saris and Satorra (1993) built on these properties and proposed a method to estimate the power of the chi-square test statistic. Their method requires that the researcher provide an alternative model structure complete with the values of the parameters. Using this alternative, the researcher generates the implied covariance matrix and mean vector that correspond to this alternative structure. The analysis of these implied moments with the original, incorrect model provides a noncentrality parameter value that permits estimation of the power of the original chi-square test. See Bollen (1989b, pp. 338–349) for a more detailed description of their method and Muthén and Curran (1997) for an application to LCM. The main challenge to implementing the Satorra and Saris approach is formulating the alternative model and its parameter values. The power of the chi-square test will depend on both the alternative structure and the values of the parameters. MacCallum et al. (1996) provide an alternative method to estimate the power of the chi-square test that is less dependent on formulating an alternative structure and that relies on the RMSEA fit index.

A less direct way of dealing with the tendency for the chi-square test statistic to be large in big samples is to supplement the chi-square test statistic with alternative fit indices. Examples of alternative fit indices are described in the chapter. However, these fit indices are not "corrections" for excess statistical power. They are alternative ways to gauge model fit.

CHAPTER THREE

Missing Data and Alternative Metrics of Time

In Chapter 2 we made the restrictive assumption that there were no missing data for any cases over time. However, it is rare to find a social science data set that does not have at least some values of its variables missing. Cross-sectionally, responses may be missing for a subset of subjects given an individual's refusal to answer certain questions, running out of time during an interview, or even simple errors in questionnaire administration. Longitudinally, missingness is commonly encountered when an individual is assessed at an earlier time point but drops out of the study and is thus not assessed at a later time point. Finally, missingness may be introduced by design when only a subset of a full sample is asked certain questions or in cohort sequential design (e.g., Schafer and Graham, 2002). Regardless of source, missingness is the norm in social science research.

Given the ubiquitous missing data in the estimation of latent curve models, we explore this topic in detail here. We begin by defining the various types of missing data that can occur within or across time. Next, we present two methods that allow for the estimation of LCMs in the presence of missing data: direct maximum likelihood and multiple imputation. We follow this with a comparison of these approaches using an empirical example. We then address the important topic of distinguishing between alternative metrics of time from the common definition of time based on wave of assessment. As we will see, restructuring data as a function of an alternative metric of time introduces missing data even if no missing data exist due to attrition or nonresponse. We conclude with an empirical example on schooling of children where years of schooling are used in place of wave of data.

3.1 MISSING DATA

3.1.1 Types of Missing Data

We begin our discussion by reviewing Rubin's (1976, 1987) typology of missing data: missing completely at random, missing at random, and missing not at random.

Latent Curve Models: A Structural Equation Perspective, by Kenneth A. Bollen and Patrick J. Curran

The terms are now widespread and we use them to describe several different mechanisms that generate missing data. Throughout this section we assume that cases are independently sampled from the population. We use a common notation in the missing data literature (see Little and Rubin, 2002) to explain the types of missing data. Let $\mathbf{Y}$ be an $N \times K$ matrix of N cases on $k = 1, 2, \ldots, K$ variables where no values are missing. The y_{ik} refers to the value of the ith case on the kth variable. Let $\mathbf{M}$ be an $N \times K$ matrix of indicators of missing values such that element $m_{ik} = 1$ when y_{ik} is missing and $m_{ik} = 0$ when y_{ik} is present. Thus, $\mathbf{M}$ gives the pattern of missing data and consists exclusively of elements that are 0's or 1's. Define $f(\mathbf{M}|\mathbf{Y}, \boldsymbol{\phi})$ to be the conditional distribution of $\mathbf{M}$, given $\mathbf{Y}$ and a vector of unknown parameters, $\boldsymbol{\phi}$. Using this notation and the definitions of Little and Rubin (2002), in the next few subsections we explain the major types of missing data.

Missing Completely At Random (MCAR) Typically, MCAR is the most difficult condition of missing data to satisfy. It requires that the probability that a value is missing be unrelated to any of the data values of $\mathbf{Y}$. More formally, MCAR holds if

$$f(\mathbf{M}|\mathbf{Y}, \boldsymbol{\phi}) = f(\mathbf{M}|\boldsymbol{\phi}) \qquad \text{for all } \mathbf{Y}, \boldsymbol{\phi} \tag{3.1}$$

In other words, the values in $\mathbf{Y}$ provide no information useful in determining whether y_{ik} is missing. For example, suppose that the probability that y_{ik} is missing is 0.1 for all i, k. Since this probability is the same for all cases and is not dependent on the value of y_{ik}, this is an example of MCAR. In this situation, the conditional probability conditioning on the observed data, $\mathbf{Y}$, and parameters, $\boldsymbol{\phi}$ [i.e., $f(\mathbf{M}|\mathbf{Y}, \boldsymbol{\phi})$], would be the same as the conditional probability, conditioning on $\boldsymbol{\phi}$ alone [i.e., $f(\mathbf{M}|\boldsymbol{\phi})$].

We can further illustrate this idea by considering empirical data drawn from the National Longitudinal Survey of Youth (NLSY).[1] We extracted our initial sample data from the 1986, 1988, 1990, and 1992 assessment periods. Children were selected to be in grade 0, 1 or 2 (representing kindergarten, first, and second grade) and to be of age 5, 6, or 7 at the first assessment (1986). Although valid data were required at the first assessment period (to establish initial grade membership), a child could provide data at all or none of the following assessments. The repeated measure of interest here is the individual and time-specific score on the reading recognition subscore of the Peabody Individual Achievement Test (PIAT). This was scaled as the percentage of 84 items that were answered correctly. The same 84 items were administered at all four time points, providing a consistent scale over time. Further, the gender of child was coded where $female = 1$ denotes female and $female = 0$ denotes male. Finally, the minority status of the child has $minority = 1$ to denote self-defined minority status and $minority = 0$ to denote nonminority status.

These selection criteria resulted in a subsample of $N = 418$ cases, $N = 166$ of whom provided data at all four time periods. Of the $N = 418$ cases, 46% were

[1]We provide initial information about the NLSY sample here and comprehensive details in the appendix to this chapter.

Table 3.1 First 10 of 418 Cases Drawn from the NLSY Subsample

Case	Female	Minority	Age '86	Read '86	Read '88	Read '90	Read '92
1	1	0	5	26.19	63.10	.	71.43
2	1	1	6	13.10	16.67	.	.
3	1	0	6	19.05	29.76	28.57	45.24
4	1	0	6	21.43	32.14	42.24	57.4
5	1	0	6	21.43	40.47	.	55.95
6	1	0	6	21.43	45.24	69.05	78.57
7	0	1	7	21.43	35.71	.	.
8	0	0	5	7.14	21.43	50	59.52
9	1	0	5	17.86	29.76	.	.
10	1	0	6	21.43	52.38	.	67.19

female, 47% were self-identified as minority, and the mean age was 5.96 years in 1986. The first 10 cases are presented in Table 3.1.

A numerical value reflects that the subject provided a valid data point for that assessment, and a "." denotes that he or she did not. As is evident in Table 3.1, only four cases provided complete data at all four assessments (cases 3, 4, 6, and 8). Case 2 provided data at the first two time points and not there after, and cases 1, 5, and 10 provided data at all but the third assessment. We will explore the various types of missing data using the full sample of $N = 418$ cases.

For example, suppose that we treat female and the 1992 measure of reading as the only variables in **Y**, so that y_{i1} is the variable female, with no missing values, and y_{i2} is the variable read '92, with some missing values. For MCAR to hold, missingness on the reading data at 1992 must be unrelated to both reading at 1992 and to whether the individual is a boy or a girl. With only two missing values for the reading data, it is difficult to evaluate the MCAR condition. With a larger sample we could test the MCAR assumption by testing whether the proportion of girls in the sample with complete data is equal to the proportion of girls in the sample composed of cases that are missing on wave 4 reading. The MCAR assumption would lead to equal proportions of girls with missing and complete data. If the proportions were significantly different, this is evidence against MCAR. As with any hypothesis test, failure to reject does not prove the null hypothesis. More generally, when there are more than two variables, there are tests for MCAR. One MCAR test divides a sample into those cases that are missing on a particular variable and those cases that are not. Then the mean value for each other variable is compared across these two groups using a t-test. The main limitation of this approach is taking account of the multiple tests and maintaining an appropriate Type I error for the group of tests. Bonferroni-type adjustments are possible. Or researchers can use an alternative test such as that described in Little (1988). In any event, it is often possible to test the MCAR assumption.

The MCAR assumption is most defensible when data are missing by design. For instance, in a social survey, only a random subsample might be asked a set of questions since it is too expensive to ask the full sample all questions. Assuming adequate implementation of the survey, MCAR is a plausible assumption for the missing values for those in the sample who were not asked the questions. But in the more typical situation, the MCAR assumption is too demanding for many applications. Fortunately, there is a useful, less restrictive assumption to which we now turn.

Missing at Random (MAR) The MCAR condition required that the missing values were unrelated to **Y**. The MAR condition is less restrictive than MCAR. Separate **Y** into $\mathbf{Y}_{\text{obs}}$ and $\mathbf{Y}_{\text{mis}}$, where the former has all nonmissing entries in **Y** and the latter has the missing components of **Y** (Little and Rubin, 2002, p. 12). The data are MAR if

$$f(\mathbf{M}|\mathbf{Y}, \boldsymbol{\phi}) = f(\mathbf{M}|\mathbf{Y}_{\text{obs}}, \boldsymbol{\phi}) \qquad \text{for all } \mathbf{Y}_{\text{mis}}, \boldsymbol{\phi} \tag{3.2}$$

The MAR condition holds if the conditional distribution of **M** given **Y** and $\boldsymbol{\phi}$ equals the conditional distribution of **M** given $\mathbf{Y}_{\text{obs}}$ and $\boldsymbol{\phi}$. This condition is less restrictive than MCAR in that the missing values can be related to the observed values in **Y** as long as "missingness" is not predicted by the missing values of **Y**. In other words, nonmissing values of other variables that are in the data set can be related to the missing values and the MAR condition still holds. MAR is less restrictive than MCAR in that it permits the missingness to be related to one or more variables in the data.

We return to the 1992 reading and female variables taken from Table 3.1 to further discuss MAR. Suppose that boys were more likely than girls to have missing reading scores for the 1992 measure. This would imply that the data were not MCAR. However, if we look at boys only and there is no relation between the data being missing and the 1992 reading variable and if the same holds among girls only, we can consider the data to be MAR. The MAR assumption permits associations between missingness and values of variables in the data set as long as the missingness is unrelated to **Y** once we control for the other variables. Unfortunately, we cannot test directly whether the data are MAR when the researcher has no information on the cause of the missingness. If another source provided information on the missing values or if a random sample on the missing values was available through a follow-up study, it might be possible to gain evidence on the plausibility of the MAR assumption. But without these, the MAR assumption is not empirically testable (Schafer, 1997).

Missing Not at Random (MNAR) The MNAR condition is the least restrictive in that it permits the probability of missingness to be related to the missing values of **Y** even after conditioning on the observed values of **Y**. If data are MNAR, then

$$f(\mathbf{M}|\mathbf{Y}, \boldsymbol{\phi}) \neq f(\mathbf{M}|\mathbf{Y}_{\text{obs}}, \boldsymbol{\phi}) \qquad \text{for all } \mathbf{Y}_{\text{mis}}, \boldsymbol{\phi} \tag{3.3}$$

For example, suppose that conditional on the values of a person's education, occupational prestige, race, and gender that those people who are missing on an income variable tend to be those with higher incomes. Or suppose that we want to study the protest activities in major metropolitan areas, but data on protests are more likely to be missing when the protests are small, even after controlling for other city traits. In both of these cases, the value of the dependent variable is related to missingness even after controlling for the explanatory variables.

MNAR does lessen the restrictiveness of the assumptions about the missing data, but it leads to complications in adequately handling the data. Heckman (1979) has one of the best known procedures for handling dependent variables whose values are MNAR, which involves estimating a selection equation that predicts missingness on the dependent variable. Winship and Mare (1992) have a useful review of such selection models. Schafer and Graham (2002, pp. 171–173) also explore the estimation of models with MNAR data. Controversy still surrounds the best way of handling MNAR, so we cannot recommend a general solution. Instead, we discuss the treatment of data that are MCAR or MAR where suitable treatments are better developed.

3.1.2 Treatment of Missing Data

We now discuss briefly several traditional ways of handling missing data, and then we give more attention to two generally superior methods: direct maximum likelihood and multiple imputation methods. As we will see, the latter two methods maintain optimal properties under less restrictive assumptions than do the traditional methods.

Complete Case Analysis (Listwise Deletion) Complete case or listwise deletion analysis deletes any cases for which any of the variables in the analysis have missing values. An advantage of this technique is that the resulting data form a rectangular file that is easy to analyze using traditional methods of estimation. Consider all of the variables for the first 10 cases in Table 3.1. Applying listwise deletion leads to 60% of the cases being dropped (e.g., only 4 of 10 cases have complete data). For the complete data, only 166 of the 418 available cases (or 39.7% of the initial sample) would be retained for analysis using listwise deletion. More generally, depending on the patterning of the missing values, listwise deletion can result in a large proportion of observations being removed. This drop in cases and ignoring some of the information in the sample is one of the disadvantages of listwise deletion. The resulting smaller N leads to lower efficiency than would be true if we had the full sample of cases.

The consequences can be more serious than just a loss in efficiency. If the data are not MCAR, listwise deletion can lead to an inconsistent and biased estimator. This might not be surprising for data that are MNAR, but it also can occur if data are MAR.[2] In general, listwise deletion is not an optimal analytical strategy, and we do not recommend it as a general approach.

[2]An important exception to this are regression analyses where the probability of missing data on the exogenous variables does not depend on the values of the dependent variable. In this case unbiased and consistent estimators maintain these properties with listwise deletion. See Little (1992) and Allison (2002) for details.

Pairwise Deletion Pairwise deletion is another common method of handling missing data. This approach uses all available cases for each pair of variables to compute summary statistics such as covariances or correlations and then inputs these into an analysis. A potential advantage of pairwise deletion is that it uses more of the data than does listwise deletion and holds the potential to be more efficient. However, pairwise deletion has several serious disadvantages. First, there is ambiguity in how to fully implement pairwise deletion. For instance, a researcher constructing a covariance matrix with pairwise deletion needs to decide how to compute the means that go into the covariances. Should the means be based on all cases available for each variable alone, or should it be the mean for the cases shared in common by the two variables? In addition, in SEMs we need to declare the sample size for the covariance matrix analyzed. Given that each covariance might be calculated with a different number of cases, what is the sample size for this matrix? Since the N goes into the calculation of the chi-square test statistic and the asymptotic standard errors, the choice of N has implications for the rest of the analysis. An additional potential problem that is most likely in small samples is that pairwise deletion can lead to a covariance matrix that is not positive definite because different elements of the covariance matrix are likely to be computed from different subsets of the full sample. Nonpositive definite matrices preclude the application of the ML estimator without some type of modifications (see Wothke, 1993).

If the data are MCAR, pairwise deletion is likely to maintain the consistency of an estimator. However, in the more realistic situations of MAR or MNAR, inconsistency or biasedness is likely. This approach is rarely recommended in practice.

Other Traditional Approaches Despite wide recommendations against their use by methodologists and statisticians (e.g., Allison, 2002), listwise and pairwise deletion remain by far the most common methods of handling missing data. However, there are other traditional methods that receive some use. A missing indicator approach creates a dummy variable that is coded 1 if a value is missing on a variable and zero otherwise. The variable with missing data has its original values for the nonmissing data but an arbitrary constant value when missing. Both the dummy variable and the recoded variable with the missing data are entered into an analysis along with the other variables (see Cohen and Cohen, 1983; Allison, 2002, pp. 9–11). Jones (1996) shows that this method generally produces biased and inconsistent estimators, so this is not a method that we would generally recommend.

Imputation methods are also popular. This technique imputes a single value for the missing values of each variable and then analyzes the resulting full rectangular file as usual. Imputing the mean of a variable for all missing cases is one common approach. Haitovsky (1968) discusses the biases that are likely with this technique. Other imputation techniques are also in use, and later we discuss recent developments in the method of *multiple imputation*. In the next section we present the direct maximum likelihood approach to missing data. Both the direct maximum likelihood and the multiple imputation methods have properties superior to those of the traditional methods that we have so far reviewed.

Direct Maximum Likelihood The maximum likelihood fitting function that we used in Chapter 2 is written in a form that makes it easy to develop a likelihood ratio test of model fit where the estimated model is compared to a saturated model. Specifically,

$$F_{\mathrm{ML}} = \ln|\mathbf{\Sigma}(\boldsymbol{\theta})| - \ln|\mathbf{S}| + \mathrm{tr}[\mathbf{\Sigma}^{-1}(\boldsymbol{\theta})\mathbf{S}] - \mathbf{p}$$
$$-[\bar{\mathbf{y}} - \boldsymbol{\mu}(\boldsymbol{\theta})]'\mathbf{\Sigma}^{-1}(\boldsymbol{\theta})[\bar{\mathbf{y}} - \boldsymbol{\mu}(\boldsymbol{\theta})] \tag{3.4}$$

where $\mathbf{S}$ and $\bar{\mathbf{y}}$ are the covariance matrix and mean vector calculated on the complete case data matrix and all else is as defined earlier. However, this pooled likelihood can be rewritten in a form that holds for each individual case in the sample and does not include the saturated model (Arbuckle, 1996):[3]

$$\ln L_i(\boldsymbol{\theta}) = K_i - \tfrac{1}{2}\ln|\mathbf{\Sigma}_i(\boldsymbol{\theta})| - \tfrac{1}{2}[\mathbf{z}_i - \boldsymbol{\mu}_i(\boldsymbol{\theta})]'\mathbf{\Sigma}_i^{-1}(\boldsymbol{\theta})[\mathbf{z}_i - \boldsymbol{\mu}_i(\boldsymbol{\theta})] \tag{3.5}$$

where the i indexes the observation in the sample, $\mathbf{z}_i$ is a vector of observed variables for the ith case, and K_i is a constant unrelated to $\boldsymbol{\theta}$. The likelihood function for all cases is

$$\ln L(\boldsymbol{\theta}) = \sum_{i=1}^{N} \ln L_i(\boldsymbol{\theta}) \tag{3.6}$$

The key idea of the direct ML approach is that the likelihood function is computed for each case using only those variables that are available for that case. The total likelihood is the sum of the values of the likelihood for each case. Interestingly, when there are no missing data, the sum of the individual likelihoods in Eq. (3.6) is equal to the overall likelihood defined in Eq. 3.4).[4] Only under partially missing data do these two approaches differ.[5]

The usual chi-square test statistic compares the $\ln L(\boldsymbol{\theta})$ of the hypothesized model to that of a saturated model and forms a likelihood ratio test statistic of whether the population moment matrices equal the implied moment matrices [i.e., $\boldsymbol{\mu} = \boldsymbol{\mu}(\boldsymbol{\theta})$ and $\mathbf{\Sigma} = \mathbf{\Sigma}(\boldsymbol{\theta})$]. Equation (3.6) provides the $\ln L(\boldsymbol{\theta})$ of the hypothesized model. The $\ln L(\boldsymbol{\theta})$ of the saturated model is required to calculate the likelihood ratio chi-square test statistic. The saturated model $\ln L(\boldsymbol{\theta})$ is calculable using the direct ML approach described above, and doing so enables us to calculate the

[3]One difference is that the saturated model is not part of the log-likelihood function as it is in F_{ML}. See Bollen (1989b, pp. 131–134) for further discussion of the relation between F_{ML} and $\ln L(\boldsymbol{\theta})$ in SEMs.

[4]However, unlike the pooled ML function, the individual likelihood function does not include a component that corresponds to the saturated model. As we describe in greater detail below, sometimes we can fit a saturated model to the missing data and compare its fit to the hypothesized model to allow for the calculation of formal likelihood ratio tests to assess omnibus goodness of fit.

[5]Indeed, given this equivalence, the freeware computer program Mx uses the direct ML estimator regardless of whether or not the data matrix is characterized by missing data.

chi-square test statistic.[6] In addition, likelihood ratio tests are available for comparing other nested models of the data. Finally, Eq. (3.5) allows for the calculation of ML estimates of the means and covariance matrix of the repeated measures based on all available cases.[7]

We use the reading data in Table 3.1 to illustrate this method further. Suppose that we want to estimate an unconditional linear latent curve model of reading recognition over the four repeated assessments. However, instead of using listwise deletion, we will use all available data provided by each observed case (here, $N = 418$). The likelihood function for the first case in Table 3.1 has three nonmissing measures of reading so that the dimensions of $\boldsymbol{\Sigma}_i(\boldsymbol{\theta})$ are 3×3 and the dimensions of $\mathbf{z}_i$ and $\boldsymbol{\mu}_i(\boldsymbol{\theta})$ are 3×1. That is, we take only those elements of the implied covariance matrix and the implied mean vector for which we have data and ignore the elements for the variables with missing information. The third case has no missing data so that $\boldsymbol{\Sigma}_i(\boldsymbol{\theta})$ is 4×4 and the dimensions of $\mathbf{z}_i$ and $\boldsymbol{\mu}_i(\boldsymbol{\theta})$ are 4×1. In a similar fashion we would determine the dimension of these matrices for each case in the sample. The value of $\boldsymbol{\theta}$ would be chosen that maximizes the sum of the likelihood function across all cases.

The direct ML approach has several desirable features. One is that as a ML estimator, it maintains the asymptotic properties of ML estimators such as consistency, asymptotic unbiasedness, asymptotic normality, and asymptotic efficiency, and permits the computation of asymptotic estimates of standard errors for significance testing (Arbuckle, 1996). In practice, this means that we will have parameter estimates, asymptotic standard errors, and significance tests available as usual. In addition, by fitting the saturated model with the same estimator, a researcher can form the likelihood ratio test statistic for overall model fit. This means that we can form a likelihood ratio (or chi-square) test of the overidentifying restrictions in the model as we have done before, and can also test other nested models as described above. Furthermore, the direct ML approach maintains these properties under MCAR and under the less restrictive assumption of MAR. The behavior of the direct ML approach when the distribution of the observed variables has excess kurtosis (Gold et al., 2003) or when the missingness is MNAR (Muthén and Jöreskog, 1983) is not well studied. Another positive feature of the direct ML technique is that it makes use of all available information in the data. No cases are discarded and all the values of the variables available for a case go into the calculations; the direct ML does not involve imputing the values of any variables that are missing, makes use of only those variables that are available, and no values

[6] One complication that can prevent calculation of the $\ln L(\boldsymbol{\theta})$ of the saturated model is when there are no cases available to estimate the covariance between two variables. In this case, the usual chi-square test statistic might not be available. However, this does not affect the ability to calculate likelihood ratio tests of other nested models.

[7] We later use these ML estimates of the means and covariance matrix in the empirical example so that we can examine the means of the repeated measures while taking account of missing data. This likelihood also is useful in forming the saturated model that goes into the chi-square test statistic. When estimating the hypothesized model, we do not use this mean and covariance matrix, but use the likelihood function of Eqs. (3.5) and (3.6) that corresponds to the hypothesized model.

are "filled in." We illustrate the direct ML procedure after we present the multiple imputation procedure for handling missing data.

Multiple Imputation Rubin's (1987) multiple imputation (MI) method of handling missing data imputes values for the missing values and then analyzes the resulting data with techniques that work on complete data sets. The procedure consists of four steps: (1) developing an appropriate imputation method, (2) forming a number of imputed data sets, (3) separate analysis of all imputed data sets, and (4) optimally combining the estimates from multiple imputed data sets (Allison, 2002).

Developing an appropriate imputation method requires a model that guides the generation of the imputations. The most common distributional assumption in multiple imputations is that the variables come from a multivariate normal distribution. Although other assumptions are possible, we retain this to motivate our discussion of the multiple imputation procedure. Under this assumption, we use regression techniques to impute the data. To illustrate this, return to Table 3.1. In the first row the third wave of the reading data is missing, whereas the other variables are present. We need to impute values for this missing measure, and we do so by forming a regression of the wave 3 reading score on the wave 1, 2, and 4 measures of reading for all cases in the sample that have values for these three variables. The predicted wave 3 reading score regression is thus

$$\widehat{y}_{i3} = \widehat{\pi}_0 + \widehat{\pi}_1 y_{i1} + \widehat{\pi}_2 y_{i2} + \widehat{\pi}_4 y_{i4} \tag{3.7}$$

where the i runs over all cases that have data on all variables in the equation, the y's are the reading variables, with the second subscripts indicating the time period, and the $\widehat{\pi}$'s are the estimated regression intercept ($\widehat{\pi}_0$) and coefficients ($\widehat{\pi}_1$, $\widehat{\pi}_2$, and $\widehat{\pi}_3$). A researcher can use variables in the imputation equation that do not appear in the substantive model as long as such variables help to predict the missingness, although we only used the other reading variables in this example.

Substituting in the values of the right-hand-side variables creates predicted values for the time 3 reading variable for cases where it is missing. However, this unadjusted regression imputed value is not an ideal prediction. One problem is that it implicitly suggests that the time 3 reading variable is predicted perfectly by the right-hand-side variables. It ignores the error that is virtually always present in a regression like this. A correction for this limitation is to add a random component to $\widehat{y}_{i3}$ that reflects the residual variance in the regression. For instance, we could form

$$\widetilde{y}_{i3} = \widehat{y}_{i3} + \widetilde{u}_{i3} \tag{3.8}$$

where $\widetilde{u}_{i3}$ comes from a normal distribution with a standard deviation equal to the residual standard deviation from the regression equation that generated $\widehat{y}_{i3}$.

Although using $\widetilde{y}_{i3}$ addresses one problem with the imputed data, a remaining difficulty is that our imputation treats the $\widehat{\pi}$'s and the standard deviation of $\widetilde{u}_{i3}$ as if they were population parameters rather than sample values. Ideally, each imputed sample should be based on a different set of $\widehat{\pi}$'s and different standard deviation of $\widetilde{u}_{i3}$ so as to reflect the uncertainty about these parameters. Typically, multiple imputation is motivated using Bayesian methods, where a posterior distribution represents the parameters and draws of parameter values from this posterior distribution that are part of the imputation process. Data augmentation and importance resampling (Rubin, 1987; Schafer, 1997) are two methods for sampling from the posterior distribution. We refer readers to Rubin (1987), Schafer (1997), Allison (2002), and Schafer and Graham (2002) for further description of these methods.

We follow these multiple imputation procedures for each missing value in the data. Applying these imputation methods to missing values enables us to create different imputed data sets, each of which has a complete set of values for all cases in the sample. In most cases, 10 or fewer imputed data sets are created (Allison, 2002, p. 50; Little and Rubin, 2002, p. 209). However, more data sets could be created in situations with a higher proportion of missing data or when greater precision is desired.

Each of these data sets forms a complete data set and we can estimate the latent curve model for each. If we refer to the total number of imputations for each missing value as M, multiple imputation results in $m = 1, 2, 3, \ldots, M$ complete data sets. Each data set leads to an estimate of the model parameters, say $\widehat{\boldsymbol{\theta}}_m$. We combine these multiple imputation estimates of the same parameters simply by taking the mean:

$$\overline{\overline{\boldsymbol{\theta}}} = \frac{\sum_{m=1}^{M} \widehat{\boldsymbol{\theta}}_m}{M} \tag{3.9}$$

Separate estimates of the asymptotic standard errors of each parameter are available. Consider one of the parameter estimates from $\overline{\overline{\boldsymbol{\theta}}}$, say $\overline{\overline{\theta}}_k$. Let s.e. $(\widehat{\theta}_{km})$ be the standard error of the kth parameter in the mth imputation. The combined multiple imputation estimate of the standard error for $\overline{\overline{\theta}}_k$ is

$$\text{s.e.}(\overline{\overline{\theta}}_k) = \sqrt{\frac{\sum_{m=1}^{M}[\text{s.e.}(\widehat{\theta}_{km})]^2}{M} + (1 + M^{-1})(M-1)^{-1}\sum_{m=1}^{M}(\widehat{\theta}_{km} - \overline{\overline{\theta}}_k)^2} \tag{3.10}$$

This standard error allows the formation of confidence intervals and significance testing.[8] We can generalize this to estimate the covariance matrix of the entire vector, $\overline{\overline{\boldsymbol{\theta}}}$. Define $\widehat{\Sigma}_{\widehat{\theta}\widehat{\theta}m}$ to be the estimated asymptotic covariance matrix of the parameters in the mth imputed sample. The mean of these estimated asymptotic covariance matrices across imputations is

$$\overline{\overline{\Sigma}}_{\widehat{\theta}\widehat{\theta}} = \frac{\sum_{m=1}^{M} \widehat{\Sigma}_{\widehat{\theta}\widehat{\theta}m}}{M} \tag{3.11}$$

[8]See Rubin (1987) and Schafer and Graham (2002) for a Student's t approximation.

The estimated asymptotic covariance matrix for $\overline{\overline{\boldsymbol{\theta}}}$ is

$$\widehat{\Sigma}_{\overline{\overline{\theta}}\,\overline{\overline{\theta}}} = \overline{\overline{\Sigma}}_{\widehat{\theta}\widehat{\theta}} + (1 + M^{-1})(M-1)^{-1}\sum_{m=1}^{M}(\widehat{\boldsymbol{\theta}}_m - \overline{\overline{\boldsymbol{\theta}}})(\widehat{\boldsymbol{\theta}}_m - \overline{\overline{\boldsymbol{\theta}}})' \tag{3.12}$$

The main diagonal of this matrix provides an estimate of the variance of the corresponding parameter, while the off-diagonal elements are the covariances of two parameters.

Combining the multiple imputed data for a single likelihood ratio test is more complicated. A simple method is based on the work of Li et al., (1991) and Allison (2002, p. 68). Let T_m be the likelihood ratio chi-square test statistic from the mth imputed data set with degrees of freedom df. Define

$$\overline{T} = \frac{\sum_{m=1}^{M} T_m}{M} \tag{3.13}$$

and

$$s_{T_m}^2 = (M-1)^{-1}\sum_{m=1}^{M}(\sqrt{T_m} - \sqrt{\overline{T}})^2 \tag{3.14}$$

The MI version of the test statistic is

$$T_{MI} = \frac{\overline{T}/\text{df} - (1 - M^{-1})s_{T_m}^2}{1 + (1 + M^{-1})s_{T_m}^2} \tag{3.15}$$

One difference from the usual likelihood ratio test statistic is that T_{MI} approximates an F-distribution under the null hypothesis. Its numerator degrees of freedom is df and its denominator degrees of freedom are

$$\frac{M-1}{\text{df}^{3/M}}\left(1 + \frac{M}{(M + M^{-1})s_{T_m}^2}\right)^2$$

Although it is useful to have a test statistic with MI, there are several aspects of it to keep in mind. First, this is an approximation that has not been well studied, particularly in the SEM context. Second, although we can estimate a p-value for the null hypothesis of perfect fit for our model, the optimal way to form the other fit indices (e.g., RMSEA, IFI) is less clear.

3.1.3 Empirical Example

We will now fit a linear latent curve model to the four repeated measures of reading recognition drawn from the NLSY data set that was introduced earlier. We begin by fitting a linear LCM to the complete case data using listwise deletion. We then use direct ML and multiple imputation procedures to fit the same LCM but using all available cases.

Linear LCM of Complete Case Data with Maximum Likelihood Estimation We start by estimating a linear latent curve model for the four repeated measures of reading recognition using the sample of $N = 166$ cases with complete data across all four waves of assessment. Because there are no missing data, we can use the full information maximum likelihood estimator described in Chapter 2 and presented in Eq. (3.4). The sample covariance matrix and mean vector calculated for the $N = 166$ complete cases is the sole unit of analysis, and these are

$$\mathbf{S} = \begin{pmatrix} 33.09 & & & \\ 31.53 & 115.68 & & \\ 24.62 & 108.70 & 177.42 & \\ 22.65 & 112.61 & 159.34 & 215.07 \end{pmatrix} \qquad \bar{\mathbf{y}} = \begin{pmatrix} 20.21 \\ 40.64 \\ 55.80 \\ 66.51 \end{pmatrix} \tag{3.16}$$

For a linear latent curve model, we set the factor loadings relating the four repeated measures to the intercept factor equal to 1.0 and set the factor loadings relating the four repeated measures to the slope factor to $\lambda_t = 0, 1, 2, 3$. Given that the repeated assessments were separated by 24-month increments, a 1-unit change in time reflects a two-year interval. Finally, because we have defined the first measure of reading recognition to be $\lambda_1 = 0$, the intercept factor represents the model implied mean of the trajectory at the initial time of measure. A path diagram of this linear latent curve model is presented in Figure 3.1.

This model was fitted to the sample data for the $N = 166$ complete cases and was found to fit the data poorly. The likelihood ratio test statistic was $\chi^2(5, N = 166) = 153.57$, $p < 0.0001$, indicating an extremely poor fit of the model to the data. Other fit indices reflected equally poor fit [TLI = 0.53; IFI = 0.61; RMSEA = 0.42 CI = (0.37, 0.48)]. A more detailed examination of the model results indicated

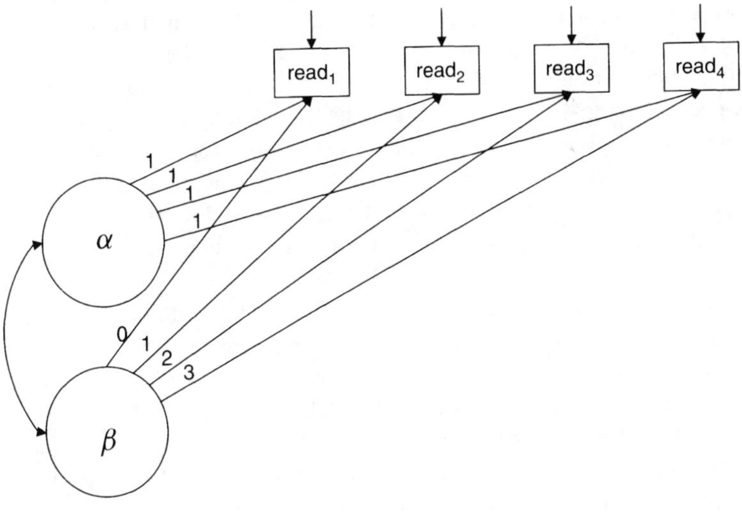

FIGURE 3.1 Path diagram of latent curve model organized on a wave of assessment.

Table 3.2 ML Parameter Estimates and Asymptotic Standard Errors for Fixed and Random Effects for Repeated Reading Measures, Listwise Deletion ($N = 166$)

Parameter	Parameter Estimate	Asymptotic Standard Error	Critical Ratio
$\hat{\mu}_{\alpha}$	21.27	0.46	45.93
$\hat{\mu}_{\beta}$	15.97	0.38	41.83
$\hat{\psi}_{\alpha\alpha}$	19.93	5.69	3.50
$\hat{\psi}_{\beta\beta}$	17.79	2.91	6.10
$\hat{\psi}_{\alpha\beta}$	5.07	2.93	1.73

that the model converged properly, and there were no Heywood cases (negative variances) or other problems in estimation.

The fixed effects reflect a linear increase in reading recognition over time, and the random effects reflects substantial individual variability around these mean values. However, caution is warranted in interpreting these values given the poor fit of the model. Despite this poor fit, we present the parameter estimates and standard errors in Table 3.2 to compare these results with subsequent results that make use of direct ML and multiple imputation.

There are a variety of reasons why this model may fit the observed data so poorly. A key issue relates to the likely nonlinearity observed in the means of the repeated measures. Whereas the linear LCM implies equal change in reading between each time point, the pattern of the sample means suggests that changes in reading are larger between the initial two time points, but the magnitude of the increase decreases as time progresses. In Chapter 4 we examine a variety of ways to model nonlinear trajectories, so we do not explore this option further here.

However, a second hypothesis for the source of poor fit lies in the fact that we omitted a large proportion of cases by using listwise deletion to allow for the complete case analysis. Indeed, fully 60% of the available cases were dropped so that we could estimate the model using the complete case maximum likelihood estimator. The omission of such a large number of cases may have introduced significant biases into the estimation of the LCM. We now turn to two alternative methods of estimation that will allow for the retention of the full sample of $N = 418$ cases: direct maximum likelihood and multiple imputation.

Linear LCM of Missing Data with Direct Maximum Likelihood Estimation To examine empirically the potential effects of using listwise deletion, we next estimated the same linear LCM as above (see Figure 3.1), but now we use the direct maximum likelihood estimator described earlier and retain all $N = 418$ cases. Whereas in the complete case analysis the sample covariance matrix and mean vector were sufficient statistics from which to estimate the model parameters, in the presence of partially present data this no longer holds. Instead, we must consider the raw data matrix that contains all of the data available for each case in

Table 3.3 ML Parameter Estimates, Asymptotic Standard Errors, and Critical Ratios for Fixed and Random Effects for Repeated Reading Measures, Direct ML with Missing Data ($N = 418$)

Parameter	Parameter Estimate	Asymptotic Standard Error	Critical Ratio
$\hat{\mu}_\alpha$	22.81	0.38	60.54
$\hat{\mu}_\beta$	16.35	0.30	54.94
$\hat{\psi}_{\alpha\alpha}$	27.67	5.53	5.00
$\hat{\psi}_{\beta\beta}$	15.82	2.57	6.17
$\hat{\psi}_{\alpha\beta}$	16.39	2.96	5.53

the sample. We then fit an individual likelihood to each case [as in Eq. (3.5)], and compute an overall likelihood by summing across all individual likelihoods [as in Eq. (3.6)].

This model also fit the data poorly. The likelihood ratio test statistic was $T_{\text{ML}} = 192.58$, df $= 5$, $p < 0.0001$, indicating a poor fit of the model to the data. Other fit indices reflected equally poor fit [TLI = 0.58; IFI = 0.65; RMSEA = 0.30 (0.27, 0.34)]. As before, the model converged and there were no improper parameter estimates. There was also evidence of linear increases in reading recognition over time and of substantial individual variability around these mean trajectory values. Again, for comparison purposes, we present the parameter estimates and standard errors in Table 3.3.

The direct ML results suggest that the omission of partially missing cases is not the source of the substantial model misfit. We next compare the direct ML results with those obtained under multiple imputation.

Linear LCM of Partially Missing Data with Multiple Imputation Estimation Finally, we estimated the same linear LCM using the method of multiple imputation. For these analyses we conducted 10 replications and pooled the model results using the equations presented earlier. As with the other linear LCMs of the same data, this model had inadequate fit as gauged using the F-statistic ($T_{\text{MI}} = 30.25$, $p < 0.0001$). For comparison purposes, we present the parameter estimates and standard errors in Table 3.4.

As with the prior models, all fixed and random effects are statistically significant at $p < 0.05$ and reflect an overall linear increase in reading over the four time periods, with significant individual variability around both starting point and change over time. However, as with the prior models, great care is needed in the interpretation of these parameters given the poor fit of the model to the observed data.

Model Comparisons It is interesting to compare the results from listwise deletion, direct ML, and multiple imputation. First, all three models fit the observed data poorly by all criteria of fit. For instance, the model chi-square was consistently highly significant and the RMSEA was nearly six times as large as the typically

Table 3.4 ML Parameter Estimates, Asymptotic Standard Errors, and Critical Ratios for Fixed and Random Effects for Repeated Reading Measures, Multiple Imputation for Missing Data ($N = 418$)

Parameter	Parameter Estimate	Asymptotic Standard Error	Critical Ratio
$\hat{\mu}_\alpha$	23.71	0.43	54.91
$\hat{\mu}_\beta$	15.32	0.37	41.54
$\hat{\psi}_{\alpha\alpha}$	35.47	5.78	6.13
$\hat{\psi}_{\beta\beta}$	13.49	2.50	5.39
$\hat{\psi}_{\alpha\beta}$	10.24	3.56	2.88

recommended cutoff. The linear LCM thus offers a poor representation of the mechanism that may have given rise to the observed data. However, for purposes of comparison across estimation methods, we can consider the parameter estimates and standard errors from each of these models.

Comparing values in Tables 3.2, 3.3, and 3.4, there is much overlap in both parameter estimates and the general substantive conclusions that would be drawn from these results. All three suggest a mean linear increase in reading over time and individual variability in both the intercept and slope factors. Interestingly, the lower efficiency resulting from listwise deletion is visible in the larger standard errors for the model based on the $N = 166$ compared to the $N = 418$. Sometimes large discrepancies exist (e.g., $\hat{\psi}_{\alpha\alpha}$, listwise 19.93, direct ML 27.67, and MI 35.47); however, all three estimates were significant well beyond the $p < 0.05$ criterion. Greater confidence is associated with the direct ML and MI results, given the more realistic assumption of MAR and greater efficiency of these methods.

3.1.4 Summary

Given the pervasiveness of missing data, it is useful to know that there are several powerful and flexible approaches that allow for the inclusion of missing data in LCMs. At this time, the direct ML and MI methods are the best alternatives to the traditional methods of listwise or pairwise deletion. Perhaps most valuable in these methods is that they are valid under the less restrictive assumption of MAR rather than MCAR.

Which method should a researcher choose, direct ML or MI? One advantage of direct ML over MI is that there is no need to impute values with direct ML. Closely related is that with direct ML a researcher uses only the original data set and does not need to reestimate the model on multiple data sets as with MI. Furthermore, with MI the coefficient estimates and significance tests could differ across researchers depending on the nature and number of the imputed data sets created. With direct ML a single set of estimates and significance tests result. In the case of SEM for LCMs, direct ML estimation is available in most of the major software packages, which makes it quite accessible. Another advantage is that usual overall fit statistics (e.g., likelihood ratio tests, fit indices) are usually available with direct ML, whereas analogous statistics using MI are less well developed.

One advantage of the MI procedure is related to the MAR assumption. Key to the MAR assumption is that variables that predict missingness are part of the process, so that, conditional on these variables, the missingness is unrelated to the values of the missing variable. In multiple imputation, researchers can easily include conditioning variables in the imputation process, including some variables that do not appear in substantive models of these data. Recent work has suggested an analogous process of introducing variables relevant to the missing data process but not part of the regular analysis with the direct ML method (e.g., Graham, 2003), but this option has not been fully explored. Another positive aspect of MI is that after the imputed data sets are collected, researchers can use any standard statistical package to estimate their models since the data sets will appear as complete data.

Another issue is that both methods typically rely on an assumption that the observed variables come from a multivariate normal distribution. Given nonnormal continuous variables and the presence of dummy or ordinal variables in social science applications, this assumption is unlikely to hold perfectly. Although much work remains to be done to explore the robustness of these techniques to violations of this assumption, early indications are that the methods are relatively robust or that corrective procedures are possible (Schafer, 1997; Enders, 2001; Allison, 2002, pp. 38–40; Gold et al., 2003).

3.2 MISSING DATA AND ALTERNATIVE METRICS OF TIME

Thus far we have considered missing data that result from some form of nonresponse on the part of the subject. That is, one or more questionnaire items might be skipped intentionally or unintentionally, certain items might not be assessed due to procedural error, or data may be partially missing due to subject attrition over time. However, there is another source of missing data that can arise from using alternative metrics of time. That is, instead of limiting ourselves to modeling the passage of time in terms of *wave of assessment*, we can consider other metrics of time that might be more meaningful for a given application. However, the use of alternative metrics of time often results in the introduction of missing data, the presence of which poses the same challenges as those described above. This section provides a more detailed exploration of the mechanisms by which missing data results from the use of alternative metrics of time, and what methods we can use to estimate LCMs given these missing data.

3.2.1 Numerical Measure of Time

In Chapter 2 we presented the level 1, level 2, and reduced-form equations that define the unconditional latent curve model. For a linear LCM, the level 1 model is

$$y_{it} = \alpha_i + \beta_i \lambda_t + \epsilon_{it} \tag{3.17}$$

the level 2 model is

$$\alpha_i = \mu_\alpha + \zeta_{\alpha_i}$$
$$\beta_i = \mu_\beta + \zeta_{\beta_i} \tag{3.18}$$

and the reduced-form expression is

$$y_{it} = (\mu_\alpha + \lambda_t \mu_\beta) + (\zeta_{\alpha_i} + \lambda_t \zeta_{\beta_i} + \epsilon_{it}) \tag{3.19}$$

where the first parenthetical term represents the fixed effects and the second parenthetical term represents the random effects.

These equations show that the numerical value of time (denoted by λ_t) plays a critical role in model estimation and subsequent interpretation. In many applied research settings, the most common metric is simply the *wave of assessment*, where $\lambda_t = 0, 1, \ldots, T-1$ and T represents the total number of assessments. This was our approach with the NLSY models presented earlier. We considered time to be equivalent to wave of assessment, and thus $T = 4$ and $\lambda_t = 0, 1, 2, 3$. However, there are often other potential metrics of time besides wave of assessment.

For example, at the first assessment period there may be variability in the *chronological age* of the subjects. Although $t = 1$ for all subjects assessed at wave 1, the chronological age of the subjects can vary substantially within a wave. If there is a theoretical question involving child development, wave of assessment may be of little interest, but chronological age may be of most importance. Given that λ_t generally reflects the value of "time," this measurement might be chronological but in different metrics (e.g., seconds, minutes, or decades), some observed developmental stage (e.g., Tanner stages of pubertal development), number of hours of instruction on a task, organizational age, or grade in school. These are just a few of the alternative metrics of time that move beyond the wave of assessment.

However, as we detail below, alternative metrics of time often create missing data. This missingness is as real as that introduced by subject attrition or nonresponse, and we must work to incorporate this into the estimation of our LCMs. To explore this using a concrete example, we consider the hypothetical situation in which we would like to measure time not in terms of wave of assessment, but instead, as chronological age. Despite our focus on age in this example, all of our discussion applies to other metrics of time.

3.2.2 When Wave of Assessment and Alternative Metrics of Time Are Equivalent

In the examples thus far, we have assumed that wave of assessment is the metric of time. Thus, time 1 is equivalent to wave 1, time 2 to wave 2, and so on. In many instances, this approach is valid. In other cases, alternative metrics are superior (e.g., Mehta and West, 2000). For example, consider the estimation of trajectories of cognitive development in children. Here, chronological age is far more likely to be a theoretically appropriate metric of time than is the more arbitrary wave of

Table 3.5 Data Structure for Hypothetical Case Where All Subjects Same Age for Each Time

Age at Assessment Period			
Time 1	Time 2	Time 3	Time 4
6	7	8	9

assessment. Fortunately, we can incorporate alternative time metrics into the LCM by using the missing data techniques we described earlier.

To begin, consider the highly restrictive and highly unlikely data structure that is "fully balanced" on time with no missing data. This implies that all subjects are the same chronological age at each wave of assessment (e.g., all children are 6 years of age at $t = 1$)[9], the assessments are equally spaced for all individuals within each adjacent time period (e.g., $t = 1$ and $t = 2$ are separated by the same interval for all subjects),[10] and there is complete data for all subjects at all time points. This data structure is presented in Table 3.5 for a hypothetical case in which $T = 4$, the interval between adjacent time periods is 12 months, and all children are 6 years of age at $t = 1$.

The observed data from this design form a rectangular matrix of dimension $N \times T$, where N is total sample size and T is total number of time points. So, for $T = 4$, repeated observations on outcome y for individual $i = 1, 2, \ldots, N$, the data matrix is

$$\begin{pmatrix} y_{1,1} & y_{1,2} & y_{1,3} & y_{1,4} \\ y_{2,1} & y_{2,2} & y_{2,3} & y_{2,4} \\ y_{3,1} & y_{3,2} & y_{3,3} & y_{3,4} \\ \vdots & \vdots & \vdots & \vdots \\ y_{N,1} & y_{N,2} & y_{N,3} & y_{N,4} \end{pmatrix} \tag{3.20}$$

In this hypothetical example, the elements of the 4×4 sample covariance matrix $\mathbf{S}$ and the 4×1 sample mean vector $\bar{\mathbf{y}}$ are the usual

$$\mathbf{S} = \begin{pmatrix} s_{11} & & & \\ s_{21} & s_{22} & & \\ s_{31} & s_{32} & s_{33} & \\ s_{41} & s_{42} & s_{43} & s_{44} \end{pmatrix} \qquad \bar{\mathbf{y}} = \begin{pmatrix} \bar{y}_1 \\ \bar{y}_2 \\ \bar{y}_3 \\ \bar{y}_4 \end{pmatrix} \tag{3.21}$$

[9]Rounding plays a significant role when considering alternative metrics of time. That is, although we can consider a sample of children to be "precisely" 6 years of age, a greater degree of precision could be introduced by considering age assessed in months, weeks, or even days. For our discussion here, we are assuming that chronological age is rounded to the nearest year.

[10]This condition implies that the interval between any two time periods is equal for all individuals, but this does *not* imply that the interval must be equal for all adjacent time periods. It is thus possible for times 1 and 2 to be separated by a six-month period and times 2 and 3 by a 12-month period assuming that these intervals hold for all individuals within adjacent periods.

Consider the first diagonal element of $\mathbf{S}$, denoted s_{11}. This represents the sample variance on measure y pooling over all N individuals at the initial assessment $t = 1$. Similarly, the first vector element of $\bar{\mathbf{y}}$ is denoted $\bar{y}_1$ and represents the sample mean on measure y pooling over all N individuals at the initial assessment $t = 1$. However, because all of the subjects are the same chronological age within assessment, s_{11} and $\bar{y}_1$ also represent the sample variance and mean of measure y for all 6-year-olds. This is, of course, due to the fact that all children within $t = 1$ are 6 years of age. Further, the elements s_{22} and $\bar{y}_2$ are the sample variance and mean for all of the children at $t = 2$, of which none have attrited from $t = 1$ and all were assessed at precisely the same interval between assessments. Thus, in this completely balanced example, wave of assessment and chronological age are equivalent.

We next want to fit a linear LCM to these data. The completely balanced data structure is reflected in the selection of numerical values entered into the factor loading matrix $\mathbf{\Lambda}$ to reflect the periods of assessment. In Chapter 2 we described that it was convenient to set the first factor loading on the slope factor to equal zero so that the intercept factor is defined as the starting point of the trajectory. For our hypothetical example, we can code this in one of two equivalent ways. First, we can set $\lambda_t = t - 1$, where $t = 1, 2, 3, 4$, reflecting that there are a total of four repeated measures. Given the age homogeneity within assessment, we could equivalently set $\lambda_t = \text{age}_t - 6$, where $\text{age}_t = 6, 7, 8, 9$, reflecting the chronological age of all subjects within each time period. Both strategies result in the factor loading matrix, defined as

$$\mathbf{\Lambda} = \begin{pmatrix} 1 & 0 \\ 1 & 1 \\ 1 & 2 \\ 1 & 3 \end{pmatrix} \tag{3.22}$$

where the first column represents the loadings on the intercept factor (α) and the second column represents the loadings on the linear slope factor (β). This results in a four-indicator, two-factor LCM as in Figure 3.1.

Because all individuals within a given assessment period have the same numerical value of time (e.g., all individuals at $t = 1$ are assigned $\lambda_1 = 0$), this implies that there is not heterogeneity on any other important metric of time (e.g., there is no heterogeneity in *chronological age* within the time period). Further, because the values of the factor loadings are equally spaced, this implies that all assessments are spaced equally for all individuals within that time assessment. For example, if 12 months were to separate $t = 1$ and $t = 2$, then coding $\lambda_1 = 0$ and $\lambda_2 = 1$ implies that every individual was reassessed precisely 12 months following the initial assessment.

As can be seen, this is a highly restrictive structure to impose on the data, particularly when considering many applications of these techniques in the social sciences. If chronological age is theoretically or empirically related to the repeated measure y over time, yet chronological age is heterogeneous within time of assessment, several potential problems might arise (e.g., Mehta and West, 2000). Fortunately, we can circumvent this problem.

3.2.3 When Wave of Assessment and Alternative Metrics of Time Are Different

Consider the situation where chronological age is the primary metric of time. We extend our example so that we continue to consider $T = 4$ assessments, but within each assessment there is variability in the chronological age of the subjects. That is, instead of all subjects being 6 years of age at $t = 1$, consider subjects who are 6, 7, or 8 years of age at $t = 1$. This general design is common in many areas of social science research.

These three ages within a time period form three age *cohorts*. Cohort 1 consists of all of the 6-year-olds, cohort 2 consists of all of the 7-year-olds, and cohort 3 consists of all of the 8-year-olds. These cohorts might be an intentional aspect of the study design (e.g., one-third of each age is recruited for participation), or these might be created after initial data collection (e.g., subjects are simply assigned to a cohort based on their age at the first assessment). The general structure of the data for this design is presented in Table 3.6, where each row represents chronological age of assessment of each child within each cohort at each assessment period.

We can structure the observed data arising from this design in a data matrix in the same way as above [e.g., Eq. 3.20]. Namely, we collect the repeated observations for the $T = 4$ assessments on N individuals on variable y into a rectangular data matrix of order $N \times T$. Instead of the earlier situation in which all subjects in the sample were 6 years old at $t = 1$, here the subjects are 6, 7, or 8 years of age at $t = 1$.

This heterogeneity within wave of assessment can be a source of bias in the model, and this is highlighted by considering the sample covariance matrix and mean vector that would result from this data structure.

Specifically, even though there are six chronological ages in the data (ranging from 6 to 11 years of age over the four waves of assessment), the covariance matrix is still of dimension 4×4 and the mean vector of dimension 4×1, reflecting the $T = 4$ assessments. However, again consider the first diagonal element of $\mathbf{S}$, denoted s_{11}. As before, this reflects the sample variance of the repeated measures

Table 3.6 Data Structure for Hypothetical Case with Three Cohorts Within Each Time

	Age at Assessment Period			
Cohort	Time 1	Time 2	Time 3	Time 4
1	6	7	8	9
2	7	8	9	10
3	8	9	10	11

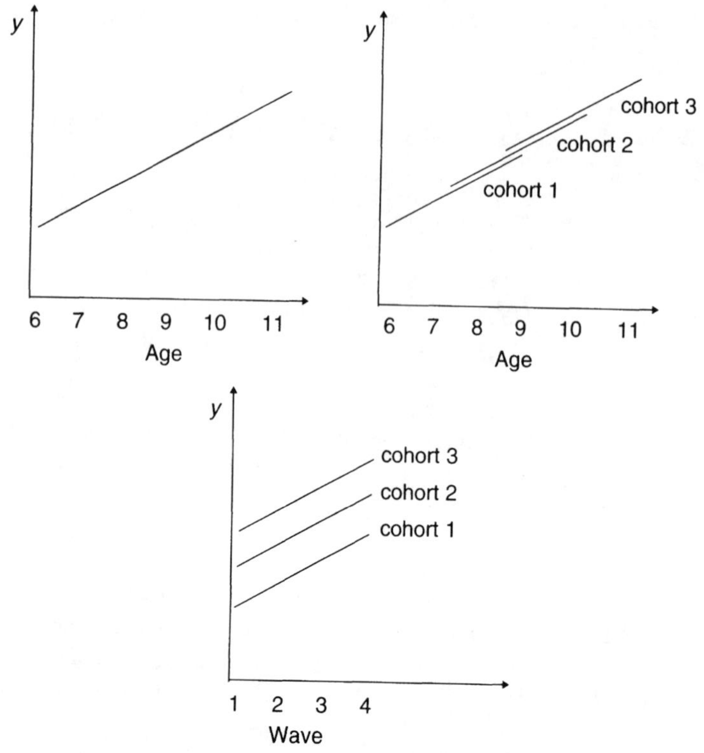

FIGURE 3.2 Pooling data over three cohorts.

for y on all N individuals at assessment $t = 1$. However, given the heterogeneity in age at the initial assessment, this sample variance pools over 6, 7, and 8 years of age. The sample mean $\overline{y}_1$ is calculated similarly by pooling over all three distinct chronological ages within $t = 1$. This pooling does not allow us to model change as a function of age, only wave of assessment.

The potential problem of pooling over age is further highlighted graphically. Consider the hypothetical trajectories presented in Figure 3.2. The upper left panel displays the mean trajectory of y as a function of chronological age ranging from 6 to 11. The upper right panel demonstrates that the single continuous trajectory is actually composed of three overlapping trajectories, one for each cohort. Finally, the lower panel presents the three trajectories, but now as a function of *wave* instead of *chronological age*. That is, the continuous trajectory on the left has been " cut" between each specific age range that defines each age cohort; these three pieces of the trajectories have then been moved laterally to the left and aligned in terms of wave.

It is clear that although the slopes of the trajectories in the three panels are equal, the intercepts are not. This is despite the fact that each piece represents a particular span of a single underlying trajectory. Given the characteristics of an empirical application or theoretical question, the difference between modeling time as a function of wave of assessment versus some other alternative metric can affect the estimation and interpretation of these models.

There are two approaches to handling the confounding of wave of assessment and other metrics of time. The first approach treats *wave* as the primary metric of time, but the effect of the other time metric (e.g., chronological age at the initial assessment) is *controlled* or *covaried out* in the prediction of the intercept and slope factors.[11] This has been a rather widely used strategy, particularly in applications prior to recent developments in missing data analysis.[12] However, there are several potential limitations of this approach (e.g., Mehta and West, 2000), and other strategies now exist that are better suited for incorporating alternative metrics to wave.

3.2.4 Reorganizing Data as a Function of Alternative Metrics of Time

An alternative approach that addresses the limitations of heterogeneity in alternative metrics of time within the wave of assessment is to capitalize on the missing data analysis procedures described above. Specifically, instead of treating *wave of assessment* as the primary metric of time (and potentially ignoring important heterogeneity on an alternative metric of time within the wave), we use the same data with an alternative time metric. This ignores the wave of assessment in the estimation of the model.[13]

To begin, consider the structure of the design in Table 3.7. Each row represents data from each of three specific cohorts, where the first cohort was assessed at 6 years of age, the second at 7 years of age, and the third at 8 years of age. The symbol "×" denotes the age at which an individual was assessed and the symbol "·" denotes the age at which an individual was *not* assessed. That is, the "·" represents missing data given that a subject of that age was not evaluated at that particular wave. Given certain assumptions, we can use the methods of missing data estimation described earlier in this chapter to fit LCMs to these data that have been restructured with chronological age as the time metric.

A critical aspect of the coding of time as chronological age is that although there are only $T = 4$ repeated observations for each individual, we can arrange the data not on *wave* but on *age* and thus consider a span of six years instead of just four. However, this introduces a challenge in model estimation. Specifically, the raw data matrix has only partially complete data, due to the data reorganization.

To better understand this approach, again consider a raw data matrix, but unlike before, this matrix is now of dimension $N \times T$, where $T = 6$ to represent chronological age instead of wave of assessment.[14] Sample elements of this data matrix

[11] We do not explore this approach here, but we present a detailed examination of conditional latent curve models in Chapter 5.

[12] Indeed, we have used this strategy (e.g., Curran et al., 1997, 1998).

[13] Although it is often important to evaluate the relation between wave of assessment and the alternative metric of time prior to fitting the LCMs.

[14] We continue to use T to denote the general metric of time, whether this be wave, age, or some other metric of interest.

Table 3.7 Cohort Sequential Design to Order Data Not on Wave But on Age

	Chronological age					
Cohort	6	7	8	9	10	11
1	×	×	×	×	.	.
2	.	×	×	×	×	.
3	.	.	×	×	×	×

are as follows:

$$\begin{pmatrix} y_{1,6} & y_{1,7} & y_{1,8} & y_{1,9} & . & . \\ . & y_{2,7} & y_{2,8} & y_{2,9} & y_{2,10} & . \\ . & . & y_{3,8} & y_{3,9} & y_{3,10} & y_{3,11} \\ \vdots & \vdots & \vdots & \vdots & \vdots & \vdots \\ . & . & y_{N,8} & y_{N,9} & y_{N,10} & y_{N,11} \end{pmatrix} \tag{3.23}$$

Although any given individual (represented in a particular row) provides four repeated observations, there is information available on a total of six discrete chronological ages when considering the overlapping data points. Thus, the symbol ".", which represents missing data, is due to the reorganization of the data matrix as a function of chronological age. This is how reorganizing data to conform to a metric of time other than wave of assessment creates a missing data problem.

The next goal is to fit a LCM to the restructured data that is partially missing using either the direct ML or MI procedures described earlier. The general rule for defining the LCM is that there are as many observed indicators of the latent curve factors as there are distinct ages that were assessed. Thus, to fit a linear latent curve model to our hypothetical raw data that has been reorganized on chronological age, we would define the factor loadings for the linear slope factor to be $\lambda_t = \text{age}_t - 6$, resulting in the factor loading matrix

$$\mathbf{\Lambda} = \begin{pmatrix} 1 & 0 \\ 1 & 1 \\ 1 & 2 \\ 1 & 3 \\ 1 & 4 \\ 1 & 5 \end{pmatrix} \tag{3.24}$$

The path diagram for this linear LCM is presented in Figure 3.3.

As is evident in the figure, although each subject can provide between one and four repeated observations, we can estimate a latent curve model spanning a total of 6 years of age. This is a key strength of this analytic approach.

Despite the advantages, it is critical to remember that subjects of a given age may have not been assessed at that age within the same time period (or at whatever time

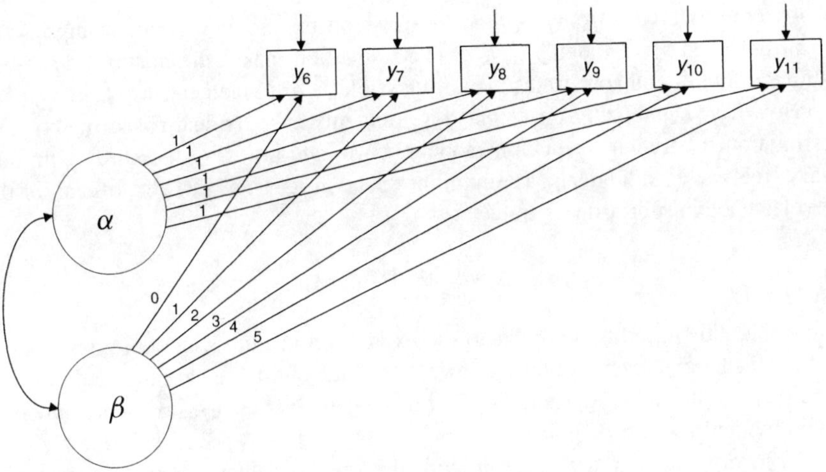

FIGURE 3.3 Latent curve model when data are organized by age rather than wave.

metric is under consideration). For example, subjects in cohort 1 may have been 9 years of age at the fourth assessment, subjects in cohort 2 may have been 9 years of age at the third assessment, and subjects in cohort 3 maybe have been 9 years of age at the second assessment. In some circumstance, it might be quite important to empirically evaluate the appropriateness of treating all subjects equitably within a given age even though they may have been assessed within different years. There are methods for testing for such "cohort effects," but we do not explore these further here. See Baltes (1968), Glenn (1976), and Adam (1978) for classic treatments of these issues.

3.2.5 Individually Varying Values of Time

Up to this point, we have defined time to be invariant over individual i within time point t. That is, regardless of whether we organize the repeated measures on wave of assessment or age at assessment (or any other relevant metric of interest), the variable of time has been denoted λ_t, indicating that all individuals assessed at time t have the same value of time λ. This strategy allows us to use as many indicators to define the latent growth factors as there are discrete time-specific measures available. Subsequently, all of the possible values for time are encoded in a single factor loading matrix $\mathbf{\Lambda}$.

One potential limitation of this strategy is that as the number of time-specific measures increases, the definition and estimation of the model becomes increasingly complex. For example, if we were interested in developmental trajectories of infants, it may be inappropriate to round to the nearest year and age would best be coded in months. There would thus be a single indicator on the latent curve factors for every possible age in months of all of the infants assessed over all time points. This may lead to a high number of indicators to define the latent growth factors. At the extreme, it is possible that there would be time-specific measures that were observed for only a *single* observation (e.g., only one infant might have been observed at 11 months of age). This situation makes the methods described above difficult, and sometimes even impossible,[15] to implement in practice.

However, capitalizing on recent developments in individual-based methods of estimation in structural equation modeling, we can allow the measure of time to vary freely over individuals. To highlight this, consider a slight reexpression of the level 1 equation for a linear trajectory:

$$y_{it} = \alpha_i + \beta_i \lambda_{it} + \epsilon_{it} \tag{3.25}$$

Note that although this is similar to the expression in Eq. (3.17) there is a critical difference here: namely, we have allowed the value of time to vary over individuals, and this is reflected in λ_{it}, denoting that the specific value of λ varies as a function of individual i and time point t.

To incorporate this method of individually varying values of time, we must turn to *definition variable methodology*. In this approach, age is treated as a variable that is used to define the fixed values of some portion of the factor loading matrix, and thus the term definition variable (Neale et al., 2002). The observed values of time for each individual are coded into an individual-specific factor loading matrix, and these individual matrices are then summed in the estimation of the overall likelihood function. This is done in much the same way as the individual likelihoods are combined in the direct ML procedures for missing data described above. Importantly, if a definition variable approach is used for an application in which the observations are balanced on time (e.g., $\lambda_{it} = \lambda_t$ for all i at time t), then the individual loading matrices simplify to the overall factor loading matrix $\mathbf{\Lambda}$ [e.g., Eq. (3.22)]. We do not detail this approach further here, but see Mehta and West (2000) and Neale et al. (2002) for demonstrations of this powerful approach.

3.2.6 Summary

Although not exhaustive, we have reviewed several issues that arise when fitting LCMs to metrics of time other than the wave of assessment. Our example focused on distinguishing between wave of assessment and chronological age, although these same issues apply regardless of what specific alternative metric of time is of interest. Importantly, moving from wave of assessment to another time metric typically creates missing data. Thus, we are fortunate that we can bring to bear the same methodologies for missing data estimation as those used for subject attrition or nonresponse. We did not discuss the situation of having missing data due to both changing the time metric *and* nonresponse or subject attrition. Fortunately, the same missing value techniques apply to these problems individually or in combination, so no new technique is needed. To demonstrate these techniques in practice, we conclude this chapter with an empirical example in which we move from wave of

[15]For example, it would be impossible to calculate a variance for a repeated measure for which there was only a single observation made at a given point in time. The covariance matrix would not be positive definite given the zero diagonal element.

assessment to enrolled grade in school when studying developmental trajectories of childhood reading achievement.

3.2.7 Empirical Example: Reading Achievement

Earlier in this chapter we explored two available options for estimating LCMs in the presence of missing data: direct maximum likelihood estimation and multiple imputation. To demonstrate these two approaches, we considered empirical data drawn from the NLSY. As we described earlier, we initially selected subjects who were first assessed in 1986, who were enrolled in grade 0, 1, or 2, and who were between 5 and 7 years of age.[16] There were $N = 166$ children who met these inclusion criteria and provided complete data in 1986, 1988, 1990, and 1992. Further, there were $N = 418$ children who met these inclusion criteria and provided complete data in 1986, but not all of whom provided complete data at all four time periods.

The sample means of the time-specific measures of reading achievement are in the top panel of Table 3.8 for the $N = 166$ complete cases, and the direct ML estimates of the means that allows MAR missing data are in the bottom panel of Table 3.8 for the $N = 418$ partially complete cases.[17] These means suggest a monotonically increasing pattern in reading achievement scores over time, but these also suggest that the rate of increase is slowing with time. Earlier, we fitted a linear LCM based on the $T = 4$ repeated assessments arranged on calendar year and found support for both fixed and random effects in developmental trajectories of reading achievement over time. However, these models fit the data poorly and raised serious doubts about the adequacy of our fitted linear LCM.

Recall that there was heterogeneity among the cases in school grade and chronological age within each year of assessment. That is, within the 1986 assessment, children were in grade 0, 1, or 2 and were of age 5, 6, or 7. This heterogeneity then naturally extends to the other years of assessment as well (e.g., in 1992 children are in grade 5, 6, or 7 and of age 11, 12, or 13). Methods by which we could "covary" the effects of grade or age at the initial assessment period using conditional LCMs are presented in Chapter 5. However, these methods consider only the grade and age of the subjects at the initial time period and do not allow us to consider differential change in grade or age over all assessment periods. Given that our focus is on reading achievement, a skill that is clearly influenced by years of schooling, school grade is an alternative metric of time to consider. Thus, one hypothesis that explains the poor fit of our linear LCM is that we were considering the wrong metric of time.

To test this hypothesis, we arrange the data in terms of *grade* instead of wave of assessment (as measured in terms of calendar year). We simply arrange each individual child's data not in terms of the calendar year in which they were assessed,

[16]We continue to refer to kindergarten as grade 0.

[17]The overall N for the direct ML estimates for missing data is 418. That is, there are 418 components that enter the overall likelihood function, one for each individual. The bottom row of Table 3.8 reports the number of cases available for each year rather than the single overall N of 418 that is part of the direct ML procedure.

Table 3.8 Year of Assessment Means of Reading Achievement for Listwise Deletion and for Direct ML Estimation with Missing Data

	Calendar Year of Assessment			
	1986	1988	1990	1992
Sample mean	20.21	40.64	55.79	66.51
Sample size by year	166	166	166	166
ML estimated mean	21.50	41.76	56.82	66.80
Sample size by year	418	356	236	224

Table 3.9 Means of Reading Achievement by Grade Using Direct ML Estimation with Missing Data ($N = 418$)

	Grade Enrolled at Time of Assessment						
	0	1	2	3	4	5	6
ML estimated mean	19.33	32.88	40.26	48.06	54.68	60.87	67.20
Sample size by grade	346	145	168	146	112	158	159

but instead, in terms of the grade in which they were enrolled when they provided the reading achievement score. This would allow us to fit a linear LCM to the reading scores organized on a more meaningful metric of time.

For example, the ML estimated means of the grade-specific scores for reading achievement based on the same $N = 418$ partially complete cases are presented in Table 3.9. This rearrangement highlights that even though any given child could contribute from one to four repeated measures on reading achievement, we can use a cohort-sequential approach to model a trajectory that spans a total of seven grades.[18]

We could now fit a latent curve model based on these $N = 418$ cases but as defined by $T = 7$ repeated assessments of reading achievement as opposed to the $T = 4$ values by year of assessment. However, given the transition in metric of time from assessment year to enrolled grade, we can reconsider how we selected the subsample from the NLSY in the first place.

Recall that we considered children who were enrolled in grades 0, 1, or 2 in 1986, and we then followed these children forward. There were $N = 418$ who met the inclusion criteria in 1986, and $N = 166$ of these were assessed at all four time points. However, consider a child who was not assessed in 1986 but provided valid data when enrolled in grade 2 in 1990. This child was not included in our original selection criteria because we equated 1986 to be equivalent to "time 1," and the

[18]Under some conditions we might want to empirically evaluate the presence of cohort effects, but we do not explore this here.

Table 3.10 Means of Reading Achievement by Grade Using Direct ML Estimation with Missing Data, Including New Cases ($N = 1767$)

	Grade Enrolled at Time of Assessment						
	0	1	2	3	4	5	6
ML estimated mean	19.82	30.49	41.30	48.14	56.18	61.53	67.04
Sample size by grade	885	645	622	524	394	370	245

child enrolled in grade 2 in 1990 would not have been in grade 0, 1, or 2 in 1986. Given the defined selection criteria, it was logical to omit this child.

However, when we move from year of assessment to grade as the metric of time, the calendar year of assessment becomes much less important. Specifically, if we are interested in the reading achievement scores of children across grade, whether a child was in grade 2 in 1986 or in grade 2 in 1990 is inconsequential.[19] What is important is that we recognize that the child was in grade 2 at any point in the study period. Our original selection procedure is thus ignoring a large proportion of the sample who was not in grade 0, 1, or 2 in 1986, but were within the grade range over the full four assessment periods between 1986 and 1992.

To better capitalize on these data, we expanded our inclusion criteria in terms of grade *at time* of assessment instead of grade *within a given year* of assessment. Our modified inclusion criteria were simple: We considered any child who was in grades 0 through 6 and ages 5 through 13 within *any* year of assessment. We chose these ranges to mimic what might be common in a typical elementary school setting. With these new inclusion criteria, a total of $N = 1767$ children could be included in the longitudinal analysis. The direct ML estimated means taking account of missing data are presented in Table 3.10.

There are several advantages to this approach. First, this data arrangement is more consistent with our theoretical model of developmental trajectories of reading achievement. That is, we do not believe these trajectories to be necessarily tied to the calendar year, but instead, to the progression through elementary school. Second, there is clear evidence of nonlinearity in the mean change over time. As we describe in greater detail in Chapter 4, having additional repeated measures allows for the testing of more complex nonlinear functions over time. Third, there is a decrease in the likelihood that we are introducing bias into our model results as a function of restricting the entry of such a large number of subjects into the sample. Finally, there is a marked increase in statistical power given that we are moving from a sample size of $N = 418$ to a sample size of $N = 1767$. All of these advantages combine to make this an important analytical approach to consider.[20]

[19]It may be an overstatement to claim that year of assessment is "inconsequential." Although this is true for our example here, there may be important cohort or period effects that must be considered in other empirical applications.

[20]Here we are focusing on grade as the primary metric of time and are not considering chronological age within grade. This could easily be accomplished using the methods that we describe here, and could even be extended to test the interaction between age and grade.

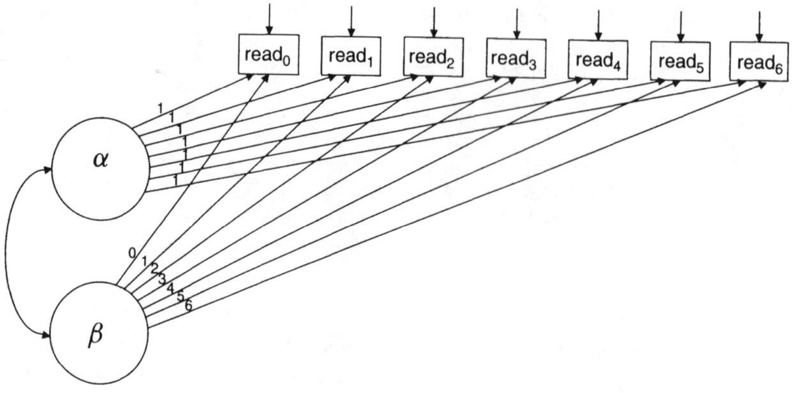

FIGURE 3.4 Latent curve model for repeated reading measures by grade.

Table 3.11 ML Parameter Estimates, Asymptotic Standard Errors, and Critical Ratios for Repeated Reading Measures by Grade Using Direct ML to Take Account of Missing Data ($N = 1767$)

	Parameter Estimate	Asymptotic Standard Error	Critical Ratio
$\hat{\mu}_\alpha$	20.78	0.18	113.95
$\hat{\mu}_\beta$	8.68	0.09	92.31
$\hat{\psi}_\alpha$	13.00	3.44	3.78
$\hat{\psi}_\beta$	4.98	0.47	10.59
$\hat{\psi}_{\alpha\beta}$	4.27	1.05	4.08

We will now reestimate the unconditional latent curve model for reading achievement scores, but instead of the sample of $N = 418$ children over $T = 4$ assessments, we now consider a sample of $N = 1767$ children of $T = 7$ grades. We will again use the direct ML estimator that we described above given the presence of missing data. The sample is characterized by discrete grade cohorts that are assessed at equal intervals, and there is missing data due to attrition over time.[21] This LCM is presented in Figure 3.4.

The linear LCM of the $T = 7$ repeated measures of reading achievement based on the sample of $N = 1767$ children fit quite poorly [$\chi^2(23) = 410.8$, $p < 0.0001$, RMSEA $= 0.10$ (0.09, 0.11), TLI $= 0.64$, IFI $= 0.61$]. This provides strong evidence that the poor fit of the linear LCM fitted to the reading data organized on wave of assessment was probably not due solely to the consideration of the wrong metric of time. We present the parameter estimates and standard errors in Table 3.11

[21] We could also productively apply the multiple imputation procedures here as well. We simply focus on direct ML for expediency.

with the realization that these should be interpreted with substantial caution, given the poor fit of the model.

Examination of these parameter estimates reveals that there is evidence for both fixed and random effects underlying the linear latent curve model of reading. Specifically, the model-implied mean starting point of reading in kindergarten is 20.78, and reading increases 8.68 units per year. Further, there is significant variability around both the starting point and rate of change. However, given the poor fit of the model, additional analyses are necessary prior to drawing any firm theoretical inferences. Indeed, we revisit this example in Chapter 4 when we explore alternatives for modeling nonlinear change over time.

3.3 CONCLUSIONS

In this chapter we have examined the challenges of missing data. Although the traditional ML estimator for complete data has many significant strengths, one limitation is the necessity to analyze data with no missing values. However, in any applied studies of repeated measures, data are generally missing. This in turn undermines use of the traditional ML estimator and prompts us to consider alternatives. Two such methods are direct ML and multiple imputation, each of which is a tool to estimate a variety of LCMs. These methods are asymptotically equivalent, although there is the expected sample-to-sample variation in results, as was evident in our empirical example. Direct ML and MI each have certain strengths and limitations, but either is a substantial improvement over listwise or pairwise deletion approaches. Indeed, we believe that rarely is the use of listwise or pairwise deletion appropriate for LCMs, given the existence of direct ML and MI. We thus strongly recommend the use of these missing data estimators.

Whereas missing data are often thought of as arising from subject nonresponse or attrition over time, this can also be introduced through the restructuring of the data as a function of a metric of time other than wave of assessment. One such mechanism that we examined closely here related to data that are not structured on wave of assessment but on chronological age. Although we explored age and grade as motivating examples, these same issues apply to any alternative metric of time. As with the powerful developments in missing data estimation, we again strongly recommend that alternative metrics of time beyond simple wave of assessment be considered.

CHAPTER FOUR

Nonlinear Trajectories and the Coding of Time

In Chapter 2 we focused exclusively on the *linear latent curve model*, in which the repeated observations were governed by an intercept and a linear slope component. This parameterization resulted in a model-implied rate of change in the repeated measures that was constant across all periods of time. That is, a 1-unit increase in time is associated with an expected β-unit increase in y regardless of specifically what point is considered on the scale of time. However, there is often theoretical interest or empirical necessity in modeling more complex trajectories of change.

For example, a particular outcome of interest may be hypothesized to develop more quickly earlier in time, but the rate of development slows with the passage of time. Alternatively, theory might predict the development of an outcome that decreases more slowly early in time, but the rate of decrease becomes larger with increasing time. Or a theory might predict that a variable grows quickly but then levels off approaching an asymptote. Regardless of the specific form, there are many situations in which a linear model of change does not correspond to the theoretical model of interest. Fortunately, a variety of approaches are available for modeling nonlinear functions within the latent curve framework. Some of these approaches directly define more complex functional forms, some approximate more complex functions with combinations of simpler functions, and some model functions that are influenced directly by the characteristics of the data.

The chapter begins by expanding the linear function to higher-order polynomials that use powers of time that result in well-known functions such as the quadratic and cubic polynomials. We then explore a creative approach in which we fix some of the factor loadings to define the metric, but freely estimate others to optimally fit the empirical data. Discussed next is a piecewise linear trajectory that approximates nonlinear trajectories by combining two linear trajectories. We follow this with an exploration of alternative nonlinear functions, including the exponential and sine functions. A section on linear and nonlinear transformations to the metric of time follows. Empirical examples demonstrate many of these approaches.

4.1 MODELING NONLINEAR FUNCTIONS OF TIME

We begin by extending the linear latent curve model to the broader class of polynomial functions.

4.1.1 Polynomial Trajectories: Quadratic Trajectory Model

One of the most common approaches to modeling nonlinearity within the standard linear regression model is to add powered terms of the explanatory variables to the model (e.g., Cohen, 1978).[1] We can use precisely the same approach to model certain types of nonlinear trajectories over time. Specifically, we can incorporate squared, cubed, or higher-power terms of time and test whether the inclusion of these additional terms leads to a significant improvement in model fit. We first explore the quadratic trajectory model and then extend this to the cubic trajectory model.

As we described in Chapter 2, a linear trajectory is defined by the fixed factor loadings on the slope factor. A common approach codes time as $\lambda_t = t - 1$, which allows for the interpretation of the intercept factor as the starting point of the trajectory.[2] To extend this model to include a nonlinear quadratic component, we simply modify the level 1 model to include a quadratic (i.e., time-squared) term. The level 1 model equation becomes

$$y_{it} = \alpha_i + \lambda_t \beta_{1i} + \lambda_t^2 \beta_{2i} + \epsilon_{it} \tag{4.1}$$

where λ_t^2 is simply the squared value of time at assessment t and β_{2i} is the individually varying value for the quadratic component of the curve. Thus, α_i defines the intercept, β_{1i} the linear slope, and β_{2i} the curvature, and these three components combine additively to reproduce the value of y for individual i at time t.

As before, treat all three of these components as random variables, and we can write expressions for these in our level 2 equations such that

$$\alpha_i = \mu_\alpha + \zeta_{\alpha i} \tag{4.2}$$

$$\beta_{1i} = \mu_{\beta_1} + \zeta_{\beta_1 i} \tag{4.3}$$

$$\beta_{2i} = \mu_{\beta_2} + \zeta_{\beta_2 i} \tag{4.4}$$

[1] It is important to note that the term *linear regression* refers to linearity *in the parameters*, not linearity in the relation between the independent and dependent variables.

[2] There are many alternative coding schemes for defining time, and we explore these in greater detail later in this chapter.

These equations are similar to the linear latent trajectory equations with assumptions about the disturbances having means of zero and being uncorrelated with the random coefficients. The main difference here is the addition of the $\lambda_t^2\beta_{2i}$ term in Eq. (4.1) and the additional level 2 equation [i.e., Eq. (4.4)] for β_{2i}. The $\lambda_t^2\beta_{2i}$ term captures the nonlinear (i.e., squared) component of time in the trajectory for our measure y_{it}. The separate equation for β_{2i} highlights that the coefficient for the squared term varies randomly over individuals in precisely the same way as did the intercept and slope component from the linear trajectory model.

As with the linear model, assume that $E(\epsilon_{it}) = 0$ for all i and t. Also assume that all of the right-hand-side random "variables" (i.e., α_i, β_{1i}, and β_{2i}) are uncorrelated with the equation disturbance ϵ_{it}. More formally, $\text{COV}(\epsilon_{it}, \alpha_i) = 0$, $\text{COV}(\epsilon_{it}, \beta_{1i}) = 0$, and $\text{COV}(\epsilon_{it}, \beta_{2i}) = 0$ for all i and $t = 1, 2, 3, \ldots, T$. It is also assumed that $E(\epsilon_{it}, \epsilon_{jt}) = 0$ for all t and $i \neq j$, and $E(\epsilon_{it}, \epsilon_{jt}) = \theta_{\epsilon_{it}} = \text{VAR}(\epsilon_{it})$ for each t and $i = j$. As we noted in Chapter 2, it is common to assume that all cases have the same variance for each time period [i.e., $E(\epsilon_{it}^2) = \text{VAR}(\epsilon_t)$]. Finally, we assume that $\text{COV}(\epsilon_{it}, \epsilon_{i,t+s}) = 0$ for $s \neq 0$ so that the errors are not correlated over time and that the disturbances for different individuals are uncorrelated.[3]

Next, again substitute the level 2 equations into the level 1 equation to define the reduced-form expression such that

$$y_{it} = \left(\mu_\alpha + \zeta_{\alpha i}\right) + (\mu_{\beta_1} + \zeta_{\beta_1 i})\lambda_t + (\mu_{\beta_2} + \zeta_{\beta_2 i})\lambda_t^2 + \epsilon_{it} \tag{4.5}$$

Simple rearrangement of this reduced form expression results in

$$y_{it} = (\mu_\alpha + \mu_{\beta_1}\lambda_t + \mu_{\beta_2}\lambda_t^2) + (\zeta_{\alpha i} + \zeta_{\beta_1 i}\lambda_t + \zeta_{\beta_2 i}\lambda_t^2 + \epsilon_{it}) \tag{4.6}$$

where the fixed effects are contained in the first parenthetical term and random effects in the second.

As with the linear model, we are interested in the estimation of parameters that define both central tendency and variability. Regarding the former, there are three fixed effects for the quadratic model: the mean of the individual intercepts (μ_α), the mean of the linear trajectory components (μ_{β_1}), and the mean of the quadratic trajectory components (μ_{β_2}). Further, we can estimate parameters that reflect the variability of individuals around the mean intercept ($\psi_{\alpha\alpha}$), the mean linear component ($\psi_{\beta_1\beta_1}$), and the mean quadratic component ($\psi_{\beta_2\beta_2}$). Finally, there is a random effect from level 1 [$\text{VAR}(\epsilon_t)$] that may vary over time.

Structural Equation Latent Curve Model We now turn to the estimation of the quadratic latent curve model (LCM) within the structural equation modeling framework.[4] We can express the level 1 equation in matrix terms such that

$$\mathbf{y} = \mathbf{\Lambda}\eta + \boldsymbol{\epsilon} \tag{4.7}$$

[3]Autocorrelated disturbances for the same case over time is permitted in some situations as long as the resulting model is identified.

[4]Although we can use a case-by-case approach to estimate a quadratic model under some circumstances, we do not pursue that approach here.

where $\mathbf{y}$ is a $T \times 1$ vector of repeated measures, $\mathbf{\Lambda}$ is a $T \times m$ matrix of factor loadings, $\boldsymbol{\eta}$ is a $m \times 1$ vector of latent factors, and $\boldsymbol{\epsilon}$ is a $T \times 1$ vector of residuals. For the simple linear trajectory model fit to T repeated measures, $m = 2$ representing the intercept and the linear slope factor. However, $m = 3$ for the quadratic model given the addition of the quadratic factor. The elements of Eq. (4.7) are

$$\begin{pmatrix} y_{i1} \\ y_{i2} \\ y_{i3} \\ \vdots \\ y_{iT} \end{pmatrix} = \begin{pmatrix} 1 & 0 & 0 \\ 1 & 1 & 1 \\ 1 & 2 & 4 \\ \vdots & \vdots & \vdots \\ 1 & T-1 & (T-1)^2 \end{pmatrix} \begin{pmatrix} \alpha_i \\ \beta_{1i} \\ \beta_{2i} \end{pmatrix} + \begin{pmatrix} \epsilon_{i1} \\ \epsilon_{i2} \\ \epsilon_{i3} \\ \vdots \\ \epsilon_{iT} \end{pmatrix} \tag{4.8}$$

which corresponds to Eq. (4.6). A path diagram for a quadratic latent curve model with $T = 5$ is presented in Figure 4.1.

Each repeated observation of y for individual i at time t is a weighted combination of a random intercept, a random linear trajectory component, a random quadratic trajectory component, and an individual and time-specific disturbance. The level 2 equation expressed via $\boldsymbol{\eta}$ in terms of a mean and deviation is

$$\boldsymbol{\eta} = \boldsymbol{\mu}_{\eta} + \boldsymbol{\zeta} \tag{4.9}$$

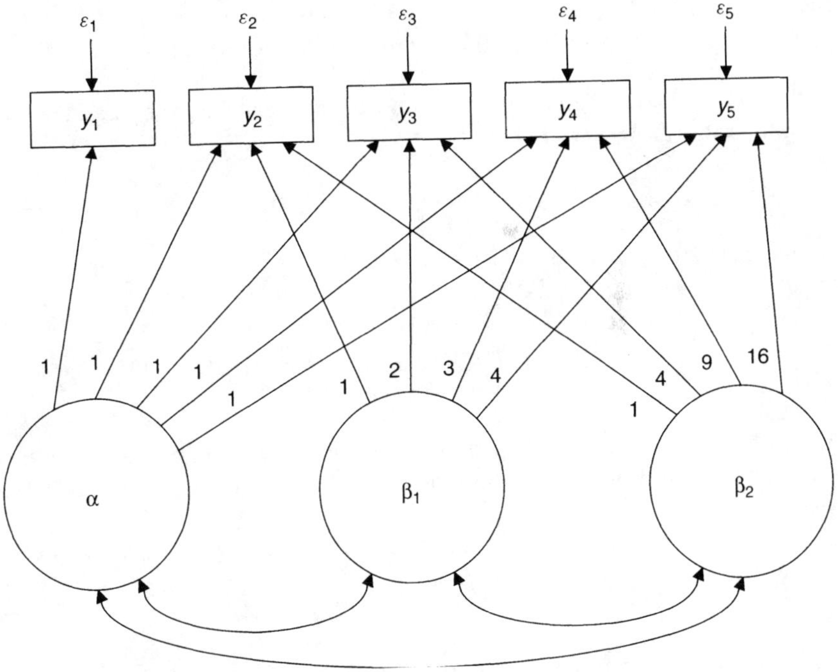

FIGURE 4.1 Path diagram of a five-wave quadratic latent curve model.

where $\boldsymbol{\mu}_\eta$ is $m \times 1$ vector of factor means and $\boldsymbol{\zeta}$ is a $m \times 1$ vector of disturbances. For the quadratic latent curve model, the matrix elements of Eq. (4.9) are

$$\begin{pmatrix} \alpha_i \\ \beta_{1i} \\ \beta_{2i} \end{pmatrix} = \begin{pmatrix} \mu_\alpha \\ \mu_{\beta 1} \\ \mu_{\beta_2} \end{pmatrix} + \begin{pmatrix} \zeta_{\alpha i} \\ \zeta_{\beta_{1i}} \\ \zeta_{\beta_{2i}} \end{pmatrix} \tag{4.10}$$

As we did with the scalar expressions, substitute Eq. (4.9) into Eq. (4.7) to create the reduced-form expression of $\mathbf{y}$ of

$$\mathbf{y} = \boldsymbol{\Lambda}\boldsymbol{\mu}_\eta + (\boldsymbol{\Lambda}\boldsymbol{\zeta} + \boldsymbol{\epsilon}) \tag{4.11}$$

The model-implied variance of the reduced form is

$$\text{VAR}\,(\mathbf{y}) = \boldsymbol{\Lambda}\boldsymbol{\Psi}\boldsymbol{\Lambda}' + \Theta_\epsilon \tag{4.12}$$

where Θ_ϵ represents the covariance structure of the residuals for the T repeated measures of y, where

$$\Theta_\epsilon = \begin{pmatrix} \text{VAR}(\epsilon_1) & 0 & 0 & \cdots & 0 \\ 0 & \text{VAR}(\epsilon_2) & 0 & 0 & 0 \\ 0 & 0 & \ddots & 0 & 0 \\ \vdots & 0 & 0 & \ddots & \vdots \\ 0 & 0 & 0 & \cdots & \text{VAR}(\epsilon_T) \end{pmatrix}$$

Further, $\boldsymbol{\Psi}$ represents the covariance matrix of the equation errors, $\boldsymbol{\zeta}$, among the latent trajectory factors such that

$$\text{VAR}\,(\boldsymbol{\zeta}) = \boldsymbol{\Psi} \tag{4.13}$$

For the quadratic model the elements of $\boldsymbol{\Psi}$ are

$$\boldsymbol{\Psi} = \begin{pmatrix} \psi_{\alpha\alpha} & \psi_{\beta_1\alpha} & \psi_{\beta_2\alpha} \\ \psi_{\beta_1\alpha} & \psi_{\beta_1\beta_1} & \psi_{\beta_2\beta_1} \\ \psi_{\beta_2\alpha} & \psi_{\beta_1\beta_2} & \psi_{\beta_2\beta_2} \end{pmatrix}$$

Because we are not yet considering predictors of the latent curve factors (a topic that we explore in Chapter 5), the variance of $\boldsymbol{\eta}$ is equal to the variance of $\boldsymbol{\zeta}$ [i.e., VAR $(\boldsymbol{\eta}) =$ VAR $(\boldsymbol{\zeta})$].

Finally, the expected value of the reduced-form trajectory model is simply

$$\boldsymbol{\mu}_y = \boldsymbol{\Lambda}\boldsymbol{\mu}_\eta \tag{4.14}$$

where $\boldsymbol{\mu}_y$ is the $T \times 1$ vector of the means of $\mathbf{y}$, $\boldsymbol{\mu}_\eta$ represents a $m \times 1$ vector of factor means, and $\boldsymbol{\Lambda}$ is defined as before. For the $m = 3$ quadratic model, the elements of Eq. (4.14) for T assessments are

$$\boldsymbol{\mu}_y = \begin{pmatrix} \mu_{y1} \\ \mu_{y2} \\ \mu_{y3} \\ \vdots \\ \mu_{yT} \end{pmatrix} = \begin{pmatrix} 1 & 0 & 0 \\ 1 & 1 & 1 \\ 1 & 2 & 4 \\ \vdots & \vdots & \vdots \\ 1 & T-1 & (T-1)^2 \end{pmatrix} \begin{pmatrix} \mu_\alpha \\ \mu_{\beta_1} \\ \mu_{\beta_2} \end{pmatrix} \tag{4.15}$$

where μ_{y_t} represents the mean of y at time t.

Identification The additional parameters estimated in this model raise questions about the number of waves of data required for model identification. Recall that in the linear trajectory model, the unknown model parameters were μ_α, μ_{β_1}, VAR(ϵ_t), $\psi_{\alpha\alpha}$, $\psi_{\beta_1\beta_1}$, and $\psi_{\alpha\beta_1}$. To these we must add the mean and variance of the quadratic factor (μ_{β_2}, $\psi_{\beta_2\beta_2}$) and the covariance of the quadratic factor with the intercept factor ($\psi_{\alpha\beta_2}$) and with the linear slope factor ($\psi_{\beta_1\beta_2}$). This results in a total of $9 + T$ model parameters where there are T error variances for y_{it} when allowing VAR(ϵ_t) to vary over time.

One necessary condition for model identification is that we must have at least as many known to be identified parameters as we have unknown parameters. In the case of three waves of data (e.g., $T = 3$), the quadratic trajectory has 12 parameters to estimate from a total of $(3)(4)/2 + 3 = 9$ identified means, variances, and covariances of the observed variables. Without further information, three waves of data are insufficient to identify the model. This is true even if it is reasonable to assume that the error variances are equal over time [VAR(ϵ_1) = VAR(ϵ_2) = VAR(ϵ_3)] since we still have 10 unknown parameters and only 9 knowns. With four or more waves of data we do have sufficient information to identify a model with a squared term added to the trajectory equation. This is because there are $(4)(5)/2 + 4 = 14$ observed variances, covariances, and means resulting in 12 parameters to estimate from a total of 14 knowns. This counting rule is a necessary but not a sufficient condition for identification.

We know of no proof of identification of the quadratic model. The appendix to this chapter contains a proof that the model is identified using the variances, covariances, and means of at least four repeated measures. For instance, $\boldsymbol{\mu} = \boldsymbol{\mu}(\boldsymbol{\theta})$ is an equality linking the means of the observed variables to the model-implied means. In the case of the quadratic LCM with four waves of data,

$$\begin{bmatrix} \mu_{y_1} \\ \mu_{y_2} \\ \mu_{y_3} \\ \mu_{y_4} \end{bmatrix} = \begin{bmatrix} \mu_\alpha \\ \mu_\alpha + \mu_{\beta_1} + \mu_{\beta_2} \\ \mu_\alpha + 2\mu_{\beta_1} + 4\mu_{\beta_2} \\ \mu_\alpha + 3\mu_{\beta_1} + 9\mu_{\beta_2} \end{bmatrix} \tag{4.16}$$

This equation immediately reveals that $\mu_\alpha = \mu_{y_1}$ and that μ_α is identified. With a little algebra, this equation also yields $\mu_{\beta_1} = 2\mu_{y_2} - \frac{3}{2}\mu_{y_1} - \frac{1}{2}\mu_{y_3}$ and $\mu_{\beta_2} = \frac{1}{2}\mu_{y_1} - \mu_{y_2} + \frac{1}{2}\mu_{y_3}$ (see the chapter appendix). Furthermore, using the covariance matrix of the repeated measures, $\boldsymbol{\Sigma}$, and the equality of $\boldsymbol{\Sigma} = \boldsymbol{\Sigma}(\boldsymbol{\theta})$ permits the identification of the remaining parameters (e.g., $\psi_{\alpha\alpha} = 3\sigma_{12} - 3\sigma_{13} + \sigma_{14}$), as shown in the chapter appendix. So with four repeated measures, the quadratic LCM is identified. In general, higher-order polynomial functions of order d require a minimum of $d + 2$ repeated observations where d represents the highest- order polynomial function in the model. A linear trajectory is a polynomial of order 1 and thus requires at least $d + 2 = 1 + 2 = 3$ repeated observations; a quadratic trajectory is a polynomial of order 2 and thus requires at least four repeated observations, and so on.[5]

Interpretation One advantage of the linear LCM was that the interpretation of the model parameters was straightforward. The linear intercept described the model implied value of the repeated measure at the initial assessment (when $\lambda_t = t - 1$), and the linear slope described the rate of change in the repeated measures per unit change in time. Importantly, this rate of change was equal between any two equally spaced time periods; this is implicit in a linear change model.

However, it is sometimes more challenging to interpret the model parameters for the quadratic LCM. The intercept continues to reflect the model-implied value at the initial assessment (again assuming that $\lambda_t = t - 1$). However, the linear component of the quadratic model described the *instantaneous rate of change* at the initial assessment. In other words, the linear component of the trajectory function is equal to the slope of the tangent line of the curve, where time is equal to zero. A 1-unit increase in time will lead to a *change* in the slope of this tangent line; this rate of change per unit time is reflected in the quadratic component of the trajectory function. Thus, larger (absolute) values of the linear component reflect steeper curvature at the initial time point, and larger (absolute) values of the quadratic component reflect more rapid change in the curvature per unit change of time. At the extreme, if the quadratic component for all cases is equal to zero, there is no change in the rate of change of the trajectory, and this defines a linear function.

Empirical Example of Quadratic LCM: Child Weight To demonstrate the estimation and interpretation of the quadratic latent curve model, we consider repeated measures data on child weight drawn from the National Longitudinal Survey of Youth [see Biesanz et al. (2004) for details]. A total of $N = 155$ children were followed for $T = 5$ time points every other year from age 5 to age 13. Initial examination of the time- specific means suggests that a linear latent curve model may not be appropriate for these data (see Table 4.1). For example, the difference in means between the first and second assessments is 15.77, between the second and third assessments is 17.02, between the third and fourth assessments is 23.92, and between the fourth and final assessments is 22.85. It is clear that the time adjacent increases of the means are not equal over time (as is implied by the linear trajectory) but are larger in magnitude for the later years than for the earlier years.

[5] As we described in Chapter 3, a subset of cases may have fewer numbers of repeated measures than are required for identification due to missing data. Here we continue to assume that a sufficient number of cases have the minimum number of repeated measures necessary to identify the LCM.

Table 4.1 Intercorrelations Between Children's Weight[a] from Ages 5 to 13 ($N = 155$)

	Weight at Time:				
	1	2	3	4	5
1. Child at age 5	—				
2. Child at age 7	0.7947	—			
3. Child at age 9	0.7264	0.8569	—		
4. Child at age 11	0.6405	0.7866	0.8651	—	
5. Child at age 13	0.6025	0.7447	0.7968	0.8981	—
Means	39.5480	55.3160	72.3350	96.2520	119.1030
Standard deviation	6.1096	11.1546	17.8567	26.9084	33.4412

[a]Weight is given in pounds (1 lb = 0.45 kg). Children at age 5 were between 55 and 60 months old when assessed in 1988 in the National Longitudinal Survey of Youth.

We began by fitting a standard linear LCM where we defined the loadings on the linear slope factor to $\lambda_t = \text{age}_t - 5$ so that the intercept represented the starting point of the trajectory. As expected, this fit the observed data poorly [$T_{\text{ML}}(10) = 159.7$, RMSEA $= 0.31$ $(0.27, 0.36)$, IFI $= 0.82$, TLI $= 0.82$]. We then fitted a quadratic latent curve model to the $T = 5$ repeated measures that corresponded to the path diagram presented above. The loadings on the linear slope factor remained $\lambda_t = \text{age}_t - 5$, and the loadings on the quadratic slope factor were set to λ_t^2. Clearly, the fit of this model is better than the linear one according to the chi-square test statistic and the incremental fit indices [$T_{\text{ML}}(6) = 26.53$, IFI $= 0.98$, TLI $= 0.96$]. However, the RMSEA raises some concern [RMSEA $=$ 0.15 $(0.09, 0.21)$]. Some caution is thus warranted when interpreting the parameter estimates and asymptotic standard errors presented in Table 4.2.

The quadratic LCM allows for several interesting insights into the characteristics of the repeated measures data. First, the mean of the intercept factor reflects that the model-implied child weight at age 5 is $\hat{\mu}_\alpha = 39.56$ pounds (because we set $\lambda_t = \text{age}_t - 5$). Given that the mean observed child weight at age 5 was 39.56 pounds, this represents an excellent model-implied reproduction of the observed initial value. Second, the mean of the *linear* latent curve factor is $\hat{\mu}_{\beta_1} = 6.98$, which represents the slope of the tangent line of the curve assessed at age 5. On average, this indicates that children have a positive linear growth component in their trajectories. The mean of the quadratic latent curve factor is $\hat{\mu}_{\beta_2} = 0.37$, indicating that, on average, the curve is increasing more steeply as age increases. Taken together, these results reflect that the developmental trajectory of child weight is increasing over time and that the magnitude of change is increasing as the children grow older. This model-implied mean curve is presented in Figure 4.2.

In addition to these significant fixed effects, there are also important significant variance components associated with all three curve components (see Table 4.2). This indicates that there is significant individual variability in starting point ($\hat{\psi}_{\alpha\alpha} = 33.91$) and rates of linear ($\hat{\psi}_{\beta_1\beta_1} = 10.75$) and quadratic ($\hat{\psi}_{\beta_2\beta_2} = 0.15$) change

Table 4.2 Parameter Estimates and Asymptotic Standard Errors of Quadratic Latent Curve Model for Weight Data ($N = 155$)

Parameter	Estimate	Standard Error
Variances		
$\psi_{\alpha\alpha}$	33.913	6.818
$\psi_{\beta_1\beta_1}$	10.749	2.277
$\psi_{\beta_2\beta_2}$	0.154	0.040
Covariances		
$\psi_{\alpha\beta_1}$	10.238	3.532
$\psi_{\alpha\beta_2}$	0.126	0.385
$\psi_{\beta_1\beta_2}$	−0.443	0.248
Means		
μ_α	39.563	0.489
μ_{β_1}	6.988	0.322
μ_{β_2}	0.373	0.042
Unique variances		
VAR(ϵ_1)	2.942	5.514
VAR(ϵ_2)	15.084	3.689
VAR(ϵ_3)	44.858	7.150
VAR(ϵ_4)	85.200	13.755
VAR(ϵ_5)	73.285	34.486
Model T_{ML}(df = 6)	26.531	

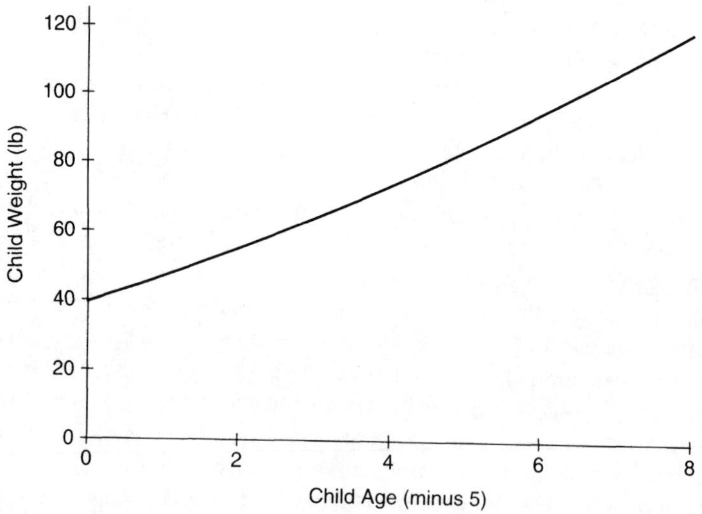

FIGURE 4.2 Model-implied means curve for weight data using the quadratic latent curve model.

over time. In sum, the results of the quadratic LCM reflect that the developmental trajectory of child weight is increasing with age, that the magnitude of weight gain is itself increasing with age, and that there are reliable individual differences across children in the starting point and the linear and quadratic components of change.

4.1.2 Polynomial Trajectories: Cubic Trajectory Models

In principle, we can estimate as high an order of polynomial function as can be identified by the number of repeated assessments. However, based on much practical experience, we have found that higher-order polynomial trajectory models become increasingly difficult to interpret when relating model results back to theory. Given a minimum of $t = 5$ time points, researchers can estimate a third-order polynomial, and this results in the well-known cubic function (remembering that a first-order polynomial is linear, and a second-order polynomial is quadratic). The estimation of the cubic trajectory model is a direct extension of the quadratic model. Specifically, a cubed term for time is introduced into the first-level equation such that

$$y_{it} = \alpha_i + \lambda_t \beta_{1i} + \lambda_t^2 \beta_{2i} + \lambda_t^3 \beta_{3i} + \epsilon_{it} \tag{4.17}$$

As before, the regression coefficient associated with the cube of time is expressed as a random variable itself. The level 2 equations are thus

$$\alpha_i = \mu_\alpha + \zeta_{\alpha i} \tag{4.18}$$

$$\beta_{1i} = \mu_{\beta_1} + \zeta_{\beta_1 i} \tag{4.19}$$

$$\beta_{2i} = \mu_{\beta_2} + \zeta_{\beta_2 i} \tag{4.20}$$

$$\beta_{3i} = \mu_{\beta_3} + \zeta_{\beta_3 i} \tag{4.21}$$

Recall that the dimensions of the factor loading matrix $\mathbf{\Lambda}$ for a polynomial function are $T \times m$, where for the linear model, $m = 2$; for the quadratic, $m = 3$; and for the cubic, $m = 4$. Thus, the factor loading matrix for a cubic trajectory with $T = 5$ is

$$\mathbf{\Lambda} = \begin{pmatrix} 1 & 0 & 0 & 0 \\ 1 & 1 & 1 & 1 \\ 1 & 2 & 4 & 8 \\ 1 & 3 & 9 & 27 \\ 1 & 4 & 16 & 64 \end{pmatrix} \tag{4.22}$$

As mentioned above, we need at least five waves of data to identify the cubic LCM.

Interpretation The interpretation of the cubic model naturally becomes more complex. The linear model implies constant change over time, the quadratic implies change in the rate of change over time, and the cubic model implies change in the change in the rate of change over time. Of course, there are many potential applications in which a cubic model might be well suited to describe the sample data, but care must be taken when interpreting these results with respect to change in the measure over time. Given that the estimation of the cubic model is a straightforward extension of the quadratic model that was described in detail above, we do not present an example of this model here.

4.1.3 Summary

In sum, we can estimate a variety of polynomial trajectory functions within the latent curve modeling framework to capture nonlinear relations. Given an adequate number of repeated measures to allow for proper model identification, we can add a squared time term to define a quadratic trajectory, a cubed time term to define a cubic trajectory, and time can be powered to whatever highest power the data will allow for identification. We can test the significance of specific parameters to allow for testing of the individual mean and variance components associated with each latent trajectory factor.

Despite the many advantages associated with the estimation of the polynomial family of functions in LCMs, there are times when this may not be the optimal nonlinear function for a given theoretical research question or given empirical data set. Our weight data example provides an instance in which the repeated measures follow a nonlinear curve, yet a fully random quadratic LCM was a less-than-optimal fit.

Keep in mind that the entire family of polynomial functions are *parametric*; that is, we define a formal function a priori and evaluate the goodness of fit of this function to our observed data. Instead of fitting predefined trajectory functions, there are alternative methods in which we can fit functions that are more reflective of the characteristics of the given empirical data set. We turn next to the estimation and interpretation of one such approach, the *completely latent* or *freed-loading model*.

4.2 NONLINEAR CURVE FITTING: ESTIMATED FACTOR LOADINGS

In their seminal work on LCMs, Meredith and Tisak (1984, 1990) explored the option of modeling curvilinear trajectories by freeing one or more of the loadings in the latent curve model. Recall that in the linear trajectory model, we fix $\lambda_t = t - 1$ across all t. Meredith and Tisak (1990) proposed that the first loading be set to zero ($\lambda_1 = 0$) and the second to 1 ($\lambda_2 = 1$) to set the metric of the latent factor, but λ_t be freely estimated for $t = 2, 3, \ldots, T$. As before, the level 1 is

$$y_{it} = \alpha_i + \lambda_t \beta_i + \epsilon_{it} \tag{4.23}$$

and the level 2 model is

$$\alpha_i = \mu_\alpha + \zeta_{\alpha i} \tag{4.24}$$

$$\beta_i = \mu_\beta + \zeta_{\beta i} \tag{4.25}$$

However, the key difference from the earlier linear models is that only λ_1 and λ_2 are fixed to predetermined values. The remaining loadings on the slope factor are freely estimated from the data.[6] For $T = 4$ the factor loading matrix is

$$\mathbf{\Lambda} = \begin{pmatrix} 1 & 0 \\ 1 & 1 \\ 1 & \lambda_3 \\ 1 & \lambda_4 \end{pmatrix} \tag{4.26}$$

where λ_3 and λ_4 are freely estimated from the sample data. The free loadings give flexibility in fitting nonlinear forms and are a type of nonlinear "spline" that best fits the data between any two time points.

Recall that for the polynomial trajectory analyses, model identification required four waves for adding a squared term, five waves for adding a cubed term, and so on. Identification of the freed-loading approach is less demanding given the estimation of fewer parameters providing that the metric of the slope factor is properly identified. We are not aware of any proofs of the identification of the freed-loading models, so we provide proof of the conditions for its identification here. As described earlier, the linear trajectory model ($\lambda_t = t - 1$) for three waves of data is overidentified with one degree of freedom. Given that $\lambda_1 = 0$ and $\lambda_2 = 1$, a three-wave model is exactly identified when estimating the third loading. To show that the freed loading is identified, consider the implied means for this model,

$$\mu_{y_1} = \mu_\alpha \tag{4.27}$$

$$\mu_{y_2} = \mu_\alpha + \mu_\beta \tag{4.28}$$

$$\mu_{y_3} = \mu_\alpha + \lambda_3 \mu_\beta \tag{4.29}$$

In Chapter 3 we used the first two equations to demonstrate that μ_α and μ_β are identified such that

$$\mu_\alpha = \mu_{y_1} \tag{4.30}$$

$$\mu_\beta = \mu_{y_2} - \mu_{y_1} \tag{4.31}$$

Given this, from Eq. (4.29) we see that

$$\lambda_3 = \frac{\mu_{y3} - \mu_\alpha}{\mu_\beta} = \frac{\mu_{y3} - \mu_{y_1}}{\mu_{y_2} - \mu_{y_1}} \tag{4.32}$$

The λ_3 will be identified as long as $\mu_{y_2} \neq \mu_{y_1}$. If μ_{y_1} and μ_{y2} were equal, the denominator of Eq. (4.32) would be zero and the right-hand side would be thus undefined. So, under the assumption of $\mu_{y_2} \neq \mu_{y_1}$, the λ_3 parameter is identified. We can use this result to establish the identification status of the remaining

[6] Note that all of the loadings remain fixed to 1.0 on the intercept factor.

parameters in the model. We make use of the general equations for the implied variances and covariances given in Chapter 3 to identify the model. To conserve space, we do not report the results here except to note that the algebraic solutions show us the necessity to make further qualifications on the values of the parameters permitted to allow identification. For instance, when solving for the COV(α, β), we get

$$\text{COV}(\alpha, \beta) = \frac{\text{COV}(y_1, y_3) - \text{COV}(y_1, y_2)}{\lambda_3 - 1} \tag{4.33}$$

In addition to the qualification that $\mu_{y_2} \neq \mu_{y_1}$, we must restrict $\lambda_3 \neq 1$ for the COV(α, β) to be identified. With these added assumptions, the three-wave one-free-loading model is just identified (i.e., df $= 0$). With four or more waves of data, the freed-loading model is overidentified given the assumptions that we just described.

The interpretation of the estimated loadings is rather straightforward. Given that $\lambda_1 = 0$ and $\lambda_2 = 1$, all subsequent change is evaluated with respect to the change observed between the first two time periods. Consider the case with $T = 4$, and say that the third and fourth loadings were estimated to be 2 and 3, respectively. This reflects that the observed change between times 1 and 3 is twice that observed between times 1 and 2 (i.e., $\lambda_3 = 2$), and the change between times 1 and 4 was three times that observed between times 1 and 2 (i.e., $\lambda_4 = 3$). Note that this is, of course, the exact interpretation resulting from the linear model: There is equal change implied between equally spaced time points.

However, say instead that the third and fourth loadings were estimated to be 1.5 and 1.8. This results in a different interpretation from that of the linear change model. Specifically, these loadings would imply that the change observed between times 1 and 3 was only 1.5 times the change observed between times 1 and 2, and not two times the change as imposed by the linear model. Similarly, the change observed between times 1 and 4 was 1.8 times that observed between times 1 and 2, not three times the change as imposed by the linear model. These results indicate that y_{it} is increasing over time, but the magnitude of this increase diminishes with the passage of time. This is a very similar interpretation to that of the quadratic model, but the freely estimated loadings model is more parsimonious given the estimation of fewer parameters. Aber and McArdle (1991) describe this approach as "stretching" the unit of time. That is, it is allowing the values of time to be estimated that linearize the relation between time and y_{it}, but achieve this linearity by transforming the metric of time. This is also what Meredith and Tisak (1990) refer to as the *completely latent trajectory model.*

Empirical Example of Freed-Loading LCM: Child Math Achievement We demonstrate the freed-loading model using the NLSY math achievement data. These are the same data that we used for modeling reading recognition among children, except here our focus is on their math performance. As we described in Chapter 3, this data set consists of as few as one and as many as four repeated measures taken on $N =$

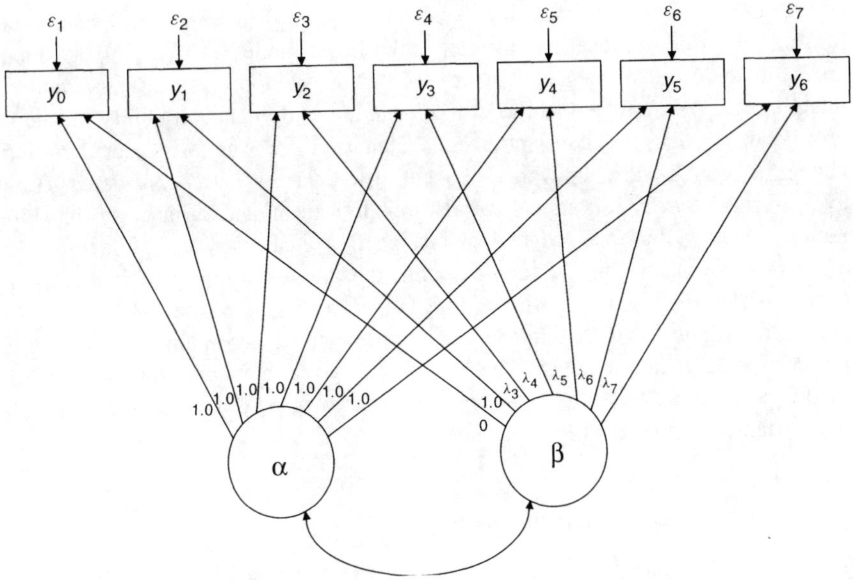

FIGURE 4.3 Freed-loading latent curve model for NLSY math achievement data ($N = 1767$)..

1767 children ranging in age from 6 to 15. The outcome of interest is the proportion of items correctly endorsed on a math achievement test at each assessment.[7]

The linear LCM did not fit, so we explored nonlinear trajectories. We first estimated a quadratic LCM, but multiple estimation problems were encountered, including negative variance estimates, correlations exceeding 1.0, and nonconverged solutions. This suggested that the quadratic model was a poor representation of the observed data. We next estimated a freed-loading LCM in which we fixed the first loading on the slope factor to 0, the second loading to 1, and freely estimated the remaining five loadings. This model is presented in Figure 4.3.

The model fit is fair, although not excellent, according to several of the fit indices [$T_{\text{ML}}(18) = 78.1$, TLI = 0.87, IFI = 0.89, RMSEA = 0.04 (0.03, 0.05)]. The sample estimates for the freely estimated factor loadings were $\hat{\lambda}_3 = 1.98$, $\hat{\lambda}_4 = 2.74$, $\hat{\lambda}_5 = 3.47$, $\hat{\lambda}_6 = 3.88$, and $\hat{\lambda}_7 = 4.36$, reflecting the nonlinear pattern observed in the means. There was also a significant mean of both the intercept ($\hat{\mu}_\alpha = 17.30$) and slope ($\hat{\mu}_\beta = 9.99$) factors, as well as significant variances in each ($\hat{\psi}_{\alpha\alpha} = 14.75$ and $\hat{\psi}_{\beta\beta} = 1.71$, respectively). These variance components reflect that there are individual differences in both starting point and nonlinear rate of change over time. We can thus conclude that math achievement changes nonlinearly over time (but does not follow either a strict linear or quadratic trajectory). Further, there are potentially important individual differences in these nonlinear trajectories.

[7]Because the same achievement test was administered at each assessment period, the metric remains the same over time.

One challenge that arises with this model is the need for appropriate interpretation of the resulting model parameters. In the linear model we were able to obtain an estimate of the mean and variance of the linear slope factor. This could be unambiguously interpreted as the group mean and individual variability in rate of change in y_{it} that was constant over t. That is, there was an expected β_i unit change in y_{it} regardless of which adjacent differences in t was considered. However, in the freed-loading model we also obtain a mean and variance of the slope factor, but now the expected rate of change in y_{it} can vary as a function of t. So, a positive mean estimate for β_i implies that the mean trajectory is increasing over time, as long as the corresponding factor loading is positive. However, the expected magnitude of the change can differ by time, depending on the magnitude of the estimated loadings. This is a perfectly reasonable interpretation of the model parameters, but care must be taken not to impose a linear interpretation inadvertently on the freed-loading model.

4.2.1 Selecting the Metric of Change

The first and second factor loadings in the freed-loading LCM described above were fixed to $\lambda_1 = 0$ and $\lambda_2 = 1$, respectively, and all later loadings were freely estimated. This means that the freely estimated factor loadings are scaled relative to the amount of change that was observed between the first two time periods. Thus, in the prior example, the model-implied change in math ability between the second and third assessments was 1.98 that observed between the first and second assessments, and so on.

However, the only general requirement to this approach is that any one factor loading be set equal to zero and any other one loading is set equal to 1, and all others are freely estimated. Given this, McArdle (1988) has suggested fixing $\lambda_1 = 0$ and the final loading $\lambda_T = 1$, and to freely estimate all of the loadings between the first and last time points. Although the overall model fit statistics will be identical to the fit obtained by setting the first and second loadings instead of the first and last, the estimated loadings have a different interpretation. Specifically, the freed loadings will reflect the *proportion* of change between two time points relative to the total change occurring from the first to the last time points.

To demonstrate this, we reestimated the freed-loading model for the sample of $N = 1767$ children assessed on math achievement. However, instead of setting the first and second loadings to 0 and 1, respectively, we set the first and last loadings to 0 and 1. Table 4.3 reports the results.

As expected, this model resulted in precisely the same fit statistics as before [$T_{\text{ML}}(18) = 78.1$, RMSEA $= 0.04$ $(0.03, 0.05)$]. However, the values of the estimated factor loadings are markedly different: $\hat{\lambda}_2 = 0.23$, $\hat{\lambda}_3 = 0.46$, $\hat{\lambda}_4 = 0.63$, $\hat{\lambda}_5 = 0.80$, and $\hat{\lambda}_6 = 0.89$.

Although these estimated loadings again reflect the nonlinear trend in math ability over time, these suggest an alternative substantive interpretation. Specifically, each value represents the *cumulative proportion of total change* that has occurred from the initial time period to that specific time period. For example, $\hat{\lambda}_2 = 0.23$

Table 4.3 Parameter Estimates and Asymptotic Standard Errors for NLSY Math Achievement Data

Parameter	Estimate	Standard Error
Loadings		
λ_0	0	
λ_1	0.230	0.009
λ_2	0.455	0.010
λ_3	0.628	0.011
λ_4	0.797	0.012
λ_5	0.891	0.012
λ_6	1	
Variances		
$\psi_{\alpha\alpha}$	14.750	3.626
$\psi_{\beta_1\beta_1}$	32.397	8.401
Covariance		
$\psi_{\alpha\beta_1}$	14.942	4.974
Means		
μ_α	17.297	0.199
μ_{β_1}	43.520	0.509
Unique variances		
VAR(ϵ_0)	23.455	3.673
VAR(ϵ_1)	59.802	4.252
VAR(ϵ_2)	73.010	5.101
VAR(ϵ_3)	60.881	5.040
VAR(ϵ_4)	39.919	4.676
VAR(ϵ_5)	31.544	4.608
VAR(ϵ_6)	24.584	5.007
Model T_{ML}(df = 18)	78.1	

reflects that 23% of the total observed change in math ability occurred between the first two assessments. Similarly, $\hat{\lambda}_4 = 0.63$ reflects that 63% of the total observed change in math achievement occurred between the first and fourth assessments; and $\hat{\lambda}_4 - \hat{\lambda}_3 = 0.63 - 0.46 = 0.17$ reflects that 17% of the total change occurred between the third and fourth assessments. Again, this parameterization results in precisely the same overall fit of the model to the observed data, but the rescaled factor loadings provide an alternative metric with which to interpret change.

4.3 PIECEWISE LINEAR TRAJECTORY MODELS

Another useful method for modeling nonlinear relations over time is to approximate the nonlinear function through the use of two or more linear piecewise splines. This piecewise linear trajectory model comes from the mixed modeling framework (e.g., Bryk and Raudenbush, 1992; Snijders and Bosker, 1999) but has been less widely utilized in the SEM approach.

The general procedure is to identify some fixed transition point during the time period under study, and to fit a linear trajectory up to that transition and a linear trajectory after that transition. The transition point might be explicit and determined theoretically (e.g., the time at which subjects transitioned from middle school to high school), or it may be more data driven (e.g., the apparent inflection point of the trajectory as suggested by the empirical data). Although recent work has explored the ability to model transition points that vary randomly over individuals (e.g., individual onset of puberty), these methods are currently not well developed. We thus focus on fitting a piecewise linear latent curve model to repeated measures in which all individuals share the same transition point.

Recall that when we considered the quadratic trajectory model, the level 1 equation was

$$y_{it} = \alpha_i + \lambda_t \beta_{1i} + \lambda_t^2 \beta_{2i} + \epsilon_{it} \tag{4.34}$$

reflecting that the repeated measure y_{it} was an additive function of an intercept, a linear component, and a curvilinear component. In the piecewise model, a similar approach will be taken, but here the repeated measure y_{it} will be an additive component of an intercept, a linear component leading up to the transition point, and a linear component following the transition point. The level 1 equation is thus

$$y_{it} = \alpha_i + \lambda_{1t} \beta_{1i} + \lambda_{2t} \beta_{2i} + \epsilon_{it} \tag{4.35}$$

Note the important differences between Eqs. (4.34) and (4.35). In the former, λ_t represents the value of time at assessment t, and this value is entered as a main effect and as a squared effect. In the latter, λ_{1t} represents one value of time at assessment t, and λ_{2t} represents a *second* value of time at assessment t. It is through the manipulation of these two values of time that we will be able to combine the two linear trajectories.

As before, because the intercept and two linear pieces are treated as random variables, these can be expressed as

$$\begin{aligned} \alpha_i &= \mu_\alpha + \zeta_{\alpha i} \\ \beta_{1i} &= \mu_{\beta_1} + \zeta_{\beta_{1i}} \\ \beta_{2i} &= \mu_{\beta_2} + \zeta_{\beta_{2i}} \end{aligned} \tag{4.36}$$

indicating that each trajectory component is expressed as an additive function of a mean and individual deviation from the mean.

There are several important distinctions between Eqs. (4.34) and (4.35). First, we are now subscripting our measure of time as λ_{1t} and λ_{2t} to indicate the passage of time in the first and the passage of time in the second piece of the trajectory, respectively. Second, there is no powering of the measure of time; that is, whereas the quadratic model incorporated nonlinearity through the power of λ_t, the piecewise model is a strictly additive combination of the nonpowered influence of time.

For example, say that the measure of interest is student self-esteem, and the first three assessments were taken prior to the transition to high school, and the second three assessments were taken after transition to high school. We would like to fit one linear piece to the first three time points and fit one linear piece to the second three time points, and to connect the two pieces at the transition point.

Although the first piece will capture linear change only over the first three time points and the second piece over the second three time points, both measures of time must be coded to reflect the passage of all six time periods. Conceptually, the first piece will bring the trajectory up to the transition point and then allow the second piece to continue after the transition. Thus, the factor loading matrix will be

$$\mathbf{\Lambda} = \begin{pmatrix} 1 & 0 & 0 \\ 1 & 1 & 0 \\ 1 & 2 & 0 \\ 1 & 2 & 1 \\ 1 & 2 & 2 \\ 1 & 2 & 3 \end{pmatrix} \tag{4.37}$$

where the first column of $\mathbf{\Lambda}$ represents the intercept factor, the second column represents the first linear piece, and the third column represents the second linear piece. The coding of time in the piecewise model nicely highlights that the first piece is defining the trajectory up to the transition point, after which the first piece "turns over" the trajectory to the second piece; similarly, the second piece makes no contribution to the trajectory prior to the transition, but "picks up" the trajectory after the transition. Figure 4.4 presents the path diagram for the piecewise linear model where the transition point occurs after the third assessment.

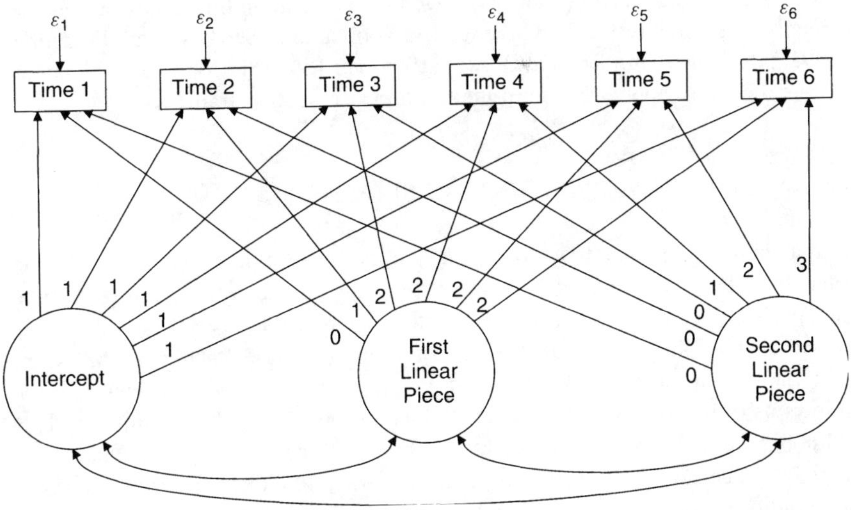

FIGURE 4.4 Path diagram of piecewise latent curve model with third-wave transition.

4.3.1 Identification

The chapter appendix provides a new proof that the piecewise linear LCM is identified with five or more waves of data. For instance, the means of the random coefficients are functions of the means of the repeated measures, such as $\mu_\alpha = \mu_{y_1}$, $\mu_{\beta_1} = \mu_{y_2} - \mu_{y_1}$, and $\mu_{\beta_2} = \mu_{y_5} - \mu_{y_4}$. Analogously, the variances and covariances of the observed variables permit identification of the variances and covariances of the random intercept and the slopes of the two pieces. For example, $\psi_{\alpha\alpha} = 2\sigma_{21} - \sigma_{31}$, $\psi_{\beta_1\beta_2} = \sigma_{31} + \sigma_{42} - \sigma_{32} - \sigma_{41}$, and $\psi_{\alpha\beta_2} = \sigma_{41} - \sigma_{31}$. Finally, the error variances are identified as well. See the appendix for details. Thus, with five or more waves of data, the piecewise linear LCM that we consider here is identified.

4.3.2 Interpretation

The interpretation of the model parameters are straightforward. As before, the means of the three latent factors represent the fixed effects for the corresponding trajectory components. Thus, the mean of the intercept factor represents the model-implied starting point of the trajectory (assuming that we begin the coding of time with zero) and the means of the two linear piece factors represent the model-implied linear change prior to and following the transition, respectively. Similarly, the variance estimates of each factor represent the degree of individual variability around each of the fixed effects. Finally, the covariances among the latent factors represent linear relations among the intercept and two linear pieces. An interesting parameter in this model is the covariance between the first and second linear pieces. This covariance (or correlation, if standardized) represents the association between individual differences in rates of change prior to and following the transition point. For example, a positive covariance between the two linear pieces would imply that, on average, individuals reporting larger slope values prior to transition also tended to report larger slope values following the transition. Inferences such as these might be of key interest in many substantive applications. See Bryk and Raudenbush (1992, p. 149) and Snijders and Bosker (1999, pp. 186–188) for a detailed example of this modeling approach within the multilevel modeling framework.

4.4 ALTERNATIVE PARAMETRIC FUNCTIONS

A standard polynomial trajectory or a piecewise specification are not the only functional forms that might be of interest. Although many interesting nonlinear functions are available (e.g., monomolecular, logistic, Gompertz), not all are easy to incorporate directly into a LCM. The reason is that the standard SEM assumes that the hypothesized model is linear in the parameters. This property is what allows us to incorporate higher-order polynomial functions directly into the LCM. Although our measure of time is squared (e.g., λ_t^2), the associated coefficient is not [e.g., β_{2i}; see Eq. (4.1)]. However, there are ways in which certain functions that are nonlinear in the parameters may be reexpressed so that they may be incorporated into the SEM [see Browne and du Toit (1991) for a more detailed discussion of this]. We will consider two examples of functions other than the polynomial that have this additive property: the exponential trajectory and the cyclic trajectory.

4.4.1 Exponential Trajectory

Exponential functions are a common function to model growth or decay that tends toward an asymptote. In these models changes are proportional to prior change. Thus, a simple model for population growth might posit that each yearly increase in population is a fixed percentage of the base population in the prior year. These same kinds of models have many potential social science applications as well. For example, du Toit and Cudeck (2001) describe methods for estimating several types of nonlinear functions within the SEM framework and place particular focus on the exponential model of change.

Modifying their notation to remain consistent with our own, the exponential trajectory is defined as

$$y_{it} = \alpha_i + \beta_i(1 - e^{-\gamma\lambda_t}) + \epsilon_{it} \tag{4.38}$$

where y_{it} is defined as before. Because $\lambda_1 = 0$ and $e^0 = 1$, α_i represents the intercept of the trajectory at the initial time period. Further, β_i represents the model-implied expected total change in y as time goes to infinity. Finally, γ affects the exponential rate of change in y over time. For $\gamma > 0$ we can think of $e^{-\gamma}$ as representing the rate in deceleration of change in y over time. That is, if we define Δ_{21} to represent the change in y between times 1 and 2, the model-implied change between times 2 and 3 is $\Delta_{32} = e^{-\gamma}\Delta_{21}$, indicating that later change is directly proportional to earlier change.

Du Toit and Cudeck (2001) demonstrate that within the SEM framework we can estimate random components for the intercept and slope (i.e., α_i and β_i, respectively), but that we need to treat γ as fixed. This is reflected in the level 2 equations in that

$$\begin{aligned} \alpha_i &= \mu_\alpha + \zeta_{\alpha_i} \\ \beta_i &= \mu_\beta + \zeta_{\beta_i} \\ \gamma &= \mu_\gamma \end{aligned} \tag{4.39}$$

This implies that individual variability is allowed in the initial starting point and in the total amount of change over time, but the rate of change is fixed for all cases in the sample. As such, we do not have a separate latent variable for γ since it is constant across cases.

The factor loadings for this exponential curve have a more complex form than the other models. Whereas in the polynomial model we could simply power our metric of time to define the curvilinear trajectories, here the nonlinear constraints are more complex and depend on the value of γ. Because the trajectory is defined by two random and one fixed component, our LCM will consist of two latent factors and an "additional" parameter. The first factor will represent the intercept component with factor loadings equal to 1.0. The second factor will represent the total change

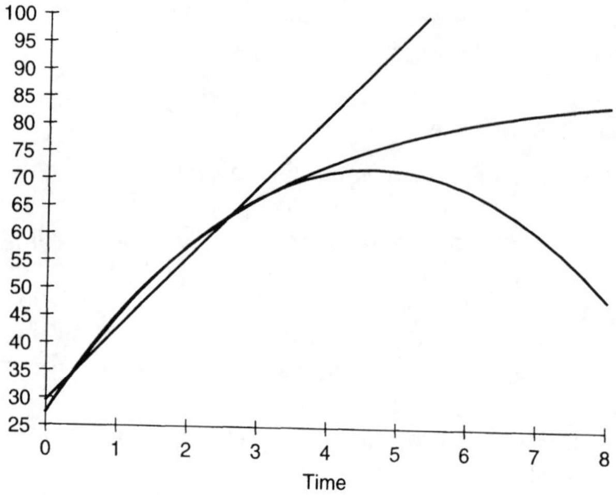

FIGURE 4.5 Comparison of linear, quadratic, and exponential latent curve models.

component with factor loadings equal to $1 - e^{-\gamma\lambda_t}$, where $\lambda_t = 0, 1, \ldots, T-1$. The factor loading for our reading example would thus be

$$\mathbf{\Lambda} = \begin{pmatrix} 1 & 0 \\ 1 & 1-e^{-\gamma(1)} \\ 1 & 1-e^{-\gamma(2)} \\ 1 & 1-e^{-\gamma(3)} \end{pmatrix} \tag{4.40}$$

The constraints on the factor loadings require a nonlinear function that depends on γ. These nonlinear constraints can either be estimated directly [as in Eq. (4.40)] or can be estimated using a higher-order polynomial (see du Toit and Cudeck, 2001, p. 268).[8] Once defined, the rest of the LCM proceeds as before. Specifically, there is a mean and variance estimated for the intercept factor, a mean and variance for the total change factor, a covariance between the intercept and total change factors, a time-specific disturbance for each repeated measure, and a single fixed rate-of-change parameter γ.

One advantage of this exponential trajectory is that unlike the polynomial family of functions, the exponential is strictly monotonic. That is, the exponential trajectory approaches an asymptote, without changing directions. In contrast, the polynomial trajectory will change directions at an inflection point and continue toward plus or minus infinity. This is highlighted in Figure 4.5, where hypothetical model-implied trajectories are plotted for a linear, quadratic, and exponential model LCM spanning eight time points. The linear trajectory tends toward positive infinity and the quadratic tends toward negative infinity. In contrast, the exponential trajectory is approaching an asymptote. If time were followed for a sufficiently longtime, the

[8]A software package is needed that will allow for nonlinear constraints on parameter estimates. At the time of this writing, Proc Calis in SAS and LISREL allow models with nonlinear constraints. See du Toit and Cudeck (2001) for a discussion of programming the exponential function in LISREL.

exponential trajectory would become even flatter. Because of this characteristic, the exponential trajectory is a *bounded function*, whereas the linear and quadratic trajectories are considered *unbounded functions*. Since most social and psychological processes are bounded, there is an advantage to using a function that does not increase or decrease without limits.

In sum, the exponential trajectory has several interesting features not available in the polynomial family of functions. Most important, this is a bounded function that tends toward an asymptote (either growing toward an asymptote or decaying toward an asymptote) and thus does not tend toward infinity as does the polynomial. A closely related method for estimating an exponential function within the SEM framework is Browne's structured latent curve model (Browne and du Toit, 1991; Browne, 1993). Although a powerful methodology, we do not explore this further here.

4.4.2 Parametric Functions with Cycles

Thus far we have explored the polynomial and exponential family of parametric functions for modeling trajectories over time. The polynomials are well known but are limited in that they are inherently unbounded functions. That is, they tend toward plus or minus infinity. Alternatively, the exponential trajectory is bounded and tends toward a growth or decay asymptote. There may be either theoretical or empirical situations in which we might like a function that allows for systematic cycling over time. For example, studying alcohol use over time might include a weekday and weekend cycle, or studying crime rates might include a seasonal cyclic pattern over time. Fortunately, there are several interesting functions that can be combined to allow us a method for estimating such cyclical data in our SEM trajectory framework (Hipp et al., 2004).

We focus here on the use of the sine and cosine functions in combination with our standard polynomial trajectory model. Recall that the cosine wave is defined as the endpoint of an arc of length x of a unit circle as measured counterclockwise from the point (1,0). We can define a function $f(x) = \cos x$, which implies the cyclical trajectory presented in Figure 4.6. This function provides us a way to incorporate a standard cycle within our general trajectory model. To this we can add an intercept-only model to raise or lower the level of the cycle, or we can include a linear or a quadratic to introduce systematic change over time. For example, consider the plot in Figure 4.7, which includes a first-order polynomial (i.e., linear growth model) combined with a cosine wave.

To estimate this model with our SEM framework, we simply need to combine the polynomial model described earlier with the cyclical sine or cosine function. Our model equations would then look like the following:

$$
\begin{aligned}
y_{it} &= \lambda_{t1}\alpha_i + \lambda_{t2}\beta_i + \lambda_{t3}\kappa_i + \epsilon_{it} \\
\alpha_i &= \mu_\alpha + \zeta_{\alpha_i} \\
\beta_i &= \mu_\beta + \zeta_{\beta_i} \\
\kappa_i &= \mu_\kappa + \zeta_{\kappa_i}
\end{aligned}
\tag{4.41}
$$

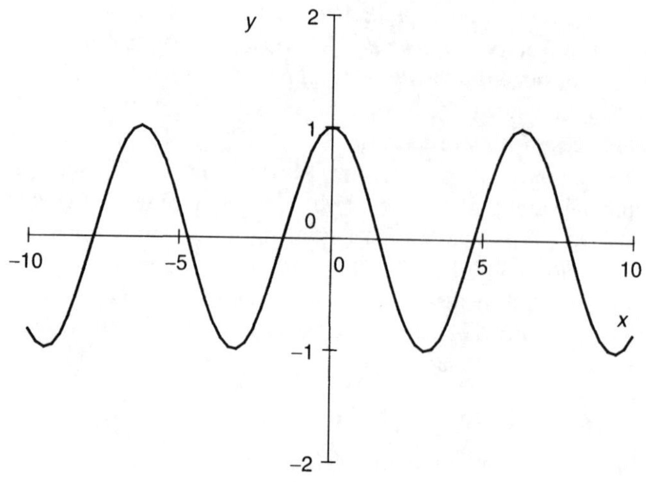

FIGURE 4.6 Cosine function for cyclical trajectories.

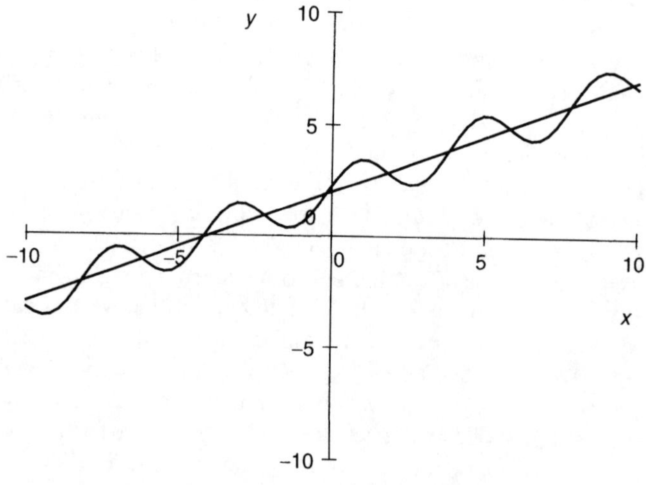

FIGURE 4.7 Cosine function trajectory combined with linear trajectory.

where κ_i is the random amplitude, $\lambda_{t1} = 1$ for all t, $\lambda_{t2} = t - 1$, and λ_{t3} is a nonlinear function of two additional parameters and is defined as

$$\lambda_{t3} = \cos[2\pi(\text{freq}(t + p))] \tag{4.42}$$

where cos denotes the cosine function. π is the well-known mathematical constant, t is time (with $t = 0$ for the initial time period), freq is the frequency of the cyclical wave (i.e., number of cycles per unit time), and p is the phase of the cyclical wave (i.e., location of the peak of the wave over a single cycle). The parameters freq and p can either be estimated from the data or can be set to predefined variables if known a priori.

We used this cyclical trajectory approach to modeling seasonal trajectories of crime over a three-year period based on data drawn from *Uniform Crime Reports* [see Hipp et al., (2004) for further details]. Monthly crime data were obtained for 9500 police reporting units across the United States. Two-month intervals were collapsed into bimonths, for a total of 18 bimonthly measures over a 36-month period. A series of cosine trajectory models were used to examine the cyclical patterns of crime over time and how these related to cycles in season and in temperature, among several other tests of similar hypotheses. A series of nested model comparisons concluded that a cosine model with a quadratic trajectory over time best characterized the sample data. The model-implied trajectory and observed sample means are presented in Figure 4.8. These results are consistent with the theory that reported crime not only followed an overall quadratic trajectory over the time period studied, but that there were consistent and predictable cycles in crime rates that coincided with seasonal cycles in temperature and weather. See Hipp et al. (2004) for further details and additional extensions to this model.

The incorporation of cyclical functions with our existing polynomial and exponential trajectories allows for a wide series of flexible and interesting models of stability and change over time. Both fixed and random effects can be estimated in these models, and parameters governing frequency and phase can either be

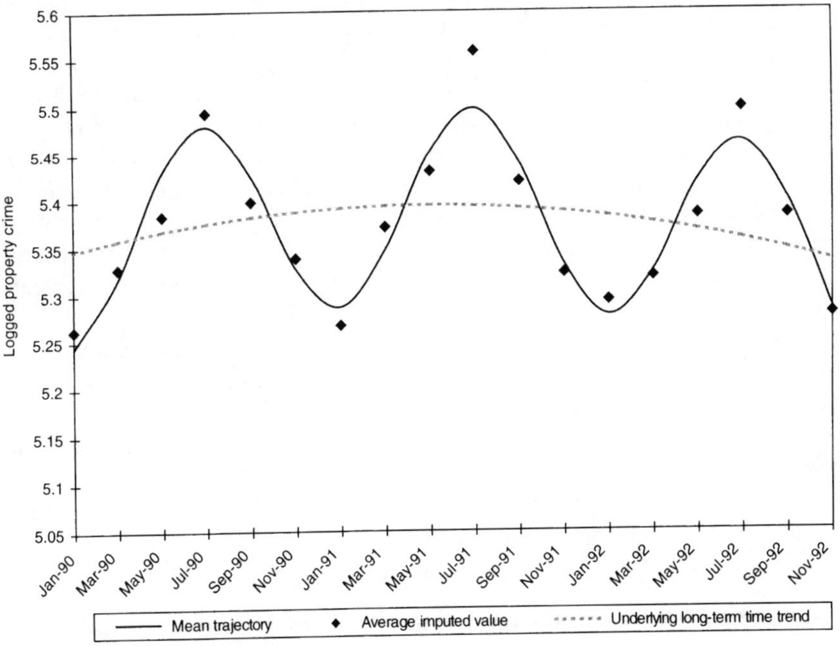

FIGURE 4.8 Model-implied crime trajectory, cosine combined with quadratic function of time ($N = 9500$).

set to predefined values or estimated from the data. The conditions to establish identification of these models are complex, and we will not explore them here. However, in general, it seems reasonable that a minimum of two full cycles occur within the observational window in order to establish estimates of frequency and phase.

Thus far we have considered members of the polynomial family of functions and the exponential family of functions, and we have suggested methods for grafting a repetitive cycle on top of these functions. These, combined with the polynomial functions, the freed-loading model, and the piecewise approach provide a flexible way to capture nonlinearity in the LCM. One other approach to nonlinearity that we have not considered is to capture nonlinearity through transformations of the time metric or of the repeated measure. We explore these briefly next.

4.4.3 Nonlinear Transformations of the Metric of Time

The freed-factor loading approach allows us to set the increment of change per unit time (i.e., change on y_{it} between the first two time points, or change between the first and last time points), and to examine change between any other two time points relative to this initial increment. The values of the freely estimated loadings depend on the characteristics of the empirical data set. Another approach that we can consider is calculating a known nonlinear transformation of the metric of time. For example, consider a situation in which we have four repeated measures taken at equal intervals and we select the coding of time as $\lambda_t = 1, 2, 3, 4$ (we will clarify why we begin the series with 1 in a moment). Now, instead of freely estimating one or more of the loadings from the data, compute a nonlinear transformation of λ_t. We will refer to the transformed metric of time as $\lambda_t^\dagger$, where $\lambda_t^\dagger = f(\lambda_t)$. That is, the newly transformed metric of time is some function of the original metric of time.

For example, consider incorporating the natural log of time as our nonlinear transformation. Thus, $\lambda_t^\dagger = f(\lambda_t) = \ln \lambda_t$ and for $\lambda_t = 1, 2, 3, 4$, $\lambda_t^\dagger = 0$, 0.693, 1.099, 1.386. The reason that we began the series with $\lambda_1 = 1$ is that the natural log of 0 is undefined. We can see that whereas the linear model fixes equal increments in the metric of time, the natural log of time introduces diminishing increments in the passage of time as time progresses. That is, times 1 and 2 are separated by 0.693 unit, times 2 and 3 by 0.496 unit, and times 3 and 4 by 0.287 unit. This induces a diminishing increment between time points with increasing values of time. Unlike the freed-factor loading model, the logged factor loadings are predetermined and thus independent of the empirical data. We could consider other nonlinear transformations of λ_t (e.g., square root or reciprocal), but we do not pursue these further here. As with many approaches to modeling change over time discussed thus far, the appropriateness of the logged loadings depends on the nature of the trajectory and whether the transformation captures these changes accurately.

4.4.4 Nonlinear Transformations of the Repeated Measures

All of the options we have explored thus far for modeling nonlinear relations between the repeated measures and time have considered the dependent measure

y_{it} in the original metric of the measure. However, an alternative option is not to transform the passage of time, but instead, to transform the metric of y_{it}. Nonlinear transformations of observed measures have long been used in other more traditional general linear models (e.g., arcsine transformations in ANOVA models and power transformations in regression analysis), and the same methods apply here. We denote a transformation of the repeated measure of y_{it} as $y_{it}^{\dagger}$, where $y_{it}^{\dagger} = f(y_{it})$ represents some function of the repeated measure. The level 1 equation of the LCM is then expressed in terms of the transformed outcome. For example, the level 1 equation for the linear LCM is simply

$$y_{it}^{\dagger} = \alpha_i + \lambda_t \beta_i + \epsilon_{it} \tag{4.43}$$

where $y_{it}^{\dagger}$ is the transformed version of y_{it}.

Although we defined a linear LCM for $y_{it}^{\dagger}$, any of the nonlinear models that we have discussed here could be used as well (e.g., a quadratic model could easily be fitted to $y_{it}^{\dagger}$ instead). Further, there are a variety of nonlinear transformations that might be considered. For example, Willett and Sayer (1994) used a natural log transformation prior to fitting a linear LCM to repeated measures of tolerance of deviant behavior. Other transformations, such as reciprocals and roots, might be considered as well. The optimal selection of a given transformation (or the use of any transformation at all) must be made on a case-by-case basis, so no general recommendations are readily available.

4.5 LINEAR TRANSFORMATIONS OF THE METRIC OF TIME

In prior sections we focused on a variety of *nonlinear* transformations of the metric of time. However, there are many situations in which *linear* transformations of time might be of interest for interpretation purposes. Within the latent curve model, the metric of time is controlled through the parameterization of the factor loading matrix, which relates the slope factor to the repeated measures. Up to this point we have parameterized the factor loading matrix such that the loading of the first repeated measure on the slope factor was always equal to zero (i.e., $\lambda_t = t - 1$, where T is the total number of assessments). This allowed the intercept factor to represent the value of the trajectory at the initial assessment period. However, this parameterization is not the only one available, and the analyst can adopt other coding schemes, depending on the goals of the research application. These transformations can affect both the meaning of the value of the zero point of time and the spacing between time assessments.[9]

For example, instead of defining the intercept term to reflect the initial period of the trajectory, researchers can define the intercept to reflect the final point of the

[9]Of course, we do not imply that these transformations affect the actual spacing of assessments in the collection of observed data. Instead, we mean that we can transform the numerical values assigned to denote the spacing between the observed assessment periods.

trajectory. Such a parameterization can be useful when evaluating group differences in the final time period. Alternatively, the intercept could be coded to reflect the midpoint of the trajectory, and this may be of use if there is not a compelling reason to focus on either the starting or ending point of the trajectory. Further, if the repeated assessments were collected at six-month intervals, the spacing of time could be transformed so that subsequent interpretations were in the metric of one-year intervals.

As we explore below, the choice of coding scheme will not affect the overall fit of the model to the data. In this sense, the coding of time is arbitrary. However, there are sometimes substantial differences in the values and meanings of certain model parameters, and these must be fully understood to properly interpret the model results.

4.5.1 Logic of Recoding the Metric of Time

Recall that in a standard multiple regression model, the intercept term of the regression equation represents the model-implied mean of the dependent variable when all predictor variables equal zero. Because we are modeling our repeated measures as a function of time, this same interpretation holds for the intercept of the trajectory equation. Recall from Chapter 2 that the expected value and model-implied variance of the repeated measures were expressed as

$$E(y_{it}) = \mu_\alpha + \lambda_t \mu_\beta \tag{4.44}$$

and

$$\text{VAR}(y_{it}) = \psi_{\alpha\alpha} + \lambda_t^2 \psi_{\beta\beta} + 2\lambda_t \psi_{\alpha\beta} + \text{VAR}(\epsilon_t) \tag{4.45}$$

respectively, where time (e.g., parameterized as the factor loadings λ_t) is treated as a predictor of the repeated measures. This parameterization implies that the intercept of the trajectory equation (e.g., μ_α) represents the model-implied mean of y_{it} when $\lambda_t = 0$. Thus, the expected value and variance of y_{it} when $\lambda_t = 0$ simplify to μ_α and $\psi_{\alpha\alpha} + \text{VAR}(\epsilon_t)$, respectively. This highlights that the interpretation of the intercept factor in the latent curve model depends directly on the meaning of the specific point in the time series corresponding to $\lambda_t = 0$, and we have control of this depending on how we parameterize our factor loading matrix. We refer to this as the *zero point* of time.

The second transformation of time to consider is altering the spacing between adjacent codings of time. Equation (4.44) highlights that the mean of the intercept factor represents the mean of y when $\lambda_t = 0$. However, this equation also highlights that the mean of the slope factor (i.e., μ_β) represents the change in the mean of y per unit change in λ_t. That is, a 1-unit change in time is associated with a μ_β-unit change in the mean of y. Given this relation, we are able to transform the spacing of adjacent time values in λ_t subsequently to alter the metric of change of μ_β. We refer to this as the *spacing* of time.

We now present a general expression for modifying the meaning of zero in the measure of time, the spacing of values of time, or both. Although we review the major issues here, see Biesanz et al. (2004) for a comprehensive treatment of this subject.

4.5.2 General Framework for Transforming Time

Because in the latent curve model we use the parameterization of $\mathbf{\Lambda}$ to reflect the coding of time, we can transform $\mathbf{\Lambda}$ in particular ways to reflect whatever coding scheme for time that we like. First, we can modify the point in the trajectory that is denoted by a value of zero. Second, we can modify the spacing between the time points. We can transform either the zero point or the spacing with the use of a simple linear equation.

Staying consistent with the prior notation, λ_t represents the value of time at time point t. Because the factor loadings for the intercept factor are all set to 1.0, the structure of $\mathbf{\Lambda}$ for T repeated assessments is

$$\mathbf{\Lambda} = \begin{pmatrix} 1 & \lambda_1 \\ 1 & \lambda_2 \\ 1 & \vdots \\ 1 & \cdot \\ 1 & \lambda_T \end{pmatrix} \tag{4.46}$$

Given that λ_t is the value of time at t, we can define λ_t^* to represent our transformed measure of time at t such that

$$\lambda_t^* = a_\lambda + b_\lambda \lambda_t \tag{4.47}$$

Here, λ_t represents the original metric of time, b_λ represents the transformation of the spacing of time, and a_λ represents the transformation of the zero point of time. We can thus transform our original metric of time by selection of particular values of a_λ and b_λ.

A more parsimonious matrix representation of the scalar expression in Eq. (4.47) is

$$\mathbf{\Lambda}^* = \mathbf{\Lambda P} \tag{4.48}$$

where $\mathbf{\Lambda}$ is our original factor loading matrix for the trajectory factors, $\mathbf{P}$ is a matrix of transformation weights, and $\mathbf{\Lambda}^*$ is the transformed factor loading matrix. The transformation weights in $\mathbf{P}$ are

$$\mathbf{P} = \begin{pmatrix} 1 & a_\lambda \\ 0 & b_\lambda \end{pmatrix} \tag{4.49}$$

where the first column is always set to 1 and 0, respectively, and a_λ transforms the zero point and b_λ transforms the spacing. More specifically,

$$\mathbf{\Lambda}^* = \mathbf{\Lambda P} = \begin{pmatrix} 1 & \lambda_1 \\ 1 & \lambda_2 \\ 1 & \lambda_3 \\ 1 & \lambda_4 \end{pmatrix} \begin{pmatrix} 1 & a_\lambda \\ 0 & b_\lambda \end{pmatrix} = \begin{pmatrix} 1 & a_\lambda + b_\lambda \lambda_1 \\ 1 & a_\lambda + b_\lambda \lambda_2 \\ 1 & a_\lambda + b_\lambda \lambda_3 \\ 1 & a_\lambda + b_\lambda \lambda_4 \end{pmatrix} \tag{4.50}$$

This general expression gives us complete control in determining both the zero point of time and spacing between time assessments.

Transforming the Initial Value of Time We will first consider transforming the point in time associated with $\lambda_t = 0$. Consider a four-time-point linear trajectory model for which the original factor loading matrix is

$$\boldsymbol{\Lambda} = \begin{pmatrix} 1 & 0 \\ 1 & 1 \\ 1 & 2 \\ 1 & 3 \end{pmatrix} \tag{4.51}$$

where $\lambda_t = t - 1$. These values reflect the fact that we are defining the intercept factor to reflect the initial measure and that the slope factor is measured in terms of 1-unit changes in time. We would like to transform $\boldsymbol{\Lambda}$ so that the mean of the trajectory represents the expected value of y_{it} at the final assessment period. To accomplish this, we define $a_\lambda = 1 - T = 1 - 4 = -3$ and we leave $b_\lambda = 1$ given that we are not yet interested in changing the spacing of time. The new factor loading matrix is thus

$$\boldsymbol{\Lambda}^* = \begin{pmatrix} 1 & 0 \\ 1 & 1 \\ 1 & 2 \\ 1 & 3 \end{pmatrix} \begin{pmatrix} 1 & -3 \\ 0 & 1 \end{pmatrix} = \begin{pmatrix} 1 & -3 \\ 1 & -2 \\ 1 & -1 \\ 1 & 0 \end{pmatrix} \tag{4.52}$$

where it can be seen that the spacing of the observations is retained but the intercept of the trajectory model now reflects the model-implied mean at the *final* time point.

A second transformation of time that might be of interest would be to define $\lambda_t = 0$ to represent the middle observation of the set of repeated measures. To code time in a way that defines the intercept term to reflect the mean of the repeated measure at the middle assessment, we simply set time to a first-order orthogonal polynomial from classic analysis of variance (ANOVA). To accomplish this, we define the "average" of time to be

$$\bar{t} = \frac{\sum_{t=1}^{T} t}{T} \tag{4.53}$$

Thus, setting $a_\lambda = -\bar{t}$ will transform $\boldsymbol{\Lambda}$, where the zero point of time reflects the center of the window of observations. For example, consider the same original $\boldsymbol{\Lambda}$ matrix in Eq. (4.51), but we would like to transform this so that $\lambda_t^* = 0$ does not reflect the first or last measurement, but instead, the middle measurement. In this case, $\bar{t} = (0 + 1 + 2 + 3)/4 = 1.5$, so $a_\lambda = -1.5$ and

$$\boldsymbol{\Lambda}^* = \begin{pmatrix} 1 & 0 \\ 1 & 1 \\ 1 & 2 \\ 1 & 3 \end{pmatrix} \begin{pmatrix} 1 & -1.5 \\ 0 & 1 \end{pmatrix} = \begin{pmatrix} 1 & -1.5 \\ 1 & -0.5 \\ 1 & 0.5 \\ 1 & 1.5 \end{pmatrix} \tag{4.54}$$

The transformed coding of time now defines the mean of the intercept trajectory to reflect the expected value of y at the middle assessment period.[10]

Transforming the Spacing Between Time Assessments We have demonstrated how we can transform $\mathbf{\Lambda}$ to modify the assessment period that corresponds to a value of time equal to zero. However, we can also modify the spacing between each time interval. We accomplish this by selecting differing values of b_λ. For example, again assuming that the original $\mathbf{\Lambda}$ started with zero and was spaced in 1-unit intervals as defined in Eq. (4.51), we can set $a_\lambda = 0$ to retain the original placement of zero but set $b_\lambda = 2$ to transform the metric of change from 1-unit time intervals to 2-unit intervals. The newly transformed matrix is thus

$$\mathbf{\Lambda}^* = \begin{pmatrix} 1 & 0 \\ 1 & 1 \\ 1 & 2 \\ 1 & 3 \end{pmatrix} \begin{pmatrix} 1 & 0 \\ 0 & 2 \end{pmatrix} = \begin{pmatrix} 1 & 0 \\ 1 & 2 \\ 1 & 4 \\ 1 & 6 \end{pmatrix} \tag{4.55}$$

where it can be seen that the intercept term represents the expected value of y_{it} at the initial time period, but that the slope term is defined in increments of 2-unit intervals.

Transforming Both the Initial Value and the Spacing We have demonstrated how to transform the zero point while retaining the original spacing and how to change the spacing while retaining the original zero point. Of course, both of these can be transformed simultaneously by selecting appropriate values for both a_λ and b_λ. For example, we can combine the two examples above for $T = 4$ to define the zero point of time to represent the middle of the repeated measures and also double the spacing to rid ourselves of time codings that require decimals [as in Eq. (4.54)]. To do this, we simply set $a_\lambda = -3$ and $b_\lambda = 2$ and the transformed factor loading matrix is

$$\mathbf{\Lambda}^* = \begin{pmatrix} 1 & 0 \\ 1 & 1 \\ 1 & 2 \\ 1 & 3 \end{pmatrix} \begin{pmatrix} 1 & -3 \\ 0 & 2 \end{pmatrix} = \begin{pmatrix} 1 & -3 \\ 1 & -1 \\ 1 & 1 \\ 1 & 3 \end{pmatrix} \tag{4.56}$$

reflecting that the mean of the intercept factor represents the expected value of y at the middle assessment, and the mean of the slope factor is defined in terms of 2-unit change in time.

Transformations of Time and Changes in Other Model Parameters Unfortunately, the transformation of the metric of time has been the source of some

[10] Note that this middle assessment period is actually observed for odd numbers of assessments, but is inferred from even numbers of assessments.

confusion in the literature. It is important to note that although we can transform $\mathbf{\Lambda}$ in a variety of ways, overall model fit is *invariant* across the linear transformation of time. This means that transforming time is strictly an issue of scaling. Using Eq. (4.48) for computing $\mathbf{\Lambda}^*$, no single time transformation will fit the data any better or worse than any other time transformation. The source of confusion seems to be related to the fact that although the overall model fit does not vary over time scaling, the numerical values of many model parameters and associated standard errors do change, sometimes drastically. Logically, this makes perfect sense. That is, if we change the point for which $\lambda_t = 0$, the mean and variance of the intercept factor is going to reflect fundamentally different values (e.g., the mean of the trajectory at the initial value, the middle value, or the final value). However, all of this confusion is avoided by understanding that the model parameter estimates and standard errors under $\mathbf{\Lambda}^*$ are derivable from the corresponding parameter estimates and standard errors under $\mathbf{\Lambda}$.

To better see this, say that $\mathbf{\Lambda}$ represents our original scaling of time, $\mathbf{\Psi}$ the covariance structure among the trajectory factors based on $\mathbf{\Lambda}$, and $\boldsymbol{\mu}_\eta$ the mean structure of the trajectory factors based on $\mathbf{\Lambda}$. For any given transformed scaling of time denoted $\mathbf{\Lambda}^*$, we can derive the resulting covariance and mean matrices such that

$$\mathbf{\Psi}^* = (\mathbf{\Lambda}^{*'}\mathbf{\Lambda}^*)\mathbf{\Lambda}^{*'}\mathbf{\Lambda}\mathbf{\Psi}\mathbf{\Lambda}'\mathbf{\Lambda}^*(\mathbf{\Lambda}^{*'}\mathbf{\Lambda}^*)^{-1} \tag{4.57}$$

$$\boldsymbol{\mu}_\eta^* = (\mathbf{\Lambda}^{*'}\mathbf{\Lambda}^*)^{-1}\mathbf{\Lambda}^{*'}\mathbf{\Lambda}\boldsymbol{\mu}_\eta \tag{4.58}$$

where $\mathbf{\Psi}^*$ and $\boldsymbol{\mu}_\eta^*$ represent the covariance matrix and mean vector that result from the transformed $\mathbf{\Lambda}^*$. This highlights that the scaling of time through $\mathbf{\Lambda}$ is inherently arbitrary, and the resulting model parameter estimates[11] are wholly predictable given $\mathbf{\Lambda}^*$.

In general, changing the zero point, but not the spacing, will influence the mean and variance of the intercept factor and the covariance between the intercept and slope factors but will not influence the mean and variance of the slope factor. Changing the spacing but not the zero point will influence the mean and variance of the slope factor and the covariance between the intercept and slope factors, but will not influence the mean and variance of the intercept factor. A summary of these differences is presented in Table 4.4. Note that in the body of the table, the terms *changed* and *unchanged* refer to the parameter estimates, standard errors, and critical ratios that exist for the model prior to the change in scaling. For example, for the "zero-point changed, spacing unchanged" section, "PE changed" reflects the fact that the parameter estimate derived under the transformed scaling of time (i.e., $\mathbf{\Lambda}^*$) would be changed relative to the same parameter estimate derived under the original scaling of time (i.e., $\mathbf{\Lambda}$), and so on.

[11] We can similarly predict the transformed standard errors of all parameter estimates for an given Λ^*, but we do not present these expressions here. See Biesanz et al. (2004) for additional details.

Table 4.4 Summary of Changes in Parameter Estimates, Asymptotic Standard Errors, and Critical Ratios as a Function of Changing Zero Point, Spacing, or Both

Zero Point Changed, Spacing Unchanged	
μ_α :	PE changed; SE changed; CR changed
$\psi_{\alpha\alpha}$:	PE changed; SE changed; CR changed
μ_β :	PE unchanged; SE unchanged; CR unchanged
$\psi_{\beta\beta}$:	PE unchanged; SE unchanged; CR unchanged
$\psi_{\alpha\beta}$:	PE changed; SE changed; CR changed
Spacing Changed, Zero Point Unchanged	
μ_α :	PE unchanged; SE unchanged; CR unchanged
$\psi_{\alpha\alpha}$:	PE unchanged; SE unchanged; CR unchanged
μ_β :	PE changed; SE changed; CR unchanged
$\psi_{\beta\beta}$:	PE changed; SE changed; CR unchanged
$\psi_{\alpha\beta}$:	PE changed; SE changed; CR unchanged
Zero Point Changed, Spacing Changed	
μ_α :	PE changed; SE changed; CR changed
ψ_α :	PE changed; SE changed; CR changed
μ_β :	PE changed; SE changed; CR unchanged
ψ_β :	PE changed; SE changed; CR unchanged
$\psi_{\alpha\beta}$:	PE changed; SE changed; CR changed

PE, parameter estimate; SE, asymptotic standard error; CR, critical ratio of PE to SE.

Transforming the Metric of Time in Quadratic Trajectory Models Our discussion of transforming time from $\mathbf{\Lambda}$ to $\mathbf{\Lambda}^*$ focuses solely on the linear trajectory model. That is, $\mathbf{\Lambda}$ consists of T rows and two columns. The first column relates the repeated measures to the intercept factor and the second column with the linear slope factor. However, as we described earlier in this chapter, given sufficient repeated measures to identify all parameters, we can add higher-order polynomials to this model. All of our results about transforming time described for the linear model expand logically to higher-order polynomials as well. For example, to define a quadratic trajectory model for $T = 4$, we would parameterize $\mathbf{\Lambda}$:

$$\mathbf{\Lambda} = \begin{pmatrix} 1 & 0 & 0 \\ 1 & 1 & 1 \\ 1 & 2 & 4 \\ 1 & 3 & 9 \end{pmatrix} \tag{4.59}$$

Our same transformation equation is

$$\mathbf{\Lambda}^* = \mathbf{\Lambda P} \tag{4.60}$$

where the dimensions of **P** expand accordingly. We do not explore these issues further here. However, see Biesanz et al. (2004) for a complete discussion and worked example of transforming the metric of time in quadratic trajectory models.

4.5.3 Summary

The overall fit of a latent curve model to the observed data is independent of linear transformations of the metric of time. That is, any linear transformation of $\mathbf{\Lambda}$ will result in identical likelihood ratio test statistics, fit indices, and residuals. However, the parameter estimates and standard errors of the latent trajectory factors are affected by these transformations, and care should be taken when interpreting these results. Understanding the coding is important so that the resulting interpretations are appropriate given the particular parameterization of time.

4.6 CONCLUSIONS

In Section 4.5 we discussed the consequences of recoding either the origin or the spacing for time. In essence, these recodings of time do not change the fit of the model, but researchers need to be aware of the shift in interpretation of the parameter estimates. The bulk of this chapter considered the issue of modeling nonlinear trajectories, a condition that differed from the time recodings. There are many situations in which a strictly linear model may not define the optimal relation between the repeated measures and time. Theory might predict that the rate of change on the repeated measures diminishes with the progression of time, or the unique characteristics of an empirical data set might demand some alternative to a linear functional form of growth. We explored here a variety of options for modeling such nonlinearity. One powerful approach was to extend the linear model to higher-order polynomial functions such as the quadratic or cubic curve. One advantage of the higher-order polynomials is that the functional form is predetermined through the factor loadings. However, it might be that the data follow a nonlinear function but the function is not well captured by the quadratic. In this case, we suggested alternative known functional forms such as the exponential trajectory. Alternatively, we showed how to free one or more factor loadings and estimate the nonlinear relations from the data, and the result was analogous to fitting a nonlinear spline over time. Another alternative was the piecewise LCM. Finally, we discussed the use of nonlinear transformations of either time or of the repeated measures, and then fitted the usual linear model to these transformed measures.

It is clear that there are a variety of approaches to modeling nonlinear trajectories over time, and it is often difficult to discern which approach is optimal for a given data application. The selection of a particular model depends on both the theoretical model that gave rise to the research question as well as the characteristics of the data. Often, a reasonable approach is to estimate several alternative nonlinear models and to compare their fit to the data.

APPENDIX 4A: Identification of Quadratic and Piecewise Latent Curve Models

4A.1 Quadratic LCM

This part of the appendix proves that the quadratic latent curve model is identified when there are at least four repeated measures.[12] With four repeated measures, $\mathbf{\Lambda}$, $\mathbf{\Psi}$, and $\mathbf{\Theta}_\epsilon$ are

$$\mathbf{\Lambda} = \begin{bmatrix} 1 & 0 & 0 \\ 1 & 1 & 1 \\ 1 & 2 & 4 \\ 1 & 3 & 9 \end{bmatrix} \quad \mathbf{\Psi} = \begin{bmatrix} \psi_{\alpha\alpha} & \psi_{\alpha\beta_1} & \psi_{\alpha\beta_2} \\ \psi_{\alpha\beta_1} & \psi_{\beta_1\beta_1} & \psi_{\beta_2\beta_1} \\ \psi_{\alpha\beta_2} & \psi_{\beta_2\beta_1} & \psi_{\beta_2\beta_2} \end{bmatrix}$$

$$\mathbf{\Theta}_\epsilon = \begin{bmatrix} \text{VAR}(\epsilon_1) & 0 & 0 & 0 \\ 0 & \text{VAR}(\epsilon_2) & 0 & 0 \\ 0 & 0 & \text{VAR}(\epsilon_3) & 0 \\ 0 & 0 & 0 & \text{VAR}(\epsilon_4) \end{bmatrix} \tag{4A.1}$$

This model is in the form of a confirmatory factor analysis (CFA). The implied covariance matrix, $\mathbf{\Sigma}(\boldsymbol{\theta})$, for a CFA is

$$\mathbf{\Sigma}(\boldsymbol{\theta}) = \mathbf{\Lambda}\mathbf{\Psi}\mathbf{\Lambda}' + \mathbf{\Theta}_\epsilon \tag{4A.2}$$

Substituting the values from Eq. (4A.1) into (4A.2) provides the implied covariance matrix for the quadratic model. Assuming that the model is correct, the population covariance matrix of the observed variables, $\mathbf{\Sigma}$, equals the implied covariance matrix, $\mathbf{\Sigma}(\boldsymbol{\theta})$; that is, $\mathbf{\Sigma} = \mathbf{\Sigma}(\boldsymbol{\theta})$.

Similarly, with $\mathbf{y} = \mathbf{\Lambda}\boldsymbol{\eta} + \boldsymbol{\epsilon}$, the implied mean vector, $\boldsymbol{\mu}(\boldsymbol{\theta})$, is

$$\boldsymbol{\mu}(\boldsymbol{\theta}) = \mathbf{\Lambda}\boldsymbol{\mu}_\eta \tag{4A.3}$$

where $\boldsymbol{\mu}_\eta$ is the mean of the latent variables or random coefficients of α, β_1, and β_2. If the model is correct, the means of the repeated measures, $\boldsymbol{\mu}$, equal the implied mean vector, $\boldsymbol{\mu}(\boldsymbol{\theta})$, or $\boldsymbol{\mu} = \boldsymbol{\mu}(\boldsymbol{\theta})$.

In general, the variances and covariances in $\mathbf{\Sigma}$ and the means in $\boldsymbol{\mu}$ are known to be identified. If we can write the model parameters in $\mathbf{\Psi}$ and $\mathbf{\Theta}_\epsilon$ as unique functions of elements from $\mathbf{\Sigma}$ and $\boldsymbol{\mu}$, we have established that these model parameters are

[12] If an LCM had additional restrictions (e.g., equality or additional constant restrictions), it would be possible to identify the quadratic model with fewer waves of data. In practice, it is rare to have justification for such additional restrictions, so we treat the more common case described in this appendix.

identified. Starting with $\boldsymbol{\mu} = \boldsymbol{\mu}(\boldsymbol{\theta})$ leads to

$$\begin{bmatrix} \mu_{y_1} \\ \mu_{y_2} \\ \mu_{y_3} \\ \mu_{y_4} \end{bmatrix} = \begin{bmatrix} 1 & 0 & 0 \\ 1 & 1 & 1 \\ 1 & 2 & 4 \\ 1 & 3 & 9 \end{bmatrix} \begin{bmatrix} \mu_\alpha \\ \mu_{\beta_1} \\ \mu_{\beta_2} \end{bmatrix}$$

$$= \begin{bmatrix} \mu_\alpha \\ \mu_\alpha + \mu_{\beta_1} + \mu_{\beta_2} \\ \mu_\alpha + 2\mu_{\beta_1} + 4\mu_{\beta_2} \\ \mu_\alpha + 3\mu_{\beta_1} + 9\mu_{\beta_2} \end{bmatrix} \tag{4A.4}$$

Manipulation of Eq. (4A.4) to solve for the model parameters results in

$$\begin{aligned} \mu_\alpha &= \mu_{y_1} \\ \mu_{\beta_1} &= 2\mu_{y_2} - \tfrac{3}{2}\mu_{y_1} - \tfrac{1}{2}\mu_{y_3} \\ \mu_{\beta_2} &= \tfrac{1}{2}\mu_{y_1} - \mu_{y_2} + \tfrac{1}{2}\mu_{y_3} \end{aligned} \tag{4A.5}$$

which establishes the identification of the means of the random intercepts (μ_α), linear slopes (μ_{β_1}), and quadratic slopes (μ_{β_2}). This leaves the identification of the elements in $\boldsymbol{\Psi}$ and $\boldsymbol{\Theta}_\epsilon$.

Substituting the values of (A1) into $\boldsymbol{\Lambda\Psi\Lambda}' + \boldsymbol{\Theta}_\epsilon$ leads to the implied covariance matrix. It is easiest first to work with the covariances of the repeated measures with each other and to solve for the elements of $\boldsymbol{\Psi}$ before establishing the identification of $\boldsymbol{\Theta}_\epsilon$. The lower half of the $\boldsymbol{\Lambda\Psi\Lambda}'$ matrix is

$$\begin{bmatrix} \psi_{\alpha\alpha} & & & \\ \begin{gathered}\psi_{\alpha\beta_1} + \psi_{\alpha\beta_2} \\ + \psi_{\alpha\alpha}\end{gathered} & \begin{gathered}2\psi_{\alpha\beta_1} + 2\psi_{\alpha\beta_2} \\ + 2\psi_{\beta_1\beta_2} + \psi_{\alpha\alpha} \\ + \psi_{\beta_1\beta_1} + \psi_{\beta_2\beta_2}\end{gathered} & & \\ \begin{gathered}2\psi_{\alpha\beta_1} + 4\psi_{\alpha\beta_2} \\ + \psi_{\alpha\alpha}\end{gathered} & \begin{gathered}3\psi_{\alpha\beta_1} + 5\psi_{\alpha\beta_2} \\ + 6\psi_{\beta_1\beta_2} + \psi_{\alpha\alpha} \\ + 2\psi_{\beta_1\beta_1} + 4\psi_{\beta_2\beta_2}\end{gathered} & \begin{gathered}4\psi_{\alpha\beta_1} + 8\psi_{\alpha\beta_2} \\ + 16\psi_{\beta_1\beta_2} + \psi_{\alpha\alpha} \\ + 4\psi_{\beta_1\beta_1} + 16\psi_{\beta_2\beta_2}\end{gathered} & \\ \begin{gathered}3\psi_{\alpha\beta_1} + 9\psi_{\alpha\beta_2} \\ + \psi_{\alpha\alpha}\end{gathered} & \begin{gathered}4\psi_{\alpha\beta_1} + 10\psi_{\alpha\beta_2} \\ + 12\psi_{\beta_1\beta_2} + \psi_{\alpha\alpha} \\ + 3\psi_{\beta_1\beta_1} + 9\psi_{\beta_2\beta_2}\end{gathered} & \begin{gathered}5\psi_{\alpha\beta_1} + 13\psi_{\alpha\beta_2} \\ + 30\psi_{\beta_1\beta_2} + \psi_{\alpha\alpha} \\ + 6\psi_{\beta_1\beta_1} + 36\psi_{\beta_2\beta_2}\end{gathered} & \begin{gathered}6\psi_{\alpha\beta_1} + 18\psi_{\alpha\beta_2} \\ + 54\psi_{\beta_1\beta_2} + \psi_{\alpha\alpha} \\ + 9\psi_{\beta_1\beta_1} + 81\psi_{\beta_2\beta_2}\end{gathered} \end{bmatrix} \tag{4A.6}$$

The elements that correspond to the covariances of the observed variables are below the main diagonal. Representing the covariance elements of the population covariance matrix ($\boldsymbol{\Sigma}$) as σ_{ij}, and writing the covariance elements in terms of the model parameters leads to

$$\begin{bmatrix} \sigma_{21} \\ \sigma_{31} \\ \sigma_{41} \\ \sigma_{32} \\ \sigma_{42} \\ \sigma_{43} \end{bmatrix} = \begin{bmatrix} 1 & 1 & 1 & 0 & 0 & 0 \\ 1 & 2 & 4 & 0 & 0 & 0 \\ 1 & 3 & 9 & 0 & 0 & 0 \\ 1 & 3 & 5 & 2 & 6 & 4 \\ 1 & 4 & 10 & 3 & 12 & 9 \\ 1 & 5 & 13 & 6 & 30 & 36 \end{bmatrix} \begin{bmatrix} \psi_{\alpha\alpha} \\ \psi_{\alpha\beta_1} \\ \psi_{\alpha\beta_2} \\ \psi_{\beta_1\beta_1} \\ \psi_{\beta_1\beta_2} \\ \psi_{\beta_2\beta_2} \end{bmatrix} \tag{4A.7}$$

as is verified by multiplying out the right-hand side of Eq. (4A.7) and comparing them to the corresponding elements below the main diagonal of Eq. (4A.6). Premultiplying both sides of Eq. (4A.7) by the inverse of the coefficient matrix on the right-hand side gives an expression that shows that each ψ parameter is a function of the covariances,

$$\begin{bmatrix} \psi_{\alpha\alpha} \\ \psi_{\alpha\beta_1} \\ \psi_{\alpha\beta_2} \\ \psi_{\beta_1\beta_1} \\ \psi_{\beta_1\beta_2} \\ \psi_{\beta_2\beta_2} \end{bmatrix} = \begin{bmatrix} 1 & 1 & 1 & 0 & 0 & 0 \\ 1 & 2 & 4 & 0 & 0 & 0 \\ 1 & 3 & 9 & 0 & 0 & 0 \\ 1 & 3 & 5 & 2 & 6 & 4 \\ 1 & 4 & 10 & 3 & 12 & 9 \\ 1 & 5 & 13 & 6 & 30 & 36 \end{bmatrix}^{-1} \begin{bmatrix} \sigma_{21} \\ \sigma_{31} \\ \sigma_{41} \\ \sigma_{32} \\ \sigma_{42} \\ \sigma_{43} \end{bmatrix}$$

$$= \begin{bmatrix} 3\sigma_{21} - 3\sigma_{31} + \sigma_{41} \\ -\frac{5}{2}\sigma_{21} + 4\sigma_{31} - \frac{3}{2}\sigma_{41} \\ \frac{1}{2}\sigma_{21} - \sigma_{31} + \frac{1}{2}\sigma_{41} \\ \frac{25}{12}\sigma_{21} - \frac{16}{3}\sigma_{31} + \frac{9}{4}\sigma_{32} + \frac{9}{4}\sigma_{41} - \frac{4}{3}\sigma_{42} + \frac{1}{12}\sigma_{43} \\ -\frac{5}{12}\sigma_{21} + \frac{4}{3}\sigma_{31} - \frac{3}{4}\sigma_{32} - \frac{3}{4}\sigma_{41} + \frac{2}{3}\sigma_{42} - \frac{1}{12}\sigma_{43} \\ \frac{1}{12}\sigma_{21} - \frac{1}{3}\sigma_{31} + \frac{1}{4}\sigma_{32} + \frac{1}{4}\sigma_{41} - \frac{1}{3}\sigma_{42} + \frac{1}{12}\sigma_{43} \end{bmatrix} \tag{4A.8}$$

Thus, all variances and covariances of the random intercepts, linear slopes, and quadratic slopes are identified. Since these are the elements of Ψ, this shows that Ψ is identified. The only remaining parameters whose identification status are unknown are the error variances in the main diagonal of $\mathbf{\Theta}_\epsilon$. Given that the other model parameters are identified, we have

$$\mathbf{\Theta}_\epsilon = \mathbf{\Sigma} - \mathbf{\Lambda}\mathbf{\Psi}\mathbf{\Lambda}' \tag{4A.9}$$

and the error variances of the repeated measures are identified. Since all individual parameters of the quadratic LCM are identified by the preceding results, the entire quadratic LCM is identified.

4A.2 Piecewise LCM

The proof that the piecewise LCM with a knot at $t = 3$ is identified is similar to that of the quadratic LCM. For this model,

$$\mathbf{\Lambda} = \begin{bmatrix} 1 & 0 & 0 \\ 1 & 1 & 0 \\ 1 & 2 & 0 \\ 1 & 2 & 1 \\ 1 & 2 & 2 \end{bmatrix} \qquad \mathbf{\Psi} = \begin{bmatrix} \psi_{\alpha\alpha} & \psi_{\alpha\beta_1} & \psi_{\alpha\beta_2} \\ \psi_{\alpha\beta_1} & \psi_{\beta_1\beta_1} & \psi_{\beta_2\beta_1} \\ \psi_{\alpha\beta_2} & \psi_{\beta_2\beta_1} & \psi_{\beta_2\beta_2} \end{bmatrix}$$

$$\mathbf{\Theta}_\epsilon = \begin{bmatrix} \mathrm{VAR}(\epsilon_1) & 0 & 0 & 0 & 0 \\ 0 & \mathrm{VAR}(\epsilon_2) & 0 & 0 & 0 \\ 0 & 0 & \mathrm{VAR}(\epsilon_3) & 0 & 0 \\ 0 & 0 & 0 & \mathrm{VAR}(\epsilon_4) & 0 \\ 0 & 0 & 0 & 0 & \mathrm{VAR}(\epsilon_5) \end{bmatrix} \tag{4A.10}$$

where we have five repeated measures. This piecewise model also is in the form of a CFA. As such, it has the same general form for the implied covariance matrix and the implied mean vector:

$$\mathbf{\Sigma}(\boldsymbol{\theta}) = \mathbf{\Lambda}\Psi\mathbf{\Lambda}' + \mathbf{\Theta}_\epsilon \tag{4A.11}$$

where $\mathbf{\Sigma} = \mathbf{\Sigma}(\boldsymbol{\theta})$ and

$$\boldsymbol{\mu}(\boldsymbol{\theta}) = \mathbf{\Lambda}\boldsymbol{\mu}_\eta \tag{4A.12}$$

where $\boldsymbol{\mu}_\eta$ is the mean of the latent variables or random coefficients of α_i, β_1, and β_2. If the model is correct, $\boldsymbol{\mu} = \boldsymbol{\mu}(\boldsymbol{\theta})$. The implied means are

$$\begin{bmatrix} \mu_{y_1} \\ \mu_{y_2} \\ \mu_{y_3} \\ \mu_{y_4} \\ \mu_{y_5} \end{bmatrix} = \begin{bmatrix} 1 & 0 & 0 \\ 1 & 1 & 0 \\ 1 & 2 & 0 \\ 1 & 2 & 1 \\ 1 & 2 & 2 \end{bmatrix} \begin{bmatrix} \mu_\alpha \\ \mu_{\beta_1} \\ \mu_{\beta_2} \end{bmatrix} = \begin{bmatrix} \mu_\alpha \\ \mu_\alpha + \mu_{\beta_1} \\ \mu_\alpha + 2\mu_{\beta_1} \\ \mu_\alpha + 2\mu_{\beta_1} + \mu_{\beta_2} \\ \mu_\alpha + 2\mu_{\beta_1} + 2\mu_{\beta_2} \end{bmatrix} \tag{4A.13}$$

A set of solutions for the means of the random coefficients is

$$\begin{aligned} \mu_\alpha &= \mu_{y_1} \\ \mu_{\beta_1} &= \mu_{y_2} - \mu_{y_1} \\ \mu_{\beta_2} &= \mu_{y_5} - \mu_{y_4} \end{aligned} \tag{4A.14}$$

To identify the remaining parameters, we substitute the values from Eq. (4A.10) into $\mathbf{\Sigma}(\boldsymbol{\theta}) = \mathbf{\Lambda}\mathbf{\Psi}\mathbf{\Lambda}' + \mathbf{\Theta}_\epsilon$ to get the implied covariance matrix. The covariance structure hypothesis is that $\mathbf{\Sigma} = \mathbf{\Sigma}(\boldsymbol{\theta})$, where the σ_{ij} elements from population covariance matrix ($\mathbf{\Sigma}$) are identified. Demonstrating that the remaining model parameters are unique functions of these covariances identifies these parameters. As with the quadratic LCM, we make use of the covariances rather than the variances of the observed variables to derive solutions.

Six covariances taken from $\mathbf{\Sigma}(\boldsymbol{\theta}) = \mathbf{\Lambda}\mathbf{\Psi}\mathbf{\Lambda}' + \mathbf{\Theta}_\epsilon$ are useful to establish identification of the elements in $\mathbf{\Psi}$:

$$\begin{bmatrix} \sigma_{21} \\ \sigma_{31} \\ \sigma_{41} \\ \sigma_{32} \\ \sigma_{42} \\ \sigma_{54} \end{bmatrix} = \begin{bmatrix} \psi_{\alpha\beta_1} + \psi_{\alpha\alpha} \\ 2\psi_{\alpha\beta_1} + \psi_{\alpha\alpha} \\ 2\psi_{\alpha\beta_1} + \psi_{\alpha\beta_2} + \psi_{\alpha\alpha} \\ 3\psi_{\alpha\beta_1} + \psi_{\alpha\alpha} + 2\psi_{\beta_1\beta_1} \\ 3\psi_{\alpha\beta_1} + \psi_{\alpha\beta_2} + \psi_{\beta_1\beta_2} + \psi_{\alpha\alpha} + 2\psi_{\beta_1\beta_1} \\ 4\psi_{\alpha\beta_1} + 3\psi_{\alpha\beta_2} + 6\psi_{\beta_1\beta_2} + \psi_{\alpha\alpha} + 4\psi_{\beta_1\beta_1} + 2\psi_{\beta_2\beta_2} \end{bmatrix}$$

$$= \begin{bmatrix} 1 & 1 & 0 & 0 & 0 & 0 \\ 1 & 2 & 0 & 0 & 0 & 0 \\ 1 & 2 & 1 & 0 & 0 & 0 \\ 1 & 3 & 0 & 2 & 0 & 0 \\ 1 & 3 & 1 & 2 & 1 & 0 \\ 1 & 4 & 3 & 4 & 6 & 2 \end{bmatrix} \begin{bmatrix} \psi_{\alpha\alpha} \\ \psi_{\alpha\beta_1} \\ \psi_{\alpha\beta_2} \\ \psi_{\beta_1\beta_1} \\ \psi_{\beta_1\beta_2} \\ \psi_{\beta_2\beta_2} \end{bmatrix} \tag{4A.15}$$

where the second equality expresses the covariances as a function of a coefficient matrix times the variances and covariances of α, β_1, and β_2. Taking the inverse of the coefficient matrix and premultiplying both sides of the equation by this inverse leads to

$$\begin{bmatrix} \psi_{\alpha\alpha} \\ \psi_{\alpha\beta_1} \\ \psi_{\alpha\beta_2} \\ \psi_{\beta_1\beta_1} \\ \psi_{\beta_1\beta_2} \\ \psi_{\beta_2\beta_2} \end{bmatrix} = \begin{bmatrix} 1 & 1 & 0 & 0 & 0 & 0 \\ 1 & 2 & 0 & 0 & 0 & 0 \\ 1 & 2 & 1 & 0 & 0 & 0 \\ 1 & 3 & 0 & 2 & 0 & 0 \\ 1 & 3 & 1 & 2 & 1 & 0 \\ 1 & 4 & 3 & 4 & 6 & 2 \end{bmatrix}^{-1} \begin{bmatrix} \sigma_{21} \\ \sigma_{31} \\ \sigma_{41} \\ \sigma_{32} \\ \sigma_{42} \\ \sigma_{54} \end{bmatrix}$$

$$= \begin{bmatrix} 2\sigma_{21} - \sigma_{31} \\ -\sigma_{21} + \sigma_{31} \\ -\sigma_{31} + \sigma_{41} \\ \frac{1}{2}\sigma_{21} - \sigma_{31} + \frac{1}{2}\sigma_{32} \\ \sigma_{31} - \sigma_{32} - \sigma_{41} + \sigma_{42} \\ -\sigma_{31} + 2\sigma_{32} + \frac{3}{2}\sigma_{41} - 3\sigma_{42} + \frac{1}{2}\sigma_{54} \end{bmatrix} \tag{4A.16}$$

Equation (A4.16) establishes the identification of the ψ's. Finally, by forming $\mathbf{\Theta}_\epsilon = \mathbf{\Sigma} - \mathbf{\Lambda}\mathbf{\Psi}\mathbf{\Lambda}'$ the error variances of the repeated measures are identified. Thus, the results above establish the identification of all parameters in the piecewise LCM when there are at least five repeated measures.

CHAPTER FIVE

Conditional Latent Curve Models

In earlier chapters we have provided the tools necessary for describing the trajectories of repeated measures. Whether the trajectories track math skills of children or crime rates of communities, these techniques allow us to summarize the patterns of changes of repeated measures in terms of trajectory parameters. If we can predict or control these parameters, we can predict or control the patterns of change in math skills, crime, or other variables that we observe. To do so, it is necessary to understand what variables influence the trajectory parameters, such as the random intercepts and slopes. Change these covariates and the trajectories of the individuals will change. To do so, we need to construct models that treat the trajectory parameters as outcomes. It is on this class of models that we focus here.

We start with an example that has two covariates or explanatory variables that directly influence the random intercepts and slopes. We then describe the model and assumptions in a way that easily generalizes to more complicated models. Next, we discuss the identification of the parameters in the conditional model. Once we establish identification, we discuss the SEM approach to conditional models, including ways to assess model fit and to interpret the parameter estimates. We conclude with a fully worked empirical example to demonstrate the estimation and interpretation of a conditional linear LCM using data on children's antisocial behavior.

5.1 CONDITIONAL MODEL AND ASSUMPTIONS

Our starting point for the *conditional latent curve model* is the same level 1 equation as in the unconditional model,

$$y_{it} = \alpha_i + \lambda_t \beta_i + \epsilon_{it} \tag{5.1}$$

where y_{it} is the value of the trajectory variable y for the ith case at time t, α_i is the random intercept for case i, and β_i is the random slope for case i. Further, λ_t

is a constant where a common coding convention is to have $\lambda_1 = 0$, $\lambda_2 = 1$. The remaining values of λ_t allow the incorporation of linear or nonlinear trajectories (see Chapter 4). In the case of a linear trajectory model $\lambda_t = t - 1$ for all t. We assume that $E(\epsilon_{it}) = 0$ for all i and t, $\text{COV}(\epsilon_{it}, \beta_i) = 0$ and $\text{COV}(\epsilon_{it}, \alpha_i) = 0$ for all i and $t = 1, 2, 3, \ldots, T$, and $E(\epsilon_{it}, \epsilon_{jt}) = 0$ for all t and $i \neq j$. The variance of the disturbance is $E(\epsilon_{it}, \epsilon_{jt}) = \theta_{\epsilon_{it}} = \text{VAR}(\epsilon_{it})$ for each t and $i = j$. Another assumption is that $\text{COV}(\epsilon_{it}, \epsilon_{i,t+s}) = 0$ for $s \neq 0$, so that the errors are not correlated over time. We also assume that the disturbances for different individuals at different time points are uncorrelated [$\text{COV}(\epsilon_{it}, \epsilon_{j,t+s}) = 0$ for $i \neq j$]. These assumptions match those for the unconditional model. The random intercepts (α_i) and slopes (β_i) determine the individual trajectory of math skills, crime, or other repeated measures, and these trajectories can differ over cases.

Recall that the unconditional latent curve model incorporated no covariates. It was defined by the means of the intercepts and slopes and the deviations from these means. In contrast, the conditional model includes variables that predict the intercepts or slopes. For instance, the trajectories for math skills might be influenced by the gender or race of the child, or the trajectory of crime might be affected by poverty or the density of the community. The conditional model permits us to test these potential influences on the trajectory parameters.

The unconditional model departs from the conditional in the level 2 equations for the random intercepts and slopes:

$$\alpha_i = \mu_\alpha + \gamma_{\alpha_1} x_{1i} + \gamma_{\alpha_2} x_{2i} + \zeta_{\alpha i} \tag{5.2}$$

$$\beta_i = \mu_\beta + \gamma_{\beta_1} x_{1i} + \gamma_{\beta_2} x_{2i} + \zeta_{\beta i} \tag{5.3}$$

where μ_α and μ_β are the intercepts for the equations that predict the random intercepts [Eq. 5.2] and slopes [Eq. 5.3] across all cases. Specifically, μ_α and μ_β are the mean intercepts and mean slopes when x_{1i} and x_{2i} are zero. The x_{1i} and x_{2i} are two covariates or predictors of the random intercepts and slopes, γ_{α_1} and γ_{α_2} are the covariate coefficients for x_{1i} and x_{2i} in the random intercept equation, and γ_{β_1} and γ_{β_2} are the covariate coefficients in the random slope equation. The covariate coefficients are interpreted the same as regression coefficients in that they provide the expected difference in the outcome for a 1-unit difference in the explanatory variable net of the other explanatory variable. Generally, the covariates x_{1i} and x_{2i} will be time invariant. That is, the covariates may take on different values over cases, but they do not change over time.[1] The $\zeta_{\alpha i}$ and $\zeta_{\beta i}$ are disturbances with means of zero and variances of $\psi_{\alpha\alpha}$ and $\psi_{\beta\beta}$ and covariance $\psi_{\alpha\beta}$. In contrast to the unconditional model, $\psi_{\alpha\alpha}$ and $\psi_{\beta\beta}$ are no longer variances of the random intercepts and random slopes, but are conditional variances, or disturbances. Finally, $\zeta_{\alpha i}$ and $\zeta_{\beta i}$ are uncorrelated with ϵ_{it}, x_{1i}, and x_{2i}.

We can combine the trajectory, intercept, and slope equations into a single equation by substituting the right-hand sides of the intercept and slope equations

[1] On occasion, a researcher might use a time-varying covariate in a model that occurs at or prior to the first wave of the repeated measures. In that situation, adding another subscript to the x's would help to clarify the timing of the covariate variable, and we describe this model in detail in Chapter 7.

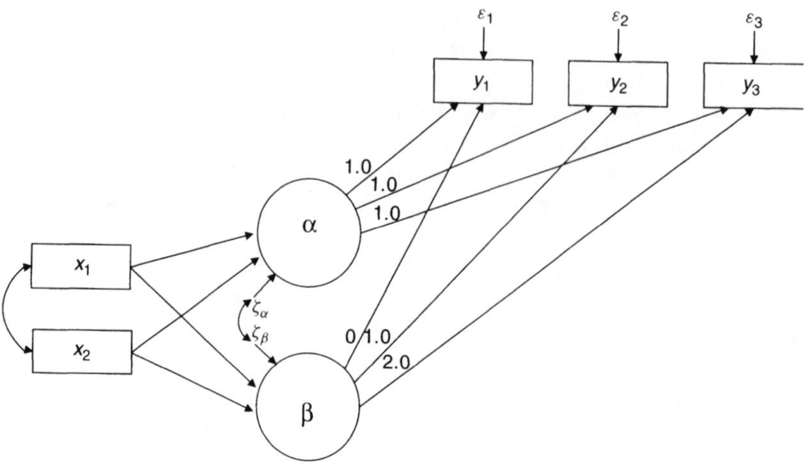

FIGURE 5.1 Linear latent curve model with three repeated measures and two covariates.

for α_i and β_i, respectively, into the trajectory equation. The result is

$$y_{it} = (\mu_\alpha + \lambda_t \mu_\beta) + (\gamma_{\alpha_1} + \lambda_t \gamma_{\beta_1}) x_{1i} \\ + (\gamma_{\alpha_2} + \lambda_t \gamma_{\beta_2}) x_{2i} + (\zeta_{\alpha_i} + \lambda_t \zeta_{\beta i} + \epsilon_{it}) \tag{5.4}$$

This is the combined model or the reduced-form equation for the latent curve model. It is a reduced form in that the endogenous random coefficients, α_i and β_i, are replaced by their exogenous determinants and disturbances. In this form of the equation we see that the trajectory of y_{it} is a function of a composite intercept, composite coefficients for x_{1i} and x_{2i} that change with λ_t, and a composite disturbance term whose variance changes with λ_t. The parenthesized terms for the intercept and coefficients for x_{1i} and x_{2i} in Eq. 5.4 are the fixed components of the latent curves, and the last term in parentheses is the random component. The fixed component represents the mean structure, and the random component represents various sources of individual variability. A hypothetical linear LCM for three repeated measures and two correlated predictors of the intercept and slope is presented in Figure 5.1.

Table 5.1 provides a summary of the model, definitions, and assumptions for the conditional latent curve model with two covariates. This model easily generalizes to any number of covariates. The main difference is the additional coefficients in the level 2 model that would correspond to the impact of the new variables on the random intercepts and slopes.

5.2 IDENTIFICATION

In Chapter 2 we established model identification of the unconditional model by showing that we could find unique solutions for all of the model parameters in

Table 5.1 Model, Definitions, and Assumptions for the Conditional Latent Curve Model with Two Covariates

Model

$y_{it} = \alpha_i + \lambda_t \beta_i + \epsilon_{it}$	*Trajectory equation*
$\alpha_i = \mu_\alpha + \gamma_{\alpha_1} x_{1i} + \gamma_{\alpha_2} x_{2i} + \zeta_{\alpha i}$	*Intercept equation*
$\beta_i = \mu_\beta + \gamma_{\beta_1} x_{1i} + \gamma_{\beta_2} x_{2i} + \zeta_{\beta i}$	*Slope equation*
$y_{it} = (\mu_\alpha + \lambda_t \mu_\beta) + (\gamma_{\alpha_1} + \lambda_t \gamma_{\beta_1}) x_{1i}$ $+ (\gamma_{\alpha_2} + \lambda_t \gamma_{\beta_2}) x_{2i} + (\zeta_{\alpha_i} + \lambda_t \zeta_{\beta i} + \epsilon_{it})$	*Combined equation*

where

$i = 1, 2, ..., N$, where N is the total number of cases
$t = 1, 2, ..., T$, where T is the total number of time points

Assumptions

$E(\epsilon_{it}) = 0$ for $i = 1, 2, \ldots, N$, $t = 1, 2, \ldots, T$
$E(\zeta_{\alpha i}) = 0$ for $i = 1, 2, \ldots, N$
$E(\zeta_{\beta i}) = 0$ for $i = 1, 2, \ldots, N$
$\text{COV}(\epsilon_{it}, \zeta_{\alpha i}) = 0$ for $i = 1, 2, \ldots, N$, $t = 1, 2, \ldots, T$
$\text{COV}(\epsilon_{it}, \zeta_{\beta i}) = 0$ for $i = 1, 2, \ldots, N$, $t = 1, 2, \ldots, T$
$\text{COV}(\zeta_{\alpha i}, \zeta_{\alpha j}) = 0$ for $i \neq j$
$\text{COV}(\zeta_{\beta i}, \zeta_{\beta j}) = 0$ for $i \neq j$
$\text{COV}(\epsilon_{it}, \epsilon_{j,t+s}) = 0$ for $i \neq j$, $t = 1, 2, \ldots, T$, all s
$\text{COV}(x_{ki}, \epsilon_{it}) = \text{COV}(x_{ki}, \zeta_{\alpha i}) = \text{COV}(x_{ki}, \zeta_{\beta i}) = 0$
for $i = 1, 2, \ldots, N$; $k = 1, 2, \ldots, K$; $t = 1, 2, \ldots, T$;
α_i and β_i are assigned scales
[$\text{COV}(\epsilon_{it}, \epsilon_{it+s}) = 0$ for $s \neq 0$ (optional assumption)]

terms of the variances, covariances, and means of the observed variables when we had at least three waves of data. As we demonstrate below, a conditional model will always be identified if (1) the unconditional latent curve model is identified and (2) all exogenous variables are manifest (e.g., there are no exogenous latent factors).[2] Below we present a detailed explication of these identifying conditions.

We start with the list of parameters that are known to be identified and those whose identification status is unknown.[3] To keep these results more general, consider the case where the number of covariates (x's) is equal to K. The known-to-be-identified parameters are the means, variances, and covariances of the y's and x's: $E(y_{it})$, $E(x_{ki})$, $\text{COV}(y_{it}, y_{i,t-s})$, $\text{COV}(x_{ki}, x_{k-m,i})$, and $\text{COV}(y_{it}, x_{ki})$. This leads to $\frac{1}{2}(T + K)(T + K + 3)$ known-to-be-identified means, variances, and covariances of the observed variables.

The unknown model parameters for a linear LCM are μ_α, μ_β, γ_{α_k}, γ_{β_k}, $\theta_{\epsilon_{it}}$, $\psi_{\alpha\alpha}$, $\psi_{\beta\beta}$, $\psi_{\alpha\beta}$, and λ_t. One necessary condition for model identification is that

[2] We can easily incorporate multiple indicator latent factors as exogenous predictors of the latent curve factors, but we must then establish the identification of these latent variable models. In general, if we have at least two indicators of each latent covariate, the model will be identified. See Chapter 8 for details.

[3] See Section 2.3 or Bollen (1989b, pp. 88–89).

we must have at least as many known-to-be-identified parameters as we have unknown parameters.[4] There will be NT parameters for $\theta_{\epsilon it}$, T parameters for λ_t, $2K$ parameters for γ_{α_k}, γ_{β_k}, and five parameters for μ_α, μ_β, $\psi_{\alpha\alpha}$, $\psi_{\beta\beta}$, and $\psi_{\alpha\beta}$ (for a linear model). The number of unknown parameters far exceeds the number of known-to-be-identified parameters, and the model is underidentified without further restrictions.

We can move closer to identification by imposing some common restrictions. First assume that the trend values captured by λ_t are known, such as in the linear trend case, where $\lambda_t = t - 1$. Second, assume that each case has the same error variances in the same time period, although the variances can differ over time; that is, $\theta_{\epsilon it} = \theta_{\epsilon t}$. Now we have $T + 2K + 5$ unknowns. Consider three waves of data and two covariates (x_{1i} and x_{2i}). This gives us 20 means, variances, and covariances of observed variables with which to work and 12 model parameters whose identification status is unknown. Since the known-to-be-identified elements exceed the unknown, we have satisfied a necessary (but not sufficient) condition of identification (i.e., the t-rule, Bollen, 1989b, p. 328) and the model might be identified.

We can use a *two-step rule* of identification (Bollen, 1989b, pp. 328–331) and the results from Chapter 2 to establish that this model is identified. The two-step rule first converts the structural equation model into a measurement model where the only direct effects are those between the latent variables and their indicators. All other associations are converted to covariances among the latent variables and exogenous x's. The first step of the two-step rule is to establish that this model would be identified as a measurement model. To establish its identification, any identification rules for confirmatory factor analysis are applicable (see Bollen, 1989b, pp. 238–254), or a researcher can use algebraic means to establish identification of the measurement model form of the model. Assuming that identification of the measurement model is established, the second step is to demonstrate that the latent variable structural model is identified while ignoring the measurement model parameters. That is, the researcher should treat the latent variable structural model as if it were a simultaneous equation model in observed variables. If the latter model is identified, the entire model is identified. The Two-Step Rule is a sufficient rule of model identification (Bollen, 1989b).

This rule is best explained further with an example. We begin with a linear latent curve model for $t = 3$ waves of data and two covariates. This was presented in Figure 5.1. Figure 5.2 illustrates the first step of converting the conditional latent curve model with two covariates and three waves of data into a measurement model. The impact of the x's on the random intercepts and slopes is replaced by covariances of the x's with the random intercepts and slopes. The first step is to establish that this measurement model would be identified. A good starting point is to examine the part of Figure 5.2 that is identical to the unconditional latent curve model. This is in the lower part of the path diagram. We can use

[4]Note that this assumes no inequality constraints that could obtain identification in some special cases.

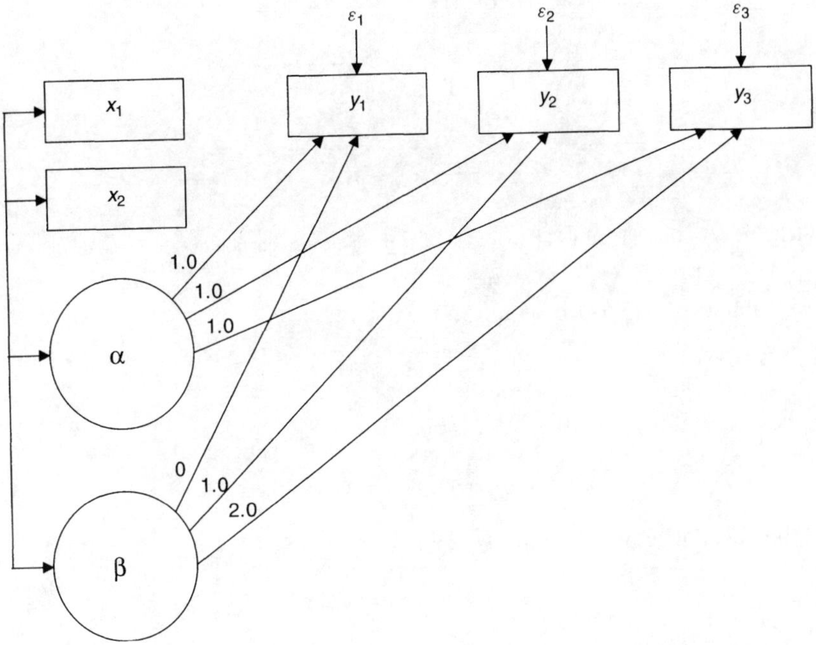

FIGURE 5.2 Conditional latent curve model converted to a measurement model.

the algebraic results from Chapter 2 for the three-wave model to establish that the variances of the errors and the variances and covariances of the random intercepts (α_i) and random slopes (β_i) are identified. Chapter 2 also showed that μ_α and μ_β are identified. This leaves the identification of the covariances of the x's with each other and with α_i and β_i, and the means of the x's to complete step 1.

First we know that the means, variances, and covariances of the x's are identified because the x's are observed variables. To obtain the covariances of the x's with α_i, consider the equation for y_{i1}:

$$y_{i1} = \alpha_i + \epsilon_{i1} \tag{5.5}$$

The covariance of x_{1i} with y_{i1} is

$$\text{COV}(x_{1i}, y_{i1}) = \text{COV}(x_{1i}, \alpha_i) \tag{5.6}$$

and a similar approach shows that $\text{COV}(x_{2i}, y_{i1}) = \text{COV}(x_{2i}, \alpha_i)$. The only remaining parameters for which we need to establish identification are $\text{COV}(x_{1i}, \beta_i)$ and $\text{COV}(x_{2i}, \beta_i)$.

To show the identification of these two covariances, consider the y_{i2} equation,

$$y_{i2} = \alpha_i + \beta_i + \epsilon_{i2} \tag{5.7}$$

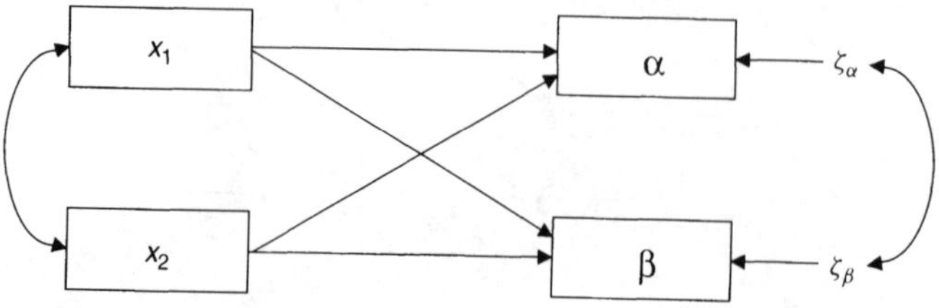

FIGURE 5.3 Conditional LCM converted to a simultaneous equation observed variable model.

If we subtract Eq. (5.5) from Eq. (5.7), we get

$$y_{i2} - y_{i1} = \beta_i + \epsilon_{i2} - \epsilon_{i1} \tag{5.8}$$

The covariance of x_{1i} with $(y_{i2} - y_{i1})$ is

$$\text{COV}((y_{i2} - y_{i1}), x_{1i}) = \text{COV}(\beta_i, x_{1i}) \tag{5.9}$$

and the covariance of x_{2i} with $(y_{i2} - y_{i1})$ is

$$\text{COV}((y_{i2} - y_{i1}), x_{2i}) = \text{COV}(\beta_i, x_{2i}) \tag{5.10}$$

This establishes that the model in Figure 5.2 is identified as a measurement model and completes the first step of the Two-Step Rule.

Figure 5.3 is the path diagram for the second step, where we treat the relation between the exogenous x's and the latent random coefficients as if all variables were observed variables. This is why boxes enclose the random intercepts and slopes rather than the ellipses in the prior diagram. In this form we can use any of the rules of identification that apply to simultaneous equation models (Bollen, 1989b, pp. 88–104). A simple rule that applies to Figure 5.3 is the Null Beta Rule. It states simply that if there are no direct relationships between the endogenous variables in the model, this is sufficient for model identification (Bollen, 1989b, pp. 94–95). These conditions are satisfied, so the second step of the Two-Step Rule is identified and hence the original model is identified.

As the reader can verify, the Two-Step Rule would also work if we had a single covariate or added many more x's. Thus, the conditional univariate latent curve model is identified with three waves of data and any number of x's. The conditional latent curve model with a freed loading for λ_3 also would be identified by using the Two-Step Rule. What happens if we have only two waves of data? In this case the conditional model as a whole is not identified (see Chapter 2 for details), although μ_α, μ_β, γ_{α_k}, and γ_{β_k} are identified.

5.3 STRUCTURAL EQUATION MODELING APPROACH

It is possible to estimate conditional latent curve models using the case-by-case approach described in Chapter 2. In brief, the case-by-case intercepts and slopes are estimated for the unconditional model as discussed in Chapter 2. Then the $\widehat{\alpha}_i$

and $\widehat{\beta}_i$ are regressed on the covariates to determine the impact of the covariates on the estimated trajectory parameters. However, this case-by-case approach shares the same limitations as listed in Chapter 2, with the additional complication of how to treat missing data. Given these limitations, this chapter concentrates on the SEM approach.

5.3.1 Implied Moment Matrices

Chapter 2 presented the structural equation model approach to the unconditional latent curve model. The implied mean structure and the implied covariance structure were an important part of understanding this approach. These two implied moment matrices also play a role in the SEM approach to the conditional latent curve model. To illustrate this, we first rewrite the equation for the repeated measures y_{it} in matrix notation such that

$$\mathbf{y}_i = \mathbf{\Lambda}\boldsymbol{\eta}_i + \boldsymbol{\epsilon}_i \tag{5.11}$$

where

$$\mathbf{y}_i = \begin{bmatrix} y_{i1} \\ y_{i2} \\ \vdots \\ y_{iT} \end{bmatrix} \tag{5.12}$$

$$\mathbf{\Lambda} = \begin{bmatrix} 1 & 0 \\ 1 & 1 \\ \vdots & \vdots \\ 1 & T-1 \end{bmatrix} \tag{5.13}$$

$$\boldsymbol{\eta}_i = \begin{bmatrix} \alpha_i \\ \beta_i \end{bmatrix} \tag{5.14}$$

$$\boldsymbol{\epsilon}_i = \begin{bmatrix} \epsilon_{i1} \\ \epsilon_{i2} \\ \vdots \\ \epsilon_{iT} \end{bmatrix} \tag{5.15}$$

Equation (5.11) is the matrix expression for the level 1 model. It is in a form that permits T time points. Similarly, a matrix expression for the level 2 model for the latent curve parameters (α_i and β_i) is

$$\boldsymbol{\eta}_i = \boldsymbol{\mu}_\eta + \mathbf{\Gamma}\mathbf{x}_i + \boldsymbol{\zeta}_i \tag{5.16}$$

where

$$\boldsymbol{\eta}_i = \begin{bmatrix} \alpha_i \\ \beta_i \end{bmatrix} \tag{5.17}$$

$$\boldsymbol{\mu}_\eta = \begin{bmatrix} \mu_\alpha \\ \mu_\beta \end{bmatrix} \tag{5.18}$$

$$\boldsymbol{\Gamma} = \begin{bmatrix} \gamma_{\alpha 1} & \gamma_{\alpha 2} & \cdots & \gamma_{\alpha K} \\ \gamma_{\beta_1} & \gamma_{\beta_2} & \cdots & \gamma_{\beta_K} \end{bmatrix} \tag{5.19}$$

$$\mathbf{x}_i = \begin{bmatrix} x_{1i} \\ x_{2i} \\ \vdots \\ x_{Ki} \end{bmatrix} \tag{5.20}$$

$$\boldsymbol{\zeta}_i = \begin{bmatrix} \zeta_{\alpha i} \\ \zeta_{\beta i} \end{bmatrix} \tag{5.21}$$

Equation (5.16) permits K exogenous x's to influence the latent curve parameters. By substituting Eq. (5.16) into Eq. (5.11) we have the combined or reduced-form model,

$$\mathbf{y}_i = \boldsymbol{\Lambda}\left(\boldsymbol{\mu}_\eta + \boldsymbol{\Gamma}\mathbf{x}_i\right) + \boldsymbol{\Lambda}\boldsymbol{\zeta}_i + \boldsymbol{\epsilon}_i \tag{5.22}$$

This equation is in a general form that permits any number of repeated measures and any number of exogenous predictors of the latent curve parameters. In this form we can easily find the implied mean structure by taking the expected value of both sides of Eq. (5.22):

$$\boldsymbol{\mu}_y(\boldsymbol{\theta}) = \boldsymbol{\Lambda}\left(\boldsymbol{\mu}_\eta + \boldsymbol{\Gamma}\boldsymbol{\mu}_\mathbf{x}\right) \tag{5.23}$$

where $\boldsymbol{\mu}_y(\boldsymbol{\theta})$ is the model implied mean vector for the y's in the model and $\boldsymbol{\theta}$ contains the unknown model parameters for which we want estimates. From this equation we can determine the implied mean structure for any of the y_{it} variables. For instance, the implied mean structure for a three-wave model with two covariates is

$$\begin{aligned}
\begin{bmatrix} \mu_{y_1}(\boldsymbol{\theta}) \\ \mu_{y_2}(\boldsymbol{\theta}) \\ \mu_{y_3}(\boldsymbol{\theta}) \end{bmatrix} &= \begin{bmatrix} 1 & 0 \\ 1 & 1 \\ 1 & 2 \end{bmatrix} \left(\begin{bmatrix} \mu_\alpha \\ \mu_\beta \end{bmatrix} + \begin{bmatrix} \gamma_{\alpha_1} & \gamma_{\alpha_2} \\ \gamma_{\beta_1} & \gamma_{\beta_2} \end{bmatrix} \begin{bmatrix} \mu_{x_1} \\ \mu_{x_2} \end{bmatrix} \right) \\
&= \begin{bmatrix} \mu_\alpha + \gamma_{\alpha_1}\mu_{x_1} + \gamma_{\alpha_2}\mu_{x_2} \\ \mu_\alpha + \mu_\beta + [\gamma_{\alpha_1} + \gamma_{\beta_1}]\mu_{x_1} + [\gamma_{\alpha_2} + \gamma_{\beta_2}]\mu_{x_2} \\ \mu_\alpha + 2\mu_\beta + [\gamma_{\alpha_1} + 2\gamma_{\beta_1}]\mu_{x_1} + [\gamma_{\alpha_2} + 2\gamma_{\beta_2}]\mu_{x_2} \end{bmatrix}
\end{aligned} \tag{5.24}$$

The implied mean structure for the x's is simply the mean values of the x's since these variables are exogenous. Combining the means of the y's with the means of the x's, the implied mean structure for all observed variables is

$$\begin{aligned}
\boldsymbol{\mu} &= \boldsymbol{\mu}(\boldsymbol{\theta}) \\
\begin{bmatrix} \boldsymbol{\mu}_\mathbf{y} \\ \boldsymbol{\mu}_\mathbf{x} \end{bmatrix} &= \begin{bmatrix} \boldsymbol{\Lambda}\left(\boldsymbol{\mu}_\eta + \boldsymbol{\Gamma}\boldsymbol{\mu}_\mathbf{x}\right) \\ \boldsymbol{\mu}_\mathbf{x} \end{bmatrix}
\end{aligned} \tag{5.25}$$

In an analogous fashion we can derive a covariance structure equation that writes the covariances and variances of the observed variables as functions of the parameters in $\boldsymbol{\theta}$:

$$\boldsymbol{\Sigma} = \boldsymbol{\Sigma}(\boldsymbol{\theta}) \tag{5.26}$$

where $\boldsymbol{\Sigma}$ is the population covariance matrix of the y's and x's and $\boldsymbol{\Sigma}(\boldsymbol{\theta})$ is the population model-implied covariance matrix written as a function of the model parameters in $\boldsymbol{\theta}$. Taking this further, we make the standard assumption that $\theta_{it} = \theta_t$ for all i.[5] Using the combined Eq. (5.22), the vector of exogenous variables $\mathbf{x}_i$, and the definition of a covariance matrix, we can find the model-implied covariance matrix. A first step in finding the covariance matrix is to deviate $\mathbf{y}_i$ and $\mathbf{x}_i$ from their means, that is, form $(\mathbf{y}_i - \boldsymbol{\mu}_\mathbf{y})$ and $(\mathbf{x}_i - \boldsymbol{\mu}_\mathbf{x})$.[6] Given that $\mathbf{x}_i$ is exogenous, we cannot simplify the deviation score, $(\mathbf{x}_i - \boldsymbol{\mu}_\mathbf{x})$, further. However, using Eq. (5.22) for $\mathbf{y}_i$ and Eq. (5.23) for $\boldsymbol{\mu}_y$, we find that

$$\begin{aligned}
\mathbf{y}_i - \boldsymbol{\mu}_\mathbf{y} &= \left[\boldsymbol{\Lambda}\left(\boldsymbol{\mu}_\eta + \boldsymbol{\Gamma}\mathbf{x}_i\right) + \boldsymbol{\Lambda}\boldsymbol{\zeta}_i + \boldsymbol{\epsilon}_i\right] - \left[\boldsymbol{\Lambda}\left(\boldsymbol{\mu}_\eta + \boldsymbol{\Gamma}\boldsymbol{\mu}_\mathbf{x}\right)\right] \\
&= \boldsymbol{\Lambda}\boldsymbol{\Gamma}(\mathbf{x}_i - \boldsymbol{\mu}_\mathbf{x}) + \boldsymbol{\Lambda}\boldsymbol{\zeta}_i + \boldsymbol{\epsilon}_i \\
&= \boldsymbol{\Lambda}\left(\boldsymbol{\Gamma}(\mathbf{x}_i - \boldsymbol{\mu}_\mathbf{x}) + \boldsymbol{\zeta}_i\right) + \boldsymbol{\epsilon}_i
\end{aligned} \tag{5.27}$$

A definition of the covariance matrix of variables is that it is the expected value of the deviation variables times the transpose of the deviation variables. Using this definition, the implied covariance matrix is

$$\begin{aligned}
\boldsymbol{\Sigma}(\boldsymbol{\theta}) &= E\left(\begin{bmatrix} \mathbf{y}_i - \boldsymbol{\mu}_y \\ \mathbf{x}_i - \boldsymbol{\mu}_\mathbf{x} \end{bmatrix}\begin{bmatrix} \mathbf{y}_i - \boldsymbol{\mu}_y \\ \mathbf{x}_i - \boldsymbol{\mu}_\mathbf{x} \end{bmatrix}'\right) \\
&= \begin{bmatrix} \boldsymbol{\Lambda}(\boldsymbol{\Gamma}\boldsymbol{\Sigma}_{xx}\boldsymbol{\Gamma}' + \boldsymbol{\Psi})\boldsymbol{\Lambda}' + \boldsymbol{\Sigma}_{\epsilon\epsilon} & \boldsymbol{\Lambda}\boldsymbol{\Gamma}\boldsymbol{\Sigma}_{xx} \\ \boldsymbol{\Sigma}_{xx}\boldsymbol{\Gamma}'\boldsymbol{\Lambda}' & \boldsymbol{\Sigma}_{xx} \end{bmatrix}
\end{aligned} \tag{5.28}$$

where $\boldsymbol{\Sigma}_{xx}$ is the population covariance matrix of the x's, and the other symbols retain their meanings from above. The upper left quadrant of the implied covariance matrix of $\boldsymbol{\Sigma}(\boldsymbol{\theta})$ in Eq. (5.28) shows the composition of the covariance matrix of the repeated measures, the y's, in terms of the structural parameters of the model. The lower left and upper right quadrants provide the functions of the model parameters that equals the covariance matrix of the x's with the y's. The lower right quadrant equals the covariance matrix of the exogenous x's. For example, in the case of two exogenous variables (x_{1i} and x_{2i}) and a linear trajectory ($\lambda_t = t - 1$), the model-implied variance for y_1 is

$$\begin{aligned}
\mathrm{VAR}(y_1) &= \gamma_{\alpha_1}^2 \mathrm{VAR}(x_1) + \gamma_{\alpha 2}^2 \mathrm{VAR}(x_2) \\
&\quad + 2\gamma_{\alpha_1}\gamma_{\alpha_2}\mathrm{COV}(x_1, x_2) + \mathrm{VAR}(\zeta_\alpha) + \mathrm{VAR}(\epsilon_1)
\end{aligned} \tag{5.29}$$

[5]In contrast to the case-by-case approach, here we permit the error variances to differ over time, but we assume that the error variances are the same at the same time point for all individuals.

[6]We remind the reader that we are using deviation scores to simplify the analytical expression of the implied covariance matrix. We are not recommending that researchers use deviation scores in their data analysis. Rather, the raw scores are more useful in latent curve analysis.

and the model-implied covariance between x_1 and y_2 is

$$\text{COV}(x_1, y_2) = (\gamma_{\alpha_1} + \gamma_{\beta_1})\text{VAR}(x_1) + (\gamma_{\alpha_2} + \gamma_{\beta_2})\text{COV}(x_1, x_2) \tag{5.30}$$

Similarly, we could find expressions for the variances and covariances of the other observed variables by using the elements in Eq. (5.28).

Chapter 2 showed the implied moment matrices, $\boldsymbol{\mu} = \boldsymbol{\mu}(\boldsymbol{\theta})$ and $\boldsymbol{\Sigma} = \boldsymbol{\Sigma}(\boldsymbol{\theta})$, to be crucial for estimation of the model parameters. The same is true in this conditional model. In the next section we describe the use of the raw data to select values of the latent curve model parameters that will maximize the natural log of the likelihood function.

5.3.2 Estimation

We use the Direct ML estimator for missing data that we discussed in Chapter 3. The maximum likelihood function for each case in the sample is (Arbuckle, 1996)[7]

$$\ln L_i(\boldsymbol{\theta}) = K_i - \tfrac{1}{2}\ln|\boldsymbol{\Sigma}_i(\boldsymbol{\theta})| - \tfrac{1}{2}[\mathbf{z}_i - \boldsymbol{\mu}_i(\boldsymbol{\theta})]'\boldsymbol{\Sigma}_i(\boldsymbol{\theta})[\mathbf{z}_i - \boldsymbol{\mu}_i(\boldsymbol{\theta})] \tag{5.31}$$

where the i indexes the observation in the sample, $\mathbf{z}_i$ is a vector of observed variables for the ith case, and K_i is a constant unrelated to $\boldsymbol{\theta}$. The likelihood function for all cases is

$$\ln L(\boldsymbol{\theta}) = \sum_{i=1}^{N} \ln L_i(\boldsymbol{\theta}) \tag{5.32}$$

As before, numerical minimization procedures are required to find the value $\widehat{\boldsymbol{\theta}}$ that maximizes the likelihood function. As a ML estimator, $\widehat{\boldsymbol{\theta}}$ is consistent, asymptotically unbiased, asymptotically normal, and asymptotically efficient. Furthermore, the inverse of the information matrix,

$$\text{ACOV}\left(\widehat{\boldsymbol{\theta}}\right) = \left(-E\left[\frac{\partial^2 \ln L(\boldsymbol{\theta})}{\partial\boldsymbol{\theta}\partial\boldsymbol{\theta}'}\right]\right)^{-1} \tag{5.33}$$

provides the asymptotic covariance matrix of $\widehat{\boldsymbol{\theta}}$. Thus, we can estimate all of the model parameters for the latent curve model and perform tests of statistical significance when we substitute the estimated values of $\widehat{\boldsymbol{\theta}}$ into the asymptotic covariance matrix.

The desirable properties of the *ML* estimator hold when the distributional assumptions hold. In the conditional latent curve model, one assumption sufficient for this is that the observed variables come from distributions with no excess multivariate kurtosis (Browne, 1984). An alternative assumption with particular

[7]The original likelihood function does not include a component that corresponds to the saturated model as does the usual ML fitting function. As we mention below, we can fit a saturated model to the missing data and compare its fit to the hypothesized model. We can form a likelihood ratio chi-square test from the values of these two log-likelihood values.

relevance to the conditional model is that the disturbances ($\zeta_{\alpha i}$, $\zeta_{\beta i}$, ϵ_i) are from a multinormal distribution. This assumption permits dummy variables and other forms of nonnormality for the $\mathbf{x}_i$ covariates as long as the conditional distribution of $\mathbf{y}_i|\mathbf{x}_i$ is multinormal (Bollen, 1989b, pp. 126–128). Failing either of these two assumptions, the disturbances would need to be independent of the $\mathbf{x}_i$ covariates for the ML estimator and significance tests to remain robust (e.g., Satorra, 1990). The consistency of the ML estimator holds even if these preceding assumptions fail, although the significance tests could well be inaccurate. Furthermore we can use bootstrapping techniques (Bollen and Stine, 1990, 1993) or corrections to the asymptotic standard errors and test statistic (Satorra and Bentler, 1994) that take account of excess kurtosis in the variables.

Another issue that arises with the SEM approach to conditional latent curve models is whether to estimate the unconditional model first and then the conditional model.[8] One advantage of estimating the unconditional model first is that the researcher can assess whether the repeated measures conform to the hypothesized latent curve structure without the possible complications of including covariates. It is this strategy we typically use in our own work. Alternatively, if prior work has demonstrated the appropriateness of a latent curve model for similar data, the researcher could choose to go right to the conditional model. Regardless of the sequence of estimation, the researcher needs to assess the fit of the model, the topic to which we now turn.

5.3.3 Model Fit

As we discussed in Chapter 2, we have measures of overall fit and component fit for latent curve models. Much of the material on model fit that we presented in Chapter 2 applies to the conditional latent curve models of this chapter. Here we briefly summarize the principal measures of fit and give more attention to the new aspects of assessing fit that arise with the conditional models.[9] One different aspect of measuring overall fit concerns a consequence of missing data. In some missing data situations, there might be no cases in the analysis that contain values for two particular variables, and this makes it impossible to calculate their covariance. The stand-alone chi-square test statistic and the fit indices that make use of it use a saturated model as the basis of comparing the hypothesized model. The saturated model is not available when some covariances are not available due to missing data. Thus, these fit indices will not be available. It still will be possible to compare other nested models with likelihood ratio tests, but any fit statistics that require the covariance matrix of the saturated model will not be usable.

Overall Fit Measures Table 5.2 lists overall fit measures for the ML estimator introduced in Chapter 2. The properties of these fit indices do not differ from those

[8]With the case-by-case approach the unconditional model is estimated prior to and separately from the conditional model. Using SEMs, the researcher can estimate the conditional model without estimating the unconditional one.

[9]Readers wishing to review the discussion of model fit should return to Chapter 2.

Table 5.2 Summary of Overall Fit Measures

Name	Definition	Ideal Fit
Chi-square test statistic	$T_{ML} = (N - 1)F_{ML}$ $df = \frac{1}{2}(T + K)(T + K + 3) - u$	Nonsignificant p-value
Tucker–Lewis index	$TLI = \frac{T_b/df_b - T_h/df_h}{(T_b/df_b) - 1}$	1
Incremental fit index	$IFI = \frac{T_b - T_h}{T_b - df_h}$	1
Relative noncentrality index	$RNI = \frac{(T_b - df_b) - (T_h - df_h)}{T_b - df_b}$	1
Root-mean-square error of approximation	$RMSEA = \sqrt{\frac{T_h - df_h}{(N-1)df_h}}$	≤ 0.05

described in Chapter 2. (As before, u is the number of unrestricted parameters that are estimated.)

Components of Fit A complete assessment of fit not only considers the overall fit, but examines the components of fit as well. Much of the discussion from Chapter 2 carries over to the conditional latent curve model, so we review only briefly those aspects of fit and devote more attention to the new methods of assessing component fit.

Researchers should examine the signs, significance, and magnitude of the parameter estimates from ML estimation. When these deviate from expectations, the analyst will need to investigate the reasons. Furthermore, the R^2 values for each of the repeated measures provides information on the degree of variance in the measure that is explained by the latent curve parameters. High values signify that the latent curve trajectory explains a considerable amount of the variation in a measure. If there are particular years that are noticeably lower in their R^2 values, it suggests that something has occurred in those years to create a departure from the trajectory pattern. The investigator should determine if there are any factors that could explain the departure in particular years, and this could lead to a reformulation of the model.

There are also R^2 values for the equations that predict the α_i's and β_i's. That is, R^2_α and R^2_β reflect the amount of variance accounted for in the latent intercept and slope factors, respectively, by the optimal linear combination of the set of exogenous predictors. These provide estimates of how well the x's can account for the random intercepts and random slopes. This gives us a sense of the degree to which our covariates account for variation in the latent curve parameters. Keep in mind that we have discussed two different R^2 values that reflect different quantities. The R^2 values for the repeated measures (y's) often will differ from the R^2 values for the latent curve parameters (α's and β's). The former tell us the variance in the y's explained by the latent trajectory parameters; the latter tell us how well our covariates explain the latent curve parameters.

Additional component fit measures are residual moment matrices defined as

$$\bar{\mathbf{z}} - \boldsymbol{\mu}(\widehat{\boldsymbol{\theta}}) \tag{5.34}$$

$$\mathbf{S} - \boldsymbol{\Sigma}(\widehat{\boldsymbol{\theta}}) \tag{5.35}$$

where $\bar{\mathbf{z}} = [\bar{\mathbf{y}}' \ \bar{\mathbf{x}}']$.

The residual moment matrices give us information on how well we are reproducing the sample means and sample covariance matrix for the observed variables. Although there is not a one-to-one correspondence between a large residual and the parameter that should be freed (see Costner and Shoenberg, 1973), these residuals can provide "hints" about problematic areas of the model. See Chapter 2 for further discussion.

5.4 INTERPRETATION OF CONDITIONAL MODEL ESTIMATES

The interpretation of the regression of the latent curve factors on a set of exogenous covariates is often complex, and great care is needed in forming an accurate understanding of these conditional effects. There are two types of effects that are often of greatest interest in a conditional LCM: the direct effects of the covariates on the latent factors, and the indirect effects of the covariates on the repeated measures as mediated by the latent curve factors. We explore each of these in turn.

5.4.1 Direct Effects of Exogenous Covariates on Latent Curve Factors

Often of greatest initial interest is the interpretation of the direct effects of the exogenous covariates on the latent curve factors (α_i, β_i). The magnitude and direction of these effects are reflected in the regression coefficients linking the covariates to each of the latent factors. These regression coefficients are interpreted in precisely the same way as is done in the traditional linear regression model.

Single Covariate For example, consider the regression of the intercept factor and linear slope factor on a single exogenous covariate x_{1i}. The level 2 model is

$$\alpha_i = \mu_\alpha + \gamma_{\alpha_1} x_{1i} + \zeta_{\alpha i} \tag{5.36}$$

$$\beta_i = \mu_\beta + \gamma_{\beta_1} x_{1i} + \zeta_{\beta i} \tag{5.37}$$

where all terms were defined earlier.

The simplest case is where x_{1i} is a dummy variable that denotes discrete group membership (e.g., $x_{1i} = 0$ or $x_{1i} = 1$ to denote gender or treatment group). Because the intercept of a regression equation is the mean of the criterion when all predictors equal 0, μ_α and μ_β represent the mean intercept and mean slope of the latent trajectory for the group coded $x_{1i} = 0$. Further, γ_{α_1} and γ_{β_1} represent the difference in the mean intercept and mean slope of the latent trajectory of the group coded $x_{1i} = 1$ compared to the group coded $x_{1i} = 0$. For example, if $\gamma_{\alpha_1} > 0$ and

$\gamma_{\beta_1} = 0$, this would reflect that relative to the group coded as 0, the group coded as 1 was characterized by a higher intercept of the latent trajectory but that there were no group differences in slopes.

The interpretation of the regression coefficients is only slightly more complex if the covariate is continuous. In this case, μ_α and μ_β continue to represent the mean intercept and mean slope of the latent trajectory when $x_{1i} = 0$. Thus, the interpretation of these values depends on the scaling of the covariate. If $x_{1i} = 0$ represents a meaningful value, μ_α and μ_β can be directly interpreted; if $x_{1i} = 0$ is outside the logical range of the covariate, these interpretations must be made with caution. Further, γ_{α_1} represents the expected change in α_i resulting from a 1-unit change in x_{1i}. Similarly, γ_{β_1} represents the expected change in β_i resulting from a 1-unit change in x_{1i}. However, whereas in the case in which the covariate is dichotomous, here these model-implied changes in the intercept and slope hold across the range of the continuous predictor.

Multiple Covariates The interpretation is only slightly more complicated when there are multiple covariates. For instance, consider the more general level 2 model

$$\alpha_i = \mu_\alpha + \gamma_{\alpha_1} x_{1i} + \gamma_{\alpha_2} x_{2i} + \cdots + \gamma_{\alpha_K} x_{Ki} + \zeta_{\alpha i} \qquad (5.38)$$

$$\beta_i = \mu_\beta + \gamma_{\beta_1} x_{1i} + \gamma_{\beta_2} x_{2i} + \cdots + \gamma_{\beta_K} x_{Ki} + \zeta_{\beta i} \qquad (5.39)$$

Here the x_{ki}'s can be a mixture of dummy and continuous variables. However, the fixed regression coefficients now represent the regression of the latent factor on the covariate net of the other covariates. For instance, the regression coefficient for x_1 in Eq. (5.39) reflects that a 1-unit change in x_1 is associated with an expected γ_{β_1}-unit change in β_i controlling for all other x_{ki}'s. These partial regression coefficients are thus interpreted in precisely the same way as those from the standard linear regression model, but here the random intercepts and slopes are the dependent variables of interest.

Interactions Among Covariates Our discussion has focused only on the inclusion of one or more exogenous predictors that do not interact with one another. Thus, it is assumed that the magnitude of the effect of any covariate is equal across all levels of other covariates. However, there may be instances in which the magnitude of the effect of one covariate varies as a function of one or more covariates. Multiplicative interactions between exogenous covariates is one way to capture such effects.

For example, consider a level 2 model with two exogenous predictors and their interaction as covariates. If we designate the first predictor as x_1, the second predictor as x_2, and the multiplicative interaction as $x_1 x_2$, the intercept and linear slope equations are

$$\alpha_i = \mu_\alpha + \gamma_{\alpha 1} x_{1i} + \gamma_{\alpha 2} x_{2i} + \gamma_{\alpha 3} x_{1i} x_{2i} + \zeta_{\alpha_i} \qquad (5.40)$$

$$\beta_i = \mu_\beta + \gamma_{\beta 1} x_{1i} + \gamma_{\beta 2} x_{2i} + \gamma_{\beta 3} x_{1i} x_{2i} + \zeta_{\beta_i} \qquad (5.41)$$

Because of the inclusion of the interaction term, $\gamma_{\alpha 1}$ and $\gamma_{\beta 1}$ represent the main effects of x_1, $\gamma_{\alpha 2}$ and $\gamma_{\beta 2}$ represent the main effects of x_2, and $\gamma_{\alpha 3}$ and $\gamma_{\beta 3}$ represent the interaction between x_1 and x_2. This implies that the magnitude of the relation between x_1 and the latent curve factors varies as a function of x_2, and vice versa. We cannot interpret the influence of one covariate without considering the other. We detail the testing and probing of these effects later in this chapter.

5.4.2 Indirect Effects of Covariates on Repeated Measures

In the section above we considered the direct effects of the covariates on the latent curve factors. This allows us to consider how a shift in the value of the covariate is associated with a corresponding shift in the latent curve intercepts and slopes. However, the latent curve factors also serve to mediate the effects of the exogenous covariates on the repeated measures. We can capitalize on this mediation to examine a variety of interesting questions.

For example, although a covariate may predict the slope factor significantly, this is not sufficient to conclude that there is a significant indirect effect between the covariate and the repeated measures at all time points. It is possible to determine the points in time that the covariate significantly affects the repeated measures as mediated by the latent intercepts and slopes. We examine these issues in this section.

To better understand these interesting questions, we can consider the specific indirect effects of the covariates on the repeated measures as mediated by the intercept factor, the specific indirect effects as mediated by the slope factor, and the total indirect effects as mediated by the set of latent curve factors. [See Bollen (1987) for a discussion of decomposition of effects in SEM.] The approach to decomposing direct and indirect effects provides a practical method for gaining a more complete understanding of the regression of the latent curve factors on one or more exogenous covariates.

The specific indirect effects of the covariates on the repeated measures as mediated via the intercept factor are quite simple. These are nothing more than the direct effect of the covariate on the intercept factor given that all of the factor loadings are set to 1.0. However, the specific indirect effects of the covariates on the repeated measures as mediated by the slope factor are more complex. This is because the degree to which the covariates influence the repeated measures varies as a function of the specific value of time (λ_t). As we demonstrate below, these indirect effects are interpretable as multiplicative interactions between the exogenous predictors and time in the prediction of the repeated measures.

The conceptualization of main effect predictors of the slope as interactions between the predictor and time is not currently widely used in conditional LCMs. However, we feel that there is much utility to this approach. The following sections draw on developments presented in Curran et al., (2004, in press).[10]

[10]Online calculators are available that calculate all of the conditional effects discussed below (quantpsy.org).

Alternative Expressions of Conditional Relations in the LCM We begin by reconsidering the simple case in which the intercept and linear slope factors are regressed on a single time-invariant exogenous variable denoted x_{1i}. As we detail later, all of the techniques applied to a single covariate generalize directly to the inclusion of multiple covariates. The scalar expressions for these regressions are

$$\alpha_i = \mu_\alpha + \gamma_{\alpha 1} x_{1i} + \zeta_{\alpha_i} \tag{5.42}$$

$$\beta_i = \mu_\beta + \gamma_{\beta 1} x_{1i} + \zeta_{\beta_i} \tag{5.43}$$

As before, these equations highlight that a 1-unit change in the exogenous predictor x_1 is associated with a $\gamma_{\alpha 1}$-unit change in the mean of the trajectory intercept and a $\gamma_{\beta 1}$-unit change in the mean of the slope of the trajectory.

However, what is less evident is that the influence of the regression of β_i on x_{1i} in Eq. (5.43) leads to an interaction between x_{1i} and time (as measured by λ_t) with respect to the effects on the repeated measures. This interaction is best seen in the reduced-form expression

$$y_{it} = \left(\mu_\alpha + \mu_\beta \lambda_t + \gamma_{\alpha 1} x_{1i} + \gamma_{\beta 1} \lambda_t x_{1i}\right) + \left(\zeta_{\alpha_i} + \zeta_{\beta_i} \lambda_t + \epsilon_{it}\right) \tag{5.44}$$

The last term in the first set of parentheses in Eq. (5.44) highlights the interaction between the covariate and time in the prediction of the repeated measures. There is thus a main effect of the covariate ($\gamma_{\alpha 1} x_{1i}$), a main effect of time ($\mu_\beta \lambda_t$), and an interaction between the covariate and time ($\gamma_{\beta 1} \lambda_t x_{1i}$). The magnitude of the regression of y_t on x_1 varies as a function of λ_t; or, given the symmetry of the interaction, the magnitude of the regression of y_t on λ_t varies as a function of x_1.

To further explicate these conditional effects, we can express the model-implied value of y at time t conditioned on x_1 and λ_t as

$$\hat{y}_t|_{c_{x_1}, c_{\lambda_t}} = \mu_\alpha + \mu_\beta c_{\lambda_t} + \gamma_{\alpha 1} c_{x_1} + \gamma_{\beta 1} c_{\lambda_t} c_{x_1} \tag{5.45}$$

where c_{x_1} and c_{λ_t} are the conditional values of x_1 and λ_t, respectively. In other words, Eq. (5.45) defines the model-implied value of y at time t conditioned on specific values of x_1 and λ_t. We can use this prediction equation to express two types of conditional effects.

First, consider the regression of y_t on x_1 conditioned on a specific value of time (c_{λ_t}). We refer to this conditional relation as the *y_t on x_1 regression* at λ_t. Second, consider the regression of y_t on λ_t conditioned on a specific value of x_1 (c_{x_1}). We refer to this as the *y_t on λ_t regression* at x_1. We now turn to exploring each of these conditional effects in more detail.

Testing the y_t on x_1 Regressions at λ_t We continue to consider a regression of the latent intercepts and linear slopes on a single predictor x_1. Simple modification of Eq. (5.45) highlights the regression of y_t on x_1 at a given value of λ_t. Specifically,

$$\hat{y}_t|_{c_{\lambda_t}} = \left(\mu_\alpha + \mu_\beta c_{\lambda_t}\right) + \left(\gamma_{\alpha 1} + \gamma_{\beta 1} c_{\lambda_t}\right) x_1 \quad - \tag{5.46}$$

where the first parenthetical term represents the intercept and the second parenthetical term represents the slope of the y_t on x_1 regression conditioned on c_{λ_t}.

The terms in Eq. (5.46) are expressed as population values. The corresponding sample estimates are

$$\hat{y}_t|_{c_{\lambda_t}} = \hat{\omega}_{0(c_{\lambda_t})} + \hat{\omega}_{1(c_{\lambda_t})}x_1 \tag{5.47}$$

where

$$\hat{\omega}_{0(c_{\lambda_t})} = \hat{\mu}_\alpha + \hat{\mu}_\beta c_{\lambda_t} \tag{5.48}$$

$$\hat{\omega}_{1(c_{\lambda_t})} = \hat{\gamma}_{\alpha 1} + \hat{\gamma}_{\beta 1} c_{\lambda_t} \tag{5.49}$$

Here $\hat{\omega}_{0(c_{\lambda_t})}$ and $\hat{\omega}_{1(c_{\lambda_t})}$ explicitly define the intercept and the slope of the y_t on x_1 regression at a specific value of λ_t (e.g., c_{λ_t}). These sample values are of key interest when interpreting conditional effects in the LCM.

Inferential tests on these simple intercept and simple slope values are possible. Standard covariance algebra enables calculation of the asymptotic standard errors (s.e.'s) for these effects. Specifically,

$$\text{s.e.}(\hat{\omega}_{0(c_{\lambda_t})}) = \left(\text{var}\left[\hat{\mu}_\alpha\right] + 2c_{\lambda_t}\text{cov}\left[\hat{\mu}_\alpha, \hat{\mu}_\beta\right] + c_{\lambda_t}^2\text{var}\left[\hat{\mu}_\beta\right]\right)^{1/2} \tag{5.50}$$

$$\text{s.e.}(\hat{\omega}_{1(c_{\lambda_t})}) = \left(\text{var}\left[\hat{\gamma}_{\alpha_1}\right] + 2c_{\lambda_t}\text{cov}\left[\hat{\gamma}_{\alpha 1}, \hat{\gamma}_{\beta 1}\right] + c_{\lambda_t}^2\text{var}\left[\hat{\gamma}_{\beta 1}\right]\right)^{1/2} \tag{5.51}$$

where var and cov represent the sample estimates of variance and covariance elements from the estimated asymptotic covariance matrix of parameter estimates [see Eq. (5.33)]. Using a z-distribution for large samples, the test statistics for these effects are then

$$z_{\hat{\omega}_{0(c_{\lambda_t})}} = \frac{\hat{\omega}_{0(c_{\lambda_t})}}{\text{s.e.}(\hat{\omega}_{0(c_{\lambda_t})})} \tag{5.52}$$

$$z_{\hat{\omega}_{1(c_{\lambda_t})}} = \frac{\hat{\omega}_{1(c_{\lambda_t})}}{\text{s.e.}(\hat{\omega}_{1(c_{\lambda_t})})} \tag{5.53}$$

Thus, Eqs. (5.48) and (5.49) provide sample point estimates for the intercept and slope of the regression of y_t on x_1 at $\lambda_t = c_{\lambda_t}$, and Eqs. (5.50) and (5.51) provide the estimated asymptotic standard errors of these estimates. This allows us to estimate and test the relation between y_t and x_1 at any value of time. Later in the chapter an empirical example will demonstrate the application of these methods.

Testing the y_t on λ_t Regressions at x_1 Given the symmetry of the interaction between the covariate and time, there also is a conditional regression of y_t on λ_t as a function of x_1. Because this conditional relation focuses on the over-time trajectory of y_t at different values of x_1, this expression is often of greatest interest in substantive research applications.

To define the y_t on λ_t regression conditioned on x_1, we can rearrange the terms in Eq. (5.46) as

$$\hat{y}_t|_{c_{x_1}} = (\mu_\alpha + \gamma_{\alpha 1} c_{x_1}) + (\mu_\beta + \gamma_{\beta 1} c_{x_1})\lambda_t \tag{5.54}$$

where again the first parenthetical term is the intercept and the second parenthetical term is the slope of λ_t at c_{x_1}. That is, Eq. (5.54) represents the model-implied trajectory of y_t over time at any given value of c_{x_1}. As before, the conditional nature of this relation is clearly evident in that the intercept and slope of this equation are affected by the specific value of c_{x_1}.

The sample estimates of the intercepts and slopes of the y_t on λ_t regression is given at a specific value of x_1 as

$$\hat{y}_t|_{c_{x_1}} = \hat{\omega}_{0(c_{x_1})} + \hat{\omega}_{1(c_{x_1})}\lambda_t \tag{5.55}$$

where

$$\hat{\omega}_{0(c_{x_1})} = \hat{\mu}_\alpha + \hat{\gamma}_{\alpha 1} c_{x_1} \tag{5.56}$$

$$\hat{\omega}_{1(c_{x_1})} = \hat{\mu}_\beta + \hat{\gamma}_{\beta 1} c_{x_1} \tag{5.57}$$

As before, covariance algebra enables calculation of the asymptotic standard errors for these sample estimates, resulting in

$$\text{s.e.}(\hat{\omega}_{0(c_{x_1})}) = \left(\text{var}\left[\hat{\mu}_\alpha\right] + 2c_{x_1}\text{cov}\left[\hat{\mu}_\alpha, \hat{\gamma}_{\alpha 1}\right] + c_{x_1}^2\,\text{var}\left[\hat{\gamma}_{\alpha 1}\right]\right)^{1/2} \tag{5.58}$$

$$\text{s.e.}(\hat{\omega}_{1(c_{x_1})}) = \left(\text{var}\left[\hat{\mu}_\beta\right] + 2c_{x_1}\text{cov}\left[\hat{\mu}_\beta, \hat{\gamma}_{\beta 1}\right] + c_{x_1}^2\,\text{var}\left[\hat{\gamma}_{\beta 1}\right]\right)^{1/2} \tag{5.59}$$

where var and cov again represent the appropriate estimates of asymptotic variance and covariance elements from the estimated asymptotic covariance matrix of parameter estimates. These standard errors enable significance tests of the intercepts and slopes in the usual way. Specifically, using a z-distribution for large samples, the test statistics are

$$z_{\hat{\omega}_{0(c_{x_1})}} = \frac{\hat{\omega}_{0(c_{x_1})}}{\text{s.e.}(\hat{\omega}_{0(c_{x_1})})} \tag{5.60}$$

$$z_{\hat{\omega}_{1(c_{x_1})}} = \frac{\hat{\omega}_{1(c_{x_1})}}{\text{s.e.}(\hat{\omega}_{1(c_{x_1})})} \tag{5.61}$$

The application of Eqs. (5.60) and (5.61) allows for standard inferential tests of the sample estimates of the intercept and slope of the simple trajectory of y_t over time at any given value of x_1.

Note the symmetry between Eqs. (5.56) and (5.57) with Eqs. (5.48) and (5.49). The y_t on x_1 regression considers the relation between y_t and x_1 at a specific value of λ_t (defined as c_{λ_t}), whereas the y_t on λ_t regression considers the relation between

y_t and λ_t at a specific value of x_1(defined as c_{x_1}). These conditional relations are all based on the sample results from the same one-predictor conditional LCM [e.g., Eq. (5.44)] but are arranged to highlight different aspects of the symmetric multiplicative interaction between the covariate and time.

Our discussion thus far has not made any assumptions about the scale of measure of the covariate. All of the methods above can be applied to a dichotomous or continuous covariate. We next briefly demonstrate the use of these methods for the y_t on λ_t regression, but also note that these interpretations generalize directly to the y_t on x_1 regression as well.

y_t on λ_t Regression for a Single Dichotomous Covariate Suppose that the predictor is a dummy variable in which $x_1 = 0$ and $x_1 = 1$ denotes membership in one of two discrete groups (e.g., gender, treatment condition). Application of Eqs. (5.56) and (5.57) for $x_1 = 0$ and $x_1 = 1$ provides the y_t on λ_t regression for each of the two groups. Thus, for group 1, the intercept and slope are

$$\hat{\omega}_{0(c_{x_1}=0)} = \hat{\mu}_\alpha \tag{5.62}$$

$$\hat{\omega}_{1(c_{x_1}=0)} = \hat{\mu}_\beta \tag{5.63}$$

and for group 2 they are

$$\hat{\omega}_{0(c_{x_1}=1)} = \hat{\mu}_\alpha + \hat{\gamma}_{\alpha 1} \tag{5.64}$$

$$\hat{\omega}_{1(c_{x_1}=1)} = \hat{\mu}_\beta + \hat{\gamma}_{\beta 1} \tag{5.65}$$

These equations highlight several important aspects of the conditional latent curve model. First, $\hat{\mu}_\alpha$ and $\hat{\mu}_\beta$ represent the intercept and slope of the y_t on λ_t regression when the predictor equals zero (i.e., the mean intercept and slope for group $x_1 = 0$). Further, $\hat{\gamma}_{\alpha 1}$ reflects the difference between the mean intercept for group $x_1 = 1$ compared to group $x_2 = 0$ and $\hat{\gamma}_{\beta 1}$ reflects the difference between the mean slope for group $x_1 = 1$ compared to group $x_1 = 0$. Finally, the compound effects $\hat{\omega}_{0(c_{x_1}=1)}$ and $\hat{\omega}_{1(c_{x_1}=1)}$ reflect the intercept and slope of the y_t on λ_t regression within group 1.

Equations (5.58) and (5.59) provide the standard errors to test these compound effects within each group. However, these point estimates and asymptotic standard errors are available using the output from any standard SEM software package. To see this, note that for $x_1 = 0$, Eqs (5.58) and (5.59) simplify to

$$\text{s.e.}(\hat{\omega}_{0(c_{x_1}=0)}) = \left(\text{var}\left[\hat{\mu}_\alpha\right]\right)^{1/2} \tag{5.66}$$

$$\text{s.e.}(\hat{\omega}_{1(c_{x_1}=0)}) = \left(\text{var}\left[\hat{\mu}_\beta\right]\right)^{1/2} \tag{5.67}$$

indicating that the point estimates and asymptotic standard errors for the y_t on λ_t regression in the group denoted $x_1 = 0$ are the intercept terms for the intercept (α_i) and slope (β_i) equations. We can capitalize on this to compute the point estimates and standard errors for group $x_1 = 1$ without recourse to the computational formulas above. To do so, we simply reverse the dummy coding of group membership and reestimate the model. The overall fit of the second model will be identical to that of the first, but the resulting point estimates for $\hat{\mu}_\alpha$ and $\hat{\mu}_\beta$ and their standard errors will represent the estimated y_t on λ_t regression of the second group, and will be equivalent to those calculated by the formulas given above. [See Aiken and West (1991) for a more detailed discussion of this approach as applied to the standard regression model.]

y_t on λ_t Regressions for a Single Continuous Covariate The only difference between categorical and continuous predictors is that in the continuous case there are typically no natural levels at which to assess y_t on λ_t regressions. That is, in the dichotomous case, x_1 unambiguously denoted group membership, and there is a y_t on λ_t regression for group 1 and a (potentially) different y_t on λ_t regression for group 2. In contrast, in the continuous case there are theoretically an infinite number of values of x_1 to consider.[11] There is subsequently an infinite number of y_t on λ_t regressions associated with each of these possible values of x_1. We must then find a strategy for selecting a small and meaningful set of values of x_1 at which to calculate the associated y_t on λ_t regressions.

In the general regression model, Cohen and Cohen (1983) suggest choosing high, medium, and low values of the predictor, often defined as one standard deviation above the mean, at the mean, and one standard deviation below the mean, respectively. Although arbitrary, there are many situations in which these values are reasonable. We can then apply Eqs. (5.56) and (5.57) to compute the sample point estimates for the y_t on λ_t regressions and Eqs. (5.58) and (5.59) to obtain the standard errors of these point estimates for any desired conditional value, c_{x_1}. These results reflect the model-implied y_t on λ_t regressions at these three specific values of x_1.

Just as in the dichotomous case, these same point estimates and standard errors are available using SEM software packages by rescaling the predictor and reestimating the model. This is facilitated by using centered predictors, since the initial model estimation will then produce the estimates and standard errors for the y_t on λ_t regression at the mean (see, e.g., Aiken and West, 1991). First, we create two new variables that are linear transformations of the original predictor such that $x_{\text{high}} = x_1 - \text{s.d.}_{x_1}$ and $x_{\text{low}} = x_1 + \text{s.d.}_{x_1}$, where s.d._{x_1} is the standard deviation of x_1.[12] Finally, we estimate two separate models in which we use x_{high} and x_{low} in place of our original x_1; the parameter estimates and standard errors for the

[11]Of course, there are only a finite number of observed values of x_1 in any given sample, but there are potentially an infinite number of values for continuous variables.

[12]As a brief aside, note that it is correct that one s.d. is *subtracted* to compute x_{high} and that one s.d. is *added* to compute x_{low}. This is because we take advantage of the fact that the intercepts of the regression equations predicting the intercept and slope factors represent the model implied mean when all predictors are equal to zero. By adding one s.d. to all x scores, a value of zero on x represents one s.d. below the mean, and vice versa. This holds *only* when using the SEM software approach to calculating these values; it does not arise when using the computational formulas presented above.

intercept terms of the intercept and slope equations are equal to the values that would be obtained using the equations above.

Probing Conditional Effects with More Than One Predictor All of our discussion above has assumed that there is just a single predictor x_1, and we have focused on probing either the conditional regression of y_t on x_1 as a function of λ_t, or the conditional regression of y_t on λ_t as a function of x_1. Of course, the inclusion of just a single exogenous covariate is a rather simple model that is rare in practice. However, all of the methods described above generalize to the probing of conditional relations in the presence of any number of correlated exogenous predictors. For example, for the condition in which there are two exogenous predictors of the intercept and slope factors, these equations are

$$\alpha_i = \mu_\alpha + \gamma_{\alpha 1}x_{1i} + \gamma_{\alpha 2}x_{2i} + \zeta_{\alpha_i} \tag{5.68}$$

$$\beta_i = \mu_\beta + \gamma_{\beta 1}x_{1i} + \gamma_{\beta 2}x_{2i} + \zeta_{\beta_i} \tag{5.69}$$

where x_{1i} and x_{2i} are two predictors that may be either categorical or continuous. All of our methods described above for probing the relation between x_{1i} and y_t would apply equally here, but these influences would be *net* the effects of x_{2i}. Thus, no new procedures are needed for examining a conditional effect in the presence of one or more additional predictors.

5.4.3 Further Exploring Conditional Effects

The procedures that we describe above allow us to calculate specific point estimates and inferential tests of the regression of the repeated measures on the covariate at specific values of time, or of the regression of the repeated measures on time at specific values of the covariate. However, there are two additional techniques that can be used to further explicate these conditional relations: regions of significance and confidence bands.

Regions of Significance Using the foregoing procedures, we can evaluate the intercept and slope of the simple regression of y_t on x_1 for specific points in time [e.g., Eqs. (5.48) and (5.49)], or we can evaluate the intercept and slope of the regression of y_t on λ_t at specific levels of x_1 [e.g., Eqs. (5.56) and (5.57)]. One limitation of these approaches is that the conditional intercepts and slopes are calculated at a specific (often arbitrary) value of the covariate or of time. However, there are methods in which we can gain even further information about the conditional effects of x_1 and time by computing regions of significance, a technique originally applied in regression models by Johnson and Neyman (1936) with subsequent extensions by Pothoff (1964) and Rogosa (1980).[13] Regions of

[13]Pothoff (1964) distinguishes between *simultaneous* and *nonsimultaneous* regions of significance. For ease of presentation we focus on nonsimultaneous regions here, although the computation of simultaneous regions are easily obtained (see Pothoff, 1964, Eq. 3.1). The same distinction applies in the subsequent section on confidence bands, and again we focus on nonsimultaneous confidence bands.

significance will allow us to assess at precisely what periods of time the intercepts and slopes of the regressions of y_t on x_1 pass from significance to nonsignificance. Similarly, regions of significance will allow us to assess at precisely what points on the scale of x_1 (when x_1 is continuous) the intercepts and slopes of the y_t on λ_t regressions pass from significance to nonsignificance. We address each of these in turn.

We begin by asking the question: Over what range of time is the effect of x_1 predicting y significant? Our interest is in finding the specific values of time, where time is treated as a continuous variable for which the intercepts and slopes of the regression of y_t on x_1 is significantly negative, nonsignificant, or significantly positive. To determine this, we reverse the procedure for computing the significance of the y_t on x_1 slopes that we described above.

When estimating y_t on x_1 regressions at a specific value of λ_t, we choose a value, c_{λ_t}, compute the point estimate and asymptotic standard error for the slope at c_{λ_t}, and then use Eqs. (5.52) and (5.53) to calculate $z_{\hat{\omega}_{0(c_{\lambda_t})}}$ and $z_{\hat{\omega}_{1(c_{\lambda_t})}}$. When computing regions of significance, we instead choose a specific critical value for the test statistic, say ± 1.96 for $\alpha = 0.05$, and then solve Eqs. (5.52) and (5.53) for the *specific* c_{λ_t} *values* that yield this critical value. These c_{λ_t} values indicate the points in time at which intercept and slope of the y_t on x_1 regression passes from significance to nonsignificance, thus demarcating the *regions of significance*.

Given that interest is most often in tests of the slope of the y_t on x_1 regression (i.e., $\hat{\omega}_{1(c_{\lambda_t})}$), we focus on this value here (see Curran et al., 2004, for extensions of this). Note, however, that regions can easily be calculated for the intercept of the y_t on x_1 regression as well. To calculate the regions of significance for $\hat{\omega}_{1(c_{\lambda_t})}$, we substitute the critical value of ± 1.96 for $z_{\hat{\omega}_{1(c_{\lambda_t})}}$ in Eq. (5.53). Squaring this expression and performing a few algebraic rearrangements, we have

$$0 = [z_{\text{crit}}^2 \text{var}(\hat{\gamma}_{\beta 1}) - \hat{\gamma}_{\beta 1}^2] c_{\lambda_t}^2 + [2(z_{\text{crit}}^2 \text{cov}(\hat{\gamma}_{\alpha 1}, \hat{\gamma}_{\beta 1}) - \hat{\gamma}_{\alpha 1}\hat{\gamma}_{\beta 1})] c_{\lambda_t} + [z_{\text{crit}}^2 \text{var}(\hat{\gamma}_{\alpha 1}) - \hat{\gamma}_{\alpha 1}^2] \quad (5.70)$$

where z_{crit} is the critical value for the z-statistic (e.g., $z_{\text{crit}} = \pm 1.96$). We can then solve for the roots of c_{λ_t} that satisfy this equality using the quadratic formula

$$c_{\lambda_t} = \frac{-b \pm \sqrt{b^2 - 4ac}}{2a} \quad (5.71)$$

where

$$a = [z_{\text{crit}}^2 \text{var}(\hat{\gamma}_{\beta 1}) - \hat{\gamma}_{\beta 1}^2] \quad (5.72)$$

$$b = [2(z_{\text{crit}}^2 \text{cov}(\hat{\gamma}_{\alpha 1}, \hat{\gamma}_{\beta 1}) - \hat{\gamma}_{\alpha 1}\hat{\gamma}_{\beta 1})] \quad (5.73)$$

$$c = [z_{\text{crit}}^2 \text{var}(\hat{\gamma}_{\alpha 1}) - \hat{\gamma}_{\alpha 1}^2] \quad (5.74)$$

The resulting roots indicate the boundaries of the regions of significance of the regression of y_t on x_1 over all possible values of time. In most cases, there is a significant relation between y_t and x_1 at values of time less than the lower bound and greater than the upper bound, but this relation is nonsignificant at values of time between the lower and upper bounds. It is possible on occasion for the significant conditional relations to fall *within* the boundaries (e.g., Rogosa, 1980), so care is needed when evaluating these values.

The values above examine the relation between y_t and x_1 across all possible values of time. However, we can also ask the question: Over what range of the predictor x_1 is the effect of time predicting y significant? This addresses the nature of the y_t on λ_t regressions and may be of key interest to many substantive applications of conditional LCMs. To answer this, we will simply apply regions of significance to the intercept and slope estimates for the y_t on λ_t regressions across all possible values of x_1. This procedure will be useful when x_1 is continuous and we are interested in the values for x_1 over which the y_t on λ_t slopes are significantly different from zero. The procedures that define the regions of significance for the y_t on λ_t slopes are precisely the same as those illustrated above for the y_t on x_1 slopes. We again focus on the y_t on λ_t slope (i.e., $\hat{\omega}_{1(c_{x_1})}$) but note that these calculations expand directly to the intercept as well.

Specifically,

$$0 = [z_{\text{crit}}^2 \text{var}(\hat{\gamma}_{\beta 1}) - \hat{\gamma}_{\beta 1}^2]c_{x_1}^2 + [2(z_{\text{crit}}^2 \text{cov}(\hat{\mu}_\beta, \hat{\gamma}_{\beta 1}) - \hat{\mu}_\beta \hat{\gamma}_{\beta 1})]c_{x_1} + [z_{\text{crit}}^2 \text{var}(\hat{\mu}_\beta) - \hat{\mu}_\beta^2] \quad (5.75)$$

where we select a value z_{crit} and solve for the values of c_{x_1} that yield z_{crit}. This will again result in a quadratic equation

$$c_{x_1} = \frac{-b \pm \sqrt{b^2 - 4ac}}{2a} \quad (5.76)$$

where

$$a = [z_{\text{crit}}^2 \text{var}(\hat{\gamma}_{\beta 1}) - \hat{\gamma}_{\beta 1}^2] \quad (5.77)$$

$$b = [2(z_{\text{crit}}^2 \text{cov}(\hat{\mu}_\beta, \hat{\gamma}_{\beta 1}) - \hat{\mu}_\beta \hat{\gamma}_{\beta 1})] \quad (5.78)$$

$$c = [z_{\text{crit}}^2 \text{var}(\hat{\mu}_\beta) - \hat{\mu}_\beta^2] \quad (5.79)$$

As with the case of the y_t on x_1 slopes, the resulting roots of Eq. (5.76) indicate the specific regions of x_1 over which the slopes of the y_t on λ_t regressions are significantly different from zero. This is often an improvement over the arbitrary selection of high, medium, and low values at which to assess the y_t on λ_t slopes, as the regions indicate the significance of the simple trajectories over all possible values of x_1.

Confidence Bands Our discussion of tests of intercepts and slopes for y_t on x_1 regressions and the y_t on λ_t regressions, along with the calculation of regions of significance, are based on traditional null hypothesis testing procedures (e.g., solving for $z_{\text{crit}} = \pm 1.96$). However, recent calls have been made for greater focus on the construction of confidence intervals in place of the simple null hypothesis tests (Wilkinson and Inference, 1999). The calculation of confidence bands provides us a powerful way of communicating the effects of a conditional LCM that moves beyond strict null hypothesis testing.

We begin with the standard normal theory formula for a confidence interval for any given parameter estimate generally denoted $\hat{\theta}$, where

$$\text{CI} = \hat{\theta} \pm z_{\text{crit}}\left[\text{s.e.}(\hat{\theta})\right] \tag{5.80}$$

In most cases we are interested in a single effect estimate (i.e., $\hat{\theta}$), so we simply compute the confidence interval for this estimate. However, in the case of conditional effects in the LCM, both the effect estimate and the associated standard error vary as a function of the moderating variable. As such, we cannot plot just one confidence interval; instead, we must plot the confidence interval at each level of the moderating variable, known as *confidence bands*.

To compute confidence bands for the slope of the regression of y_t on x_1 conditional on λ_t (i.e., $\hat{\omega}_{1(c_{\lambda_t})}$), substitute Eqs. (5.49) and (5.51) for the corresponding terms in Eq. (5.80):

$$\begin{aligned} \text{CB}_{\hat{\omega}_{1(c_{\lambda_t})}} &= (\hat{\gamma}_{\alpha 1} + \hat{\gamma}_{\beta 1} c_{\lambda_t}) \\ &\pm z_{\text{crit}}\left[\left(\text{var}\left[\hat{\gamma}_{\alpha 1}\right] + 2c_{\lambda_t}\text{cov}\left[\hat{\gamma}_{\alpha 1}, \hat{\gamma}_{\beta 1}\right] + c_{\lambda_t}^2\,\text{var}\left[\hat{\gamma}_{\beta 1}\right]\right)^{1/2}\right] \end{aligned} \tag{5.81}$$

Similar procedures yield the confidence band for the slope estimate of the y_t on λ_t regressions (i.e., $\hat{\omega}_{1(c_{x_1})}$):

$$\begin{aligned} \text{CB}_{\hat{\omega}_{1(c_{x_1})}} &= (\hat{\mu}_{\beta} + \hat{\gamma}_{\beta 1} c_{x_1}) \\ &\pm z_{\text{crit}}\left[\left(\text{var}\left[\hat{\mu}_{\beta}\right] + 2c_{x_1}\text{cov}\left[\hat{\mu}_{\beta}, \hat{\gamma}_{\beta 1}\right] + c_{x_1}^2\,\text{var}\left[\hat{\gamma}_{\beta 1}\right]\right)^{1/2}\right] \end{aligned} \tag{5.82}$$

As is the case with standard confidence intervals, confidence bands convey the same information as null hypothesis tests of y_t on x_1 slopes and/or regions of significance. Specifically, the points where the confidence bands cross zero are the boundaries of the regions of significance, so also indicate which y_t on x_1 slopes are significant. More important, the confidence bands also convey our certainty in the y_t on x_1 slope estimates and how that certainty changes across the range of the moderating variable. Typically, there is much greater precision of estimation for medium values of the moderator, with increasing imprecision at the extreme ends of the scale. Analytically, this is true because the confidence bands expand hyperbolically around the conditional effect estimate. We demonstrate the use of confidence bands later in this chapter.

5.4.4 Higher-Order Interactions

Thus far we have concentrated on the inclusion of main effect predictors of the latent curve factors, and we have probed this effect as an interaction between the covariate and time. However, these techniques extend to higher-order interactions between exogenous covariates, curvilinear relations between exogenous covariates, or main effect predictors of curvilinear trajectory functions.

Interactions Between Exogenous Predictors of the Latent Curve Factors In previous sections we described how a main effect predictor of a latent slope factor is a two-way interaction between the predictor and time with respect to the repeated measures. Given this, a two-way interaction among exogenous predictors of a slope factor is a three-way interaction with time, with respect to the repeated measures. If we designate the first predictor as x_1, the second predictor as x_2, and the multiplicative interaction as x_1x_2, the intercept and linear slope equations are

$$\alpha_i = \mu_\alpha + \gamma_{\alpha 1}x_{1i} + \gamma_{\alpha 2}x_{2i} + \gamma_{\alpha 3}x_{1i}x_{2i} + \zeta_{\alpha_i} \tag{5.83}$$

$$\beta_i = \mu_\beta + \gamma_{\beta 1}x_{1i} + \gamma_{\beta 2}x_{2i} + \gamma_{\beta 3}x_{1i}x_{2i} + \zeta_{\beta_i} \tag{5.84}$$

Further, we can write the reduced form for the conditional mean of y_t as a function of x_1 and x_2 as

$$\begin{aligned}\hat{y}_t|_{c_{x_1},c_{x_2}} &= \left(\mu_\alpha + \gamma_{\alpha 1}c_{x_1} + \gamma_{\alpha 2}x_2 + \gamma_{\alpha 3}c_{x_1}c_{x_2}\right)\\ &\quad + \left(\mu_\beta + \gamma_{\beta 1}c_{x_1} + \gamma_{\beta 2}c_{x_2} + \gamma_{\beta 3}c_{x_1}c_{x_2}\right)\lambda_t\end{aligned} \tag{5.85}$$

We have arranged this equation in the same form as Eq. (5.44) to illustrate how the intercepts and slopes of the y_t on λ_t regressions (the first and second terms above) depend on the two predictors and their interaction. Further, this highlights that the interaction x_1x_2 itself interacts with time λ_t (via $\gamma_{\beta 3}$). We could equivalently rearrange the equation to show how the effect of x_1 depends on the interaction of x_2 and λ_t, or how the effect of x_2 depends on the interaction of x_1 and λ_t.

Unlike computing the simple slopes of main effect predictors, when exogenous variables interact in the prediction of the α_i and β_i we must evaluate the time-specific effect of one predictor at various levels of the other. Specifically, the intercept and slope for the regression of y_t on x_1 at a specific point in time λ_t (c_{λ_t}) *and* a specific level of x_2 (c_{x_2}) can be expressed as

$$\hat{\omega}_{0(c_{x_2},c_{\lambda_t})} = \hat{\mu}_\alpha + \hat{\gamma}_{\alpha_2}c_{x_2} + \hat{\mu}_\beta c_{\lambda_t} + \hat{\gamma}_{\beta 2}c_{x_2}c_{\lambda_t} \tag{5.86}$$

$$\hat{\omega}_{1(c_{x_2},c_{\lambda_t})} = \hat{\gamma}_{\alpha 1} + \hat{\gamma}_{\alpha 3}c_{x_2} + \hat{\gamma}_{\beta 1}c_{\lambda_t} + \hat{\gamma}_{\beta 3}c_{x_2}c_{\lambda_t} \tag{5.87}$$

To probe this conditional effect, we must then calculate and test the effect of x_1 at various points in time within our observational window and, if x_2 is categorical, for each group defined by x_2, or, if x_2 is continuous, for high, medium, and low values of x_2. The standard errors needed to test the simple slopes are obtained by covariance algebra in much the same way as before. Since the expression for the standard error involves many terms, we do not present it here (see Curran et al., 2004). Rather, we simply note that we could again obtain these standard errors by recoding x_2 and λ_t to place their origins at the desired values and then reestimating the model, as described above for two-way interactions. The only additional complication is that to probe higher-order interactions, more models have to be estimated to evaluate each possible combination of selected values of x_2 and λ_t. These standard errors enable the formation of critical ratios of each estimate tests of statistical significance. Parallel procedures are available to evaluate the simple slopes of x_2 for specific values of x_1 and λ_t.

Although tests of the slopes of x_1 and x_2 are informative, primary interest is likely to reside in the y_t on λ_t regressions defined at various levels of the exogenous predictors. In this case we would obtain point estimates for the intercepts and slopes by selecting values of x_1 and x_2 (i.e., c_{x_1}, c_{x_2}) and then computing the conditional intercept and slope at those values:

$$\hat{\omega}_{0(c_{x_1},c_{x_2})} = \hat{\mu}_\alpha + \hat{\gamma}_{\alpha 1} c_{x_1} + \hat{\gamma}_{\alpha 2} c_{x_2} + \hat{\gamma}_{\alpha 3} c_{x_1} c_{x_2} \tag{5.88}$$

$$\hat{\omega}_{1(c_{x1},c_{x_2})} = \hat{\mu}_\beta + \hat{\gamma}_{\beta 1} c_{x_1} + \hat{\gamma}_{\beta 2} c_{x_2} + \hat{\gamma}_{\beta 3} c_{x_1} c_{x_2} \tag{5.89}$$

The standard errors again involve many terms so are not presented here [see Curran et al. (2004) for full details]. The significance of these estimates are tested in the same way as the two-way interaction model, with the exception that we would now need to evaluate the simple trajectories at different levels of x_1 and x_2. If both predictors were continuous, and we selected high, medium, and low values of each to evaluate the simple trajectories, crossing these values would result in nine simple trajectories. If one or both were categorical, we would evaluate the simple trajectory for each group present in the analysis at selected levels of the other predictor. Finally, we could recode the predictors to place their origins at the selected values and re-estimate the model to obtain the same point estimates, standard errors, and significance tests afforded by the equations.

Similar to the computation of simple slopes, the use of regions of significance and confidence bands relies on the more complex standard error expressions for conditional effects. However, unlike the computation of simple slopes, regions of significance and confidence bands are most useful when examined on a single dimension; that is, where the estimate and standard error vary as a function of a single variable at a time, as was the case with main effect predictors of the trajectory parameters.[14] When interactions between exogenous predictors are added to the model, the two variables cannot be examined in isolation of one another, and thus both dimensions have to be considered simultaneously. Regions of significance would be two-dimensional regions on the plane defined by the two interacting predictors, and confidence bands would evolve into confidence surfaces. This additional complexity quickly diminishes the appeal and interpretability of these procedures. However, it is possible to combine the regions of significance and conditional point estimate approaches to provide a useful hybrid of these two approaches (Curran et al., 2004) although we do not detail this here.

[14]This is true even if there are multiple main effect predictors, as the conditional effects can still be examined one predictor at a time, where the interest is in the unique effect of that predictor controlling for other predictors in the model.

Nonlinear Effects Among the Exogenous Variables In the preceding section we explored ways to test and probe multiplicative interactions between two exogenous covariates in the prediction of the latent intercept and slope factors. Analogous techniques extend to the nonlinear relation between a single covariate and the latent curve factors. This is because this nonlinear effect is estimated as an interaction of a covariate with itself, and thus the methods apply here as well.

For example, suppose that we wished to regress the linear latent curve factors on a covariate and its square to test a nonlinear relation between the covariate and the trajectory parameters. Our level 2 equations would be

$$\alpha_i = \mu_\alpha + \gamma_{\alpha_1} x_{1i} + \gamma_{\alpha_2} x_{1i}^2 + \zeta_{\alpha i} \tag{5.90}$$

$$\beta_i = \mu_\beta + \gamma_{\beta_1} x_{1i} + \gamma_{\beta_2} x_{1i}^2 + \zeta_{\beta i} \tag{5.91}$$

This permits a U-shaped relationship between x_{1i} and the random intercepts and random slopes rather than assuming an always-increasing or always-decreasing effect as implied by a main effect regression of the growth factors on x_1. We can rewrite this quadratic equation as

$$\alpha_i = \mu_\alpha + (\gamma_{\alpha_1} + \gamma_{\alpha_2} x_{1i}) x_{1i} + \zeta_{\alpha i} \tag{5.92}$$

$$\beta_i = \mu_\beta + (\gamma_{\beta_1} + \gamma_{\beta_2} x_{1i}) x_{1i} + \zeta_{\beta i} \tag{5.93}$$

to highlight that the squared term is nothing more than the interaction between the predictor and itself. Thus, the impact of x_{1i} on the latent curve factors depends on the specific value of x_{1i}, so we cannot summarize the expected effect of the covariate on the random intercepts and slopes with a single value. Instead, we must evaluate this relation conditional on specific values of x_{1i}.

Given that a main effect predictor of the slope factor is a two-way interaction between the predictor and time, the quadratic effect between the predictor and the slope factor must be treated as a three-way interaction between the nonlinear effect and time. This is highlighted in the reduced-form equation, given as

$$\begin{aligned} y_{it} &= \left[\mu_\alpha + (\gamma_{\alpha_1} + \gamma_{\alpha_2} x_{1i}) x_{1i} + \zeta_{\alpha i}\right] \\ &\quad + [\mu_\beta + (\gamma_{\beta_1} + \gamma_{\beta_2} x_{1i}) x_{1i} + \zeta_{\beta i}]\lambda_t + \epsilon_{it} \\ &= (\mu_\alpha + \lambda_t \mu_\beta) + (\gamma_{\alpha_1} + \gamma_{\alpha_2} x_{1i}) x_{1i} \\ &\quad + (\gamma_{\beta_1} + \gamma_{\beta_2} x_{1i}) x_{1i} \lambda_t + (\zeta_{\alpha_i} + \zeta_{\beta_i} \lambda_t + \epsilon_{it}) \end{aligned} \tag{5.94}$$

This reduced-form equation highlights that the regression coefficient γ_{β_2} represents a three-way interaction between the nonlinear effect of x_{1i} and λ_t. Because all of the procedures that we described earlier for testing and probing an interaction between two exogenous covariates in the prediction of the slope factor can be applied here equivalently, we do not present this in further detail.

Nonlinear Effects of Time Thus far, we have only considered the regression of a linear slope factor on one or more exogenous covariates. In Chapter 4 we explored a variety of latent curve models that allowed for nonlinear trajectory functions. All of the techniques that we have described above for testing and probing conditional effects for the linear LCM generalize to many types of nonlinear LCMs as well.

For example, we defined a quadratic LCM as

$$y_{it} = \alpha_i + \lambda_t \beta_{1_i} + \lambda_t^2 \beta_{2_i} + \epsilon_{it} \tag{5.95}$$

where α_i, β_{1_i}, and β_{2_i} represented the random intercept, linear component, and curvilinear component of the quadratic trajectory, respectively. We further showed that because we treat these trajectory components as random variables, these can in turn be expressed as

$$\alpha_i = \mu_\alpha + \zeta_{\alpha_i} \tag{5.96}$$

$$\beta_{1_i} = \mu_{\beta_1} + \zeta_{\beta_{1_i}} \tag{5.97}$$

$$\beta_{2_i} = \mu_{\beta_2} + \zeta_{\beta_{2_i}} \tag{5.98}$$

Whereas Eqs. (5.96), (5.97), and (5.98) are unconditional given that we are not yet considering any predictors of the latent curve factors, we could easily extend all of these to include one or more exogenous predictors.

For example, say that we wished to regress the three latent curve factors on a single covariate x_{1i}. The conditional level 2 equations would now be

$$\alpha_i = \mu_\alpha + \gamma_{\alpha_1} x_{1i} + \zeta_{\alpha_i} \tag{5.99}$$

$$\beta_{1_i} = \mu_{\beta 1} + \gamma_{\beta_{1_1}} x_{1i} + \zeta_{\beta_{1_i}} \tag{5.100}$$

$$\beta_{2_i} = \mu_{\beta_2} + \gamma_{\beta_{2_1}} x_{1i} + \zeta_{\beta_{2_i}} \tag{5.101}$$

with reduced form

$$\begin{aligned} y_{it} = & \left(\mu_\alpha + \gamma_{\alpha_1} x_{1i} + \zeta_{\alpha_i}\right) + (\mu_{\beta_1} + \gamma_{\beta_{1_1}} x_{1i} + \zeta_{\beta_{1_i}})\lambda_t \\ & + (\mu_{\beta_2} + \gamma_{\beta_{2_1}} x_{1i} + \zeta_{\beta_{2_i}})\lambda_t^2 \end{aligned} \tag{5.102}$$

We can expand this expression and collect terms to result in

$$\begin{aligned} y_{it} = & (\mu_\alpha + \lambda_t \mu_{\beta_1} + \lambda_t^2 \mu_{\beta_2} + \gamma_{\alpha_1} x_{1i} + \gamma_{\beta_{1_1}} x_{1i} \lambda_t + \gamma_{\beta_{2_1}} x_{1i} \lambda_t^2) \\ & + (\zeta_{\alpha_i} + \lambda_t \zeta_{\beta_{1_i}} + \lambda_t^2 \zeta_{\beta_{2_i}} + \epsilon_{it}) \end{aligned} \tag{5.103}$$

which highlights the interaction between the single covariate x_{1i} and the quadratic factor (i.e., $\gamma_{\beta_{2_1}} x_{1i} \lambda_t^2$). The probing and plotting of these effects is done as with the previous models.

In Chapter 4 we also described the freed-factor loading model in which one or more loadings on the slope factor were freely estimated from the sample data. All of the methods described above are applicable to this freed-loading model as well. This does become somewhat more complicated given that the estimated factor loadings are themselves characterized by a standard error, the value of which must be considered in the calculation of the indirect effects. However, all of the techniques developed for the decomposition of effects in SEM apply directly to this situation (e.g., Bollen, 1987), so we do not explore this further here.

5.5 EMPIRICAL EXAMPLE

To demonstrate the estimation and interpretation of a conditional latent curve model, we consider data drawn from a sample of children from the National Longitudinal Survey of Youth (NLSY). The sample consists of $N = 405$ children ranging from 6 to 8 years of age at the 1986 assessment. Of these 405 children, 374 were assessed again in 1988, 297 were assessed in 1990, 294 were assessed in 1992, and 221 were assessed at all four time periods. Although any single child provided between one and four repeated measures, we used the missing data methods described in Chapter 3 to reorganize the data on chronological age; this resulted in a total of 10 repeated measures ranging from 6 to 15 years of age. Additional details about the sample are presented in Curran (1997) and Curran et al. (2004).

The repeated measure of interest was the mother's report of the child's antisocial behavior having occurred within the prior 30 days. Scores could range from 0 to 10, where higher scores reflected greater levels of antisocial behavior. There were two time-invariant covariates of interest. The first was child gender, which was coded 0 for females and 1 for males. The second was a continuous measure of parental cognitive support of the child in the home at the initial time period, which ranged from 0 to 12, and higher scores reflected greater cognitive support. Cognitive support was mean centered to aid in later interpretations.

The starting point is fitting a linear unconditional latent curve model to the 10 repeated measures of antisocial behavior. The factor loadings are $\lambda_{it} = \text{age}_{it} - 6$, where 6 was the youngest child in the sample at the first assessment and age was rounded to the nearest year. This coding leads to an the intercept that is the model-implied value of antisocial behavior for the youngest age in the sample. The first model has all time-specific error variances set equal over time. The unbalanced data that resulted from using age led to some pairs of variables for the repeated measures for which there were no observed data and this precluded estimation of a saturated model. Without a saturated model, it was not possible to estimate the usual stand alone chi-square test statistic or incremental fit indices.[15] However, this does not preclude a chi-square likelihood ratio difference test between two nested models. Using this method, we first tested the adequacy of the equal error variances over time and found that there was not a significant decrement in model

[15]A similar situation occurs in the traditional hierarchical linear model.

fit associated with this restriction, and we retained the equal variances over time.[16] We next tested the improvement in model fit with the inclusion of a quadratic latent curve factor to capture potential nonlinearity over time, and this too did not result in a significant improvement in model fit.

To evaluate the overall fit of the final model, we considered the magnitude of the parameter estimates and squared multiple correlations, the presence of large and significant modification indices, and the magnitude and distribution of residuals between the observed and model-implied covariance and mean structures. The stability of the model was also checked by examining the number of iterations needed to converge, the sensitivity of the solution to variations in start values, and the potential impact of influential observations. All of these results suggested that there was an excellent and stable fit of the linear latent curve model with equal error variances.

The mean initial value of antisocial behavior was equal to 1.634 (s.e. = 0.085), and this increased linearly at a mean rate of 0.072 (s.e. = 0.018) units per yearly increase in age. Thus, both the mean intercept and mean slope differed significantly from zero ($p < 0.0001$). Further, the variance of the intercept factor was equal to 1.003 (s.e. = 0.225) and the variance of the slope factor was equal to 0.022 (s.e. = 0.009), suggesting that individuals differ in their starting points and their rates of change.

Given this, we next consider the regression of the intercept and slope factors on two covariates: gender and cognitive support (see Figure 5.4).[17] Evaluating the same criteria described above, there was again an excellent fit of the hypothesized model to the observed data. The parameter estimates and standard errors for this model are presented in Table 5.3.

First, gender significantly predicted the intercept factor ($\hat{\gamma} = 0.778$), indicating that, on average, boys reported significantly higher values of antisocial behavior at the initial age compared to girls. However, gender did not significantly predict the slope factor ($\hat{\gamma} = 0.015$), indicating that there were no significant differences between boys and girls in rates of change in antisocial behavior over time. Next, cognitive support significantly and negatively predicted both the intercept factor ($\hat{\gamma} = -0.070$) and the slope factor ($\hat{\gamma} = -0.016$), reflecting that higher initial values of cognitive support were associated with lower initial levels and a negative change in slopes in antisocial behavior over time.

The interpretation of the gender effect is straightforward given that this is only significantly associated with the intercept factor and thus does not involve changing values of the measure of time. Specifically, boys are significantly higher than girls in antisocial behavior at age 6, but the effect of gender on the random slopes is not significant. However, the interpretation of the cognitive support effect is

[16]The freed error variance model had a slight negative error variance at the final time point. A Wald test suggested that it was not statistically significantly different from zero, suggesting that this was not due to model misspecification, although we must interpret this test with caution since this is a boundary solution (Chen et al., 2001). There were no Heywood cases with the equality constraints imposed, and this imposition did not lead to decrement in model fit and was thus retained.

[17]We omit labeling the disturbance and error arrows in this diagram to simplify its presentation.

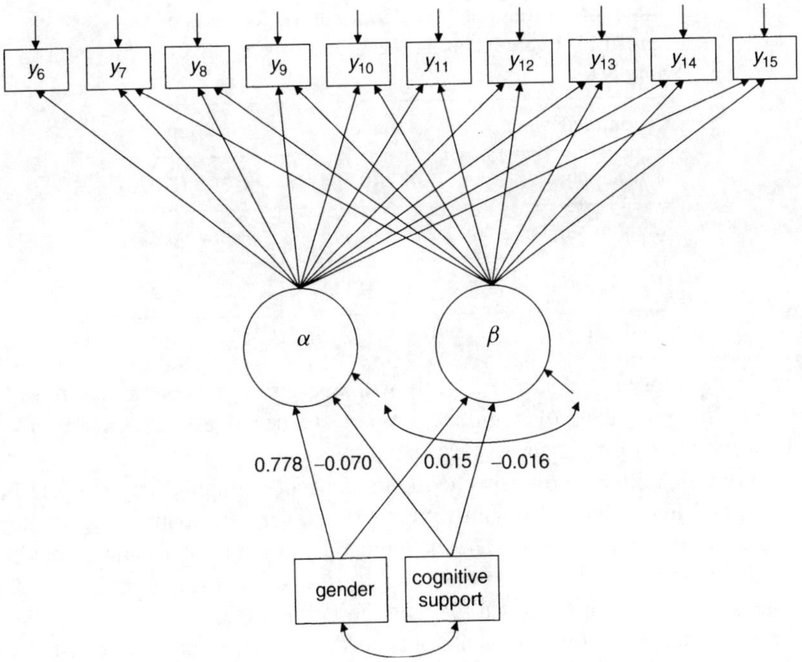

FIGURE 5.4 Conditional latent curve model of antisocial behavior regressed on gender and cognitive support ($N = 405$).

Table 5.3 Coefficient Estimates (Asymptotic Standard Errors) for Random Intercepts and Random Slopes Regressed on Gender and Cognitive Support

Predictor Variable	Intercept Factor	Slope Factor
Intercept	1.242 (0.116)	0.064 (0.024)
Child gender	0.778 (0.165)	0.015 (0.034)
Cognitive support	−0.070 (0.032)	−0.016 (0.007)

more complex given that there are significant differences associated with both the intercept and slope factor. That is, although the regression parameter reflects that a one-unit increase in cognitive is associated with a 0.016-unit decrease in the slope of the trajectory of antisocial behavior, the specific nature of this effect is unknown. We will now turn to the methods for probing conditional effects that we described above to further understand these relations.

5.5.1 y_t on λ_t Regressions

Consider first the model implied trajectories of antisocial behavior across different specific values of cognitive support. From the unconditional model we know that

Table 5.4 Intercepts and Slopes of Trajectories of Antisocial Behavior at Three Levels of Cognitive Support

Conditional Value	Intercept	Slope
Low support	1.422 (0.144)	0.105 (0.030)
Medium support	1.242 (0.117)	0.064 (0.024)
High support	1.062 (0.141)	0.023 (0.030)

the mean trajectory is increasing linearly with age. From the conditional model we know that higher values of cognitive support are negatively associated with the slope of the trajectory for antisocial behavior.

To probe this effect further, we compute the model-implied trajectory at high, medium, and low values of cognitive support.[18] There are many ways to choose these conditional values, but we use the typical definition of one standard deviation below the mean of cognitive support to define low, at the mean to define medium, and one standard deviation above the mean to define high.

We computed the intercepts and slopes of the trajectory of antisocial behavior at these three conditional values using Eqs. (5.56) and (5.57), and we computed the associated standard errors using Eqs. (5.58) and (5.59). These values are presented in Table 5.4 and are plotted in Figure 5.5. As can be seen in the table, antisocial behavior is increasing positively at all three conditional values of cognitive support. However, whereas the slope of the trajectory is significantly different from zero at low and medium values of support (i.e., $p < 0.05$), it is *not* significantly different from zero at high values of support. This is important substantive information that would not have been identified without further probing of this conditional effect.

The probing of this effect reflects that the y_t on λ_t trajectories of antisocial behavior are becoming progressively less steep with increasing values of cognitive support. Further, we know that somewhere between the mean and 1 standard deviation above the mean of cognitive support the slope becomes nonsignificant. However, this information does not identify at *exactly what value* of cognitive support the y_t on λ_t slope moves from positive and significant, to nonsignificant, to negative and significant. Computing the regions of significance for this conditional effect will allow us to determine these more precise values.

5.5.2 Regions of Significance and Confidence Bands

We calculated the regions of significance for the regression of the slope factor on cognitive stimulation and found that the lower bound of the region was equal to 0.927 and the upper bound was equal to 23.41. This means that the model-implied

[18]We used the online calculators available at quantpsy.org to compute all of the following effects.

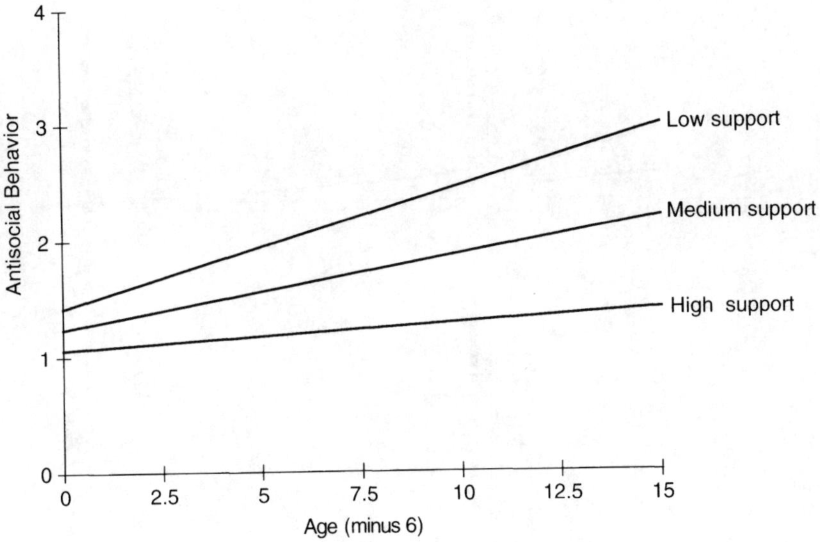

FIGURE 5.5 Model-implied trajectories of antisocial behavior at high, medium, and low levels of cognitive support.

slope of the y_t on λ_t regression of antisocial behavior is positive and significant for values of cognitive support less than 0.927, is not significantly different from zero for values between 0.927 and 23.41, and is negative and significant for values greater than 23.41. The observed range of the (centered) cognitive support variable was from −7.89 to +5.106 with a standard deviation of 2.57. Thus, the slope of the model implied y_t on λ_t trajectories of antisocial behavior were significant and positive for cognitive support values ranging up to $|-7.89/2.57| = 3.07$ standard deviations below the mean to $|0.927/2.57| = 0.36$ standard deviation above the mean but were nonsignificant for any higher scores. Although region reflected that the antisocial trajectories would become significant and negative at cognitive support values exceeding 23.41, this value was well outside the range observed in the sample, and great care is warranted when interpreting this effect.

Finally, we can compute confidence bands for this effect to visually highlight all of these values. Using Eq. (5.82), we calculated the confidence bands for the slope of the y_t on λ_t trajectory across values of cognitive support, and this is presented in Figure 5.6.

Note that the x-axis reflects values of cognitive support and that the y-axis reflects the magnitude of the linear slope of the y_t on λ_t trajectory of antisocial behavior. Further, the vertical dashed lines demarcate the region of significance; slopes to the left of the lower region are positive and significant, slopes between the two regions are not significant, and slopes to the right of the upper region are significant and negative. However, the range of cognitive support in the sample was −7.89 and 5.106, so the upper region lies outside the observed range of the predictor and should be interpreted with caution.

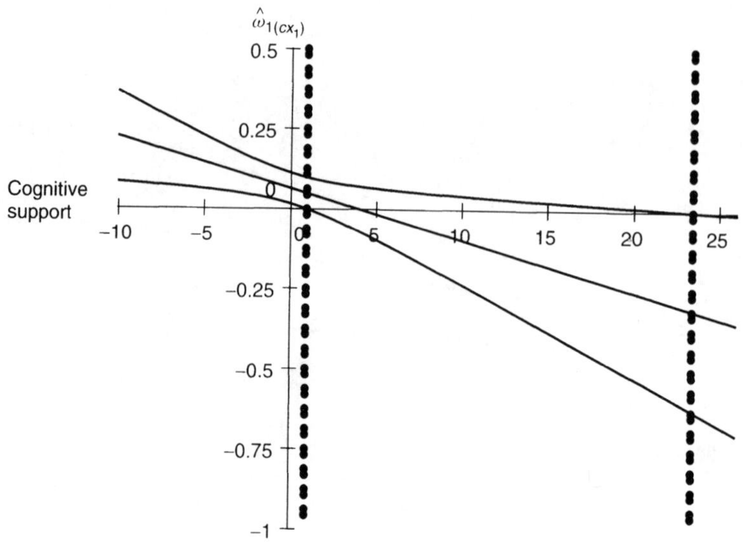

FIGURE 5.6 Slope of antisocial behavior (y-axis) trajector as a function of cognitive support (x-axis) with 95% confidence bands.

5.5.3 Summary

The unconditional LCM reflected that antisocial behavior increased linearly between ages 6 and 15, and there was evidence of significant individual variability in both the starting point and the linear rate of change. The conditional LCM showed that boys reported significantly greater antisocial behavior at age 6, but there were no gender differences in the rates of change in antisocial behavior over time. Further, higher values of cognitive support at home were associated with lower initial levels of antisocial behavior and less steep increases in antisocial behavior over time. Further probing of this effect reflected that at values of cognitive support of about one-third of a standard deviation above the mean and higher, the y_t on λ_t trajectory of antisocial behavior was not increasing significantly over time. Although the regions of significance indicated that at very high levels of cognitive support, the slope of the y_t on λ_t regression would become significantly negative, this conditional value was well outside the observed range of the data. The further probing of these conditional effects revealed much greater information than is available from simply noting that cognitive support significantly predicted the random slope factor. We recommend that researcher use these procedures to further explore their LCM results.

5.6 CONCLUSIONS

In the previous chapters we demonstrated that we could fit a latent curve model to repeated measures with linear and nonlinear trajectories and with complete and missing data. In all of these cases, we considered the unconditional LCM in which there were no predictors of the latent curve factors. In this chapter we showed that we can do more than just describe the latent curves of repeated measures. We can examine the impact of one or more covariates on the magnitude of these parameters, and this in turn will determine the trajectories. Covariates are easy to introduce in the SEM approach. Here we established the general conditions of identification. Furthermore, we illustrated that the SEM tools of assessing fit apply to the conditional latent curve model as they did for the unconditional model. Finally, we used the distinctions between direct and indirect effects from the SEM literature and the interpretation of interactions from the regression literature to interpret the effects of the covariates on both the trajectory parameters and the repeated measures. In addition, this chapter explained how to probe the effects of the covariates on the repeated measures by treating these relations as interaction effects.

All of the topics addressed in this chapter assume that the sample has been drawn from a single population. It is clear that in many areas of social science research, this may not be the case. For example, we might be interested in explicitly modeling important differences as a function of gender, race, or treatment versus control. We can draw on the many developments in the general SEM related to multiple-group analysis to aid us in this endeavor within the LCM. In the next chapter we explore the latent curve model when there is more than one group.

CHAPTER SIX

The Analysis of Groups

Social scientists often classify cases into discrete groups: experimental vs. control, women and men, or European-American, African-American, Latino-American, Asian, and others. Behind these distinctions is the assumption that the processes under study might differ across groups. So far we have not made distinctions between groups and have assumed that we had a sample from a homogeneous population where a single model and single set of parameters were appropriate. It is true that an essential aspect of our models is that individual cases can differ in their trajectories of change, but we have largely ignored the possibility that cases might systematically differ, depending on the group to which they belong.

If all parameters across groups are equal, making group distinctions within an analysis is less important. However, our substantive knowledge is rarely complete enough to know whether group distinctions are present. Ignoring group differences leads to the potential for biased and inconsistent estimators, and inappropriate predictions and hypothesis tests. Thus, it is important to assess whether group differences exist.

The purposes of this chapter are (1) to present methods to test hypotheses about group differences in latent trajectories or covariates that influence these trajectories, (2) to present two general approaches to testing for these group differences, one using dummy variables and the other by performing separate analyses of the groups, and (3) to permit group differences when group membership is unknown or latent. Although our results extend easily to any number of groups, for the sake of simplicity most of our discussion concentrates on two groups.

In the next section we focus on the dummy variable approach to analyzing different groups. This reviews material presented in Chapter 5 and includes a discussion of using dummy variables interacting with continuous variables to capture different covariate slopes in different groups. In the section we present the identification, estimation, testing, and interpretation of such models. This review then informs the next section, in which we present analyses that treat groups separately but that introduce equality constraints to test for group differences across all

Latent Curve Models: A Structural Equation Perspective, by Kenneth A. Bollen and Patrick J. Curran

model parameters. Then we address modeling groups when group membership is unknown, and conclude with a summary of our results.

6.1 DUMMY VARIABLE APPROACH

The dummy variable approach to multiple groups has the same level 1 equation as the conditional and unconditional models from the earlier chapters,

$$y_{it} = \alpha_i + \lambda_t \beta_i + \epsilon_{it} \tag{6.1}$$

where y_{it} is the value of the trajectory variable y for the ith case at time t, α_i is the random intercept, and β_i is the random slope for case i. The λ_t is a constant where a common coding convention is to have $\lambda_1 = 0$, $\lambda_2 = 1$. The remaining values of λ_t allow the incorporation of linear or nonlinear trajectories (see Chapter 4). In the case of a linear trajectory model, $\lambda_t = t - 1$ for all t. We assume that $E(\epsilon_{it}) = 0$ for all i and t, $\text{COV}(\epsilon_{it}, \beta_i) = 0$ and $\text{COV}(\epsilon_{it}, \alpha_i) = 0$ for all i and $t = 1, 2, 3, \ldots, T$, and $E(\epsilon_{it}, \epsilon_{j,t+k}) = 0$ for all k and $i \neq j$. We write the variance of the disturbance as $E(\epsilon_{it}, \epsilon_{it}) = \theta_{\epsilon_{it}} = \text{VAR}(\epsilon_{it})$ for each t and all i. For most of this chapter, we use the assumption that the variance of the error can differ over time but is the same for each individual $[E(\epsilon_{it}, \epsilon_{it}) = \text{VAR}(\epsilon_t)]$. Also, we assume that $\text{COV}(\epsilon_{it}, \epsilon_{i,t+s}) = 0$ for $s \neq 0$, so that the errors are not correlated over time, although this assumption could be loosened for some models.

These assumptions match those we have used in previous chapters. However, we would like to call attention to some implicit assumptions in this model that are relevant to the analysis of different groups. Most important, an implicit assumption is that α_i and β_i are from a single population rather than from two or more different populations with different means and variances. Furthermore, the λ_t values at time t are not different for different subsets of cases. That is, we assume that the form of the trajectories is equivalent over the groups. We also assume that the $\text{VAR}(\epsilon_t)$ is equal across groups at time t. To highlight this, we briefly review the conditional LCM first presented in Chapter 5.

The dummy variable approach manifests itself in the level 2 equations for the random intercepts and random slopes:

$$\alpha_i = \mu_\alpha + \gamma_{\alpha D_1} D_{1i} + \zeta_{\alpha i} \tag{6.2}$$

$$\beta_i = \mu_\beta + \gamma_{\beta D_1} D_{1i} + \zeta_{\beta i} \tag{6.3}$$

where μ_α and μ_β are the intercepts for the equations that predict the random intercepts [Eq. (6.2)] and random slopes [Eq. (6.3)] across all cases. The D_{1i} is a dummy variable that signifies group membership. The equation is similar to the one of Chapter 5 in an analysis of antisocial behavior data. It addresses whether the means of the intercepts and slopes differed for boys and girls. There we coded 1 for males and 0 for females where gender was the basis for group classification, but there also was a continuous variable in the analysis that we do not include here. The $\gamma_{\alpha D_1}$ and $\gamma_{\beta D_1}$ are the coefficients for D_{1i}; they give the expected difference for boys versus girls for their means of intercepts and slopes. The $\zeta_{\alpha i}$ and $\zeta_{\beta i}$ are disturbances with means of zero and variances of $\psi_{\alpha\alpha}$ and $\psi_{\beta\beta}$ and covariance of the disturbances for the intercepts and slopes of $\psi_{\alpha\beta}$. The $\zeta_{\alpha i}$ and $\zeta_{\beta i}$ are uncorrelated with ϵ_{it}. Equations (6.2) and (6.3) comprise the level 2 model. Although the dummy variable approach permits the means of the intercepts and slopes to differ by group, the level 2 equations, like the level 1 equation, have other implicit equality constraints across the groups. Specifically, we assume that $\psi_{\alpha\alpha}$, $\psi_{\beta\beta}$, and $\psi_{\alpha\beta}$ do not vary over the two groups. If they differ, the multiple group analysis that we discuss later in the chapter will be more appropriate.

Figure 6.1 is a path diagram that corresponds to this model when there are three waves of data. The pictorial similarity to the conditional model of Chapter 5 (e.g., see Figure 5.1) is more than coincidental. In fact, we could interpret the dummy variable model here as a special case of the conditional model where we have one covariate that is a dichotomous variable. But there are some subtle distinctions between these and a more typical conditional model with continuous covariates that we would like to highlight. One is the interpretation of the regression coefficients, $\gamma_{\alpha D_1}$ and $\gamma_{\beta D_1}$. The $\gamma_{\alpha D_1}$ coefficient is the shift in the mean of the random intercepts for group 2 ($D_{1i} = 1$) compared to the mean of the random intercepts for group 1 ($D_{1i} = 0$). This implies that the mean of the random intercepts for group 1 ($D_{1i} = 0$) is μ_α and the mean of the intercepts for group 2 ($D_{1i} = 1$) is $\mu_\alpha + \gamma_{\alpha D_1}$. Similarly, the $\gamma_{\beta D_1}$ coefficient is the shift in the mean of the random slopes for group 2 versus group 1. We discussed and illustrated this interpretation in Chapter 5.

We can add a continuous or interval covariate x_{1i} to the level 2 model, resulting in

$$\alpha_i = \mu_\alpha + \gamma_{\alpha D_1} D_{1i} + \gamma_{\alpha x_1} x_{1i} + \zeta_{\alpha i} \tag{6.4}$$

$$\beta_i = \mu_\beta + \gamma_{\beta D_1} D_{1i} + \gamma_{\beta x_1} x_{1i} + \zeta_{\beta i} \tag{6.5}$$

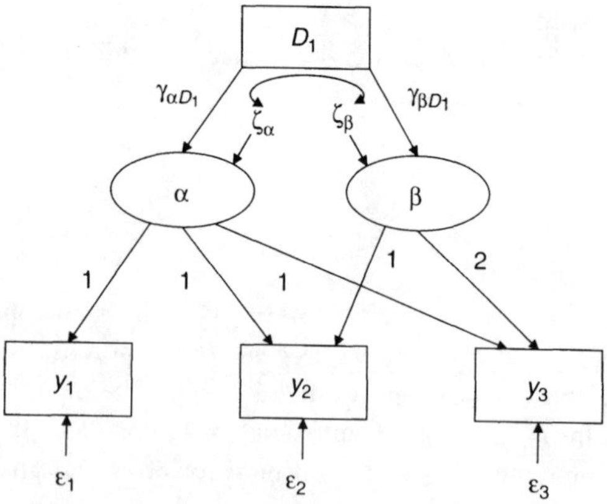

FIGURE 6.1 Three-wave latent curve model with a dummy variable for group difference.

Now we can interpret $\gamma_{\alpha D_1}$ and $\gamma_{\beta D_1}$ as the expected shifts in the intercepts of the α_i and β_i equations rather than an expected shift in their means. The $\gamma_{\alpha x_1}$ and $\gamma_{\beta x_1}$ coefficients give the expected effect on α_i and β_i for a 1-unit difference in x_{1i} net of D_{1i}. We can generalize these equations to include more groups or more covariates.

6.1.1 Various Level 2 Models

There is always a possibility that the impact of the covariates on the random intercepts or random slopes might not be the same for each group. For instance, we might want to test whether family background factors have the same impact on test performance over time for boys and girls. In the model in Eqs. (6.4) and (6.5) we assume that the impact of the x's on α_i and β_i is the same in each group. By adding interaction terms that are products of the dummy variables and the covariates, we can permit the slopes for the x's to differ across groups. When we have two groups and one covariate, the level 2 model would be

$$\alpha_i = \mu_\alpha + \gamma_{\alpha D_1} D_{1i} + \gamma_{\alpha x_1} x_{1i} + \gamma_{\alpha(D_1 x_1)} D_{1i} x_{1i} + \zeta_{\alpha i} \tag{6.6}$$

$$\beta_i = \mu_\beta + \gamma_{\beta D_1} D_{1i} + \gamma_{\beta x_1} x_{1i} + \gamma_{\beta(D_1 x_1)} D_{1i} x_{1i} + \zeta_{\beta i} \tag{6.7}$$

where the product interaction term, $D_{1i}x_{1i}$, permits a difference in the impact of x_{1i} on α_i or β_i in the two groups. The slope of x_{1i} is $\gamma_{\alpha x_1}$ or $\gamma_{\beta x_1}$ for the first group ($D_{1i} = 0$) and it is $(\gamma_{\alpha x_1} + \gamma_{\alpha(D_1 x_1)})$ or $(\gamma_{\beta x_1} + \gamma_{\beta(D_1 x_1)})$ for the second group ($D_{1i} = 1$). A test of whether $\gamma_{\alpha(D_1 x_1)}$ or $\gamma_{\beta(D_1 x_1)}$ is zero tests whether the intercepts or slopes for x_{1i} differ across groups, and if it does, the coefficients give the magnitude of the difference for the group coded 1 for D_{1i} versus the group coded 0 for D_{1i}. These results generalize easily to more than one group or more than one covariate. Products of the dummy variables and the covariates permit us to test whether the covariates' effects differ by group.

6.1.2 Identification, Estimation, and Model Fit

Identification concerns whether it is possible to find unique values for the parameters in a model. As mentioned in the preceding section, the dummy variable approach to having different groups looks the same as the conditional model of Chapter 5. This means that we can use the results on identifying a conditional model here to support the identification of the dummy variable model. In addition, we can use the identification results from Chapter 5 to show that the dummy variable approach to multiple groups remains identified when there is more than one dummy variable and additional covariates predicting the intercepts or slopes. We do not repeat that material here but refer the reader to Chapter 5.

Having an identified model, we can turn to parameter estimation. The level 1 model is still

$$y_{it} = \alpha_i + \lambda_t \beta_i + \epsilon_{it} \tag{6.8}$$

We generalize the equations for the random intercepts and random slopes to permit $G-1$ dummy variables for G groups and K covariates:

$$\alpha_i = \mu_\alpha + \sum_{g=1}^{G-1} \gamma_{\alpha D_g} D_{ig} + \sum_{k=1}^{K} \gamma_{\alpha x_k} x_{ik} + \zeta_{\alpha i} \tag{6.9}$$

$$\beta_i = \mu_\beta + \sum_{g=1}^{G-1} \gamma_{\beta D_g} D_{ig} + \sum_{k=1}^{K} \gamma_{\beta x_k} x_{ik} + \zeta_{\beta i} \tag{6.10}$$

If we want to allow the effect of the covariates on the random slopes or intercepts to differ by group, we can consider some of the x covariates to be the product of a dummy variable for the group times another x. We need to estimate the following parameters: μ_α, μ_β, $\gamma_{\alpha D_g}$, $\gamma_{\alpha x_k}$, $\gamma_{\beta D_g}$, $\gamma_{\beta x_k}$, $\theta_{\epsilon_{it}}$, $\psi_{\alpha\alpha}$, $\psi_{\beta\beta}$, $\psi_{\alpha\beta}$, VAR(x_k), COV(D_g, D_h), COV(x_k, x_m), COV(x_k, D_g), and λ_t. We assume a linear trajectory model such that $\lambda_t = t - 1$. As in earlier chapters, there are two major ways in which we could approach these models: (1) case by case and (2) using the structural equation model. In the chapter appendix we briefly discuss the case-by-case approach. Here we concentrate on the SEM approach. Since these models are essentially the same as the conditional models of Chapter 5, we can estimate the parameters the same as was explained in Chapter 5. Similarly, the assessment of overall and component fit of the model is essentially unchanged from that discussed for the conditional model in Chapter 5. Instead of repeating these ideas here, we illustrate them with an empirical example.

6.1.3 Empirical Example

In Chapters 1 and 2 we used the logarithm transformation of the total crime rate in New York for the first four bimonthly periods of 1995. Here we consider 401 police reporting units (typically cities, although sometimes counties) in New York and 756 reporting units in Pennsylvania.[1] Suppose that we want to analyze these data to see whether Pennsylvania and New York have similar trajectories. This analysis examines bimonthly mean values (mean of January–February, March–April, etc.) over the first eight months of 1995, leading to four time points. There are only two states ($G = 2$), so one dummy variable (D_{1i}) permits a test of whether the mean intercept or mean slope differs across these two states. The level 1 model is

$$y_{it} = \alpha_i + \lambda_t \beta_i + \epsilon_{it} \tag{6.11}$$

[1]These crime data come from the *Uniform Crime Reports* (UCR) covering the year 1995. We obtained these from the National Archive of Criminal Justice Data Web site, housed at the Interuniversity Consortium for Political and Social Research (www.icpsr.umich.edu/NACJD/archive.html). We use the total crime index constructed by the FBI, which is the sum of the eight major crime types: murder, robbery, assault, rape, burglary, larceny, motor vehicle theft, and arson. There are more police reporting units for New York in this analysis than in Chapters 1 and 2, because previously, cases with missing data were deleted, whereas here they are included.

and the level 2 model is

$$\alpha_i = \mu_\alpha + \gamma_{\alpha D_1} D_{1i} + \zeta_{\alpha i} \tag{6.12}$$

$$\beta_i = \mu_\beta + \gamma_{\beta D_1} D_{1i} + \zeta_{\beta i} \tag{6.13}$$

where D_1 is coded 1 for Pennsylvania and 0 for New York.

Next, we regress the random intercepts and the random slopes on the dummy variable (D_1) for state ($D_1 = 1$ for Pennsylvania; 0 for New York) to determine the magnitude and significance of the differences in intercepts and slopes across states. The ML results give us a means to assess the overall model fit with the chi-square test statistic and the other fit indices. The chi-square test statistic, T_{ML}, is 6.77 with 7 df and a p-value of 0.45. The fit indices consistently show an excellent fit (IFI = 1.00, TLI = 1.00, RMSEA = 0.00). The ML parameter estimates and asymptotic standard errors are given in Table 6.1.

The error variances are free to differ over time, but in a given time period each error variance is equal across cases. Looking first at the results for the trajectory intercepts, the initial levels of total crime rates are 5.29 for New York and 5.05 (= 5.29 – 0.24) for Pennsylvania. Although the difference seems substantively small, it is statistically significant. Comparing the random slopes across states, we see that the mean slope is 0.14 for New York and 0.10 (= 0.14 – 0.04) for Pennsylvania, and this difference is statistically significant. It suggests that over this eight-month period there is on average an upward trend in the total crime index, with the initial level and trend higher in New York than in Pennsylvania. The R^2 values of the total crime index for each time period (not shown in Table 6.1) are 0.819, 0.816, 0.846, and 0.795, suggesting that the latent curve explains most of the variance in the crime indices. Finally, the R^2 values for the random intercept and random slope are 0.024 and 0.027, respectively, suggesting that the dummy variables for Pennsylvania and New York does not explain much of the variation in the random intercepts and random slopes.

Table 6.1 Structural Equation Model ML Estimates and Asymptotic Standard Errors from Regression of Random Intercepts and Random Slopes on State (D_1 = 1 for Pennsylvania, 0 for New York) (N = 1157)

Parameter	Estimate	Asymptotic Standard Errors
μ_α	5.29	0.039
μ_β	0.14	0.010
$\gamma_{\alpha D_1}$	−0.24	0.048
$\gamma_{\beta D_1}$	−0.04	0.013
R^2_α	0.024	
R^2_β	0.027	

In sum, the conditional LCM with a single dichotomous predictor allows us to estimate the shift in the mean of the latent intercept and of the latent slope as a function of group membership. Thus, the conditional mean intercept of crime for Pennsylvania is 0.24 unit lower than that for New York, and the conditional mean linear slope for Pennsylvania is 0.04 less than that for New York. As we first presented in Chapter 5, it is possible to consider one or more continuous covariates as well. It is to this we now turn.

Adding Covariates Suppose that interests lie in the impact of percent of the population living in poverty on the initial levels and the rates of change in the violent crime indices for Pennsylvania and New York while controlling for the state dummy variable. Furthermore, we would like to determine whether the impact of poverty on the trajectory parameters is different in each of these states. A modified SEM addresses these issues. Table 6.2 provides these results for the linear model with the main effects of the dummy variable and the covariate. The statistical significance of the coefficients for the state and poverty variables show their effect on the trajectories of the total crime rates. As before, we find that New York units have a higher intercept and slope than Pennsylvania units, net their level of poverty. Furthermore, poverty has a statistically significant positive effect on the initial level, but each unit difference in poverty has a negative effect on the slope net of the state dummy variable. Because we considered only the main effects of poverty and state membership, the magnitude of the effects of one predictor does not depend on the value of the other.

A key question is whether the magnitude of the relation between poverty and the latent curve factors varies as a function of state membership. The results in Table 6.3 reflect that there is a significant interaction between poverty and state in

Table 6.2 Structural Equation Model ML Estimates and Asymptotic Standard Errors from Regressing Random Intercepts and Random Slopes on State ($D_1 = 1$ for Pennsylvania, $D_1 = 0$ for New York) and Percent of Population Living in Poverty (x_1) ($N = 1157$)

Parameter	Estimate	Asymptotic Standard Errors
μ_α	4.94	0.051
μ_β	0.16	0.014
$\gamma_{\alpha D_1}$	−0.23	0.047
$\gamma_{\alpha x_1}$	0.027	0.003
$\gamma_{\beta D_1}$	−0.04	0.013
$\gamma_{\beta x_1}$	−0.002	0.001
R^2_α	0.039	
R^2_β	0.115	

Table 6.3 Structural Equation Model ML Estimates and Asymptotic Standard Errors from Regressing Random Intercepts and Random Slopes on State ($D_1 = 1$ for Pennsylvania, $D_1 = 0$ for New York) and Percent of Population Living in Poverty (x_1) and Interaction Between State Membership and Poverty ($x_1 D_1$) ($N = 1157$)

Parameter	Estimate	Asymptotic Standard Errors
μ_α	4.84	0.074
μ_β	0.15	0.021
$\gamma_{\alpha D_1}$	−0.09	0.089
$\gamma_{\alpha x_1}$	0.035	0.005
$\gamma_{\alpha(D_1 x_1)}$	−0.012	0.006
$\gamma_{\beta D_1}$	−0.04	0.025
$\gamma_{\beta x_1}$	−0.001	0.001
$\gamma_{\beta(D_1 x_1)}$	−0.0005	0.002
R^2_α	0.119	
R^2_β	0.040	

the prediction of the intercept of crime, but not of slope. Using more significant digits than shown in the Table 6.3, the p-value is 0.053. The interaction term in the equation that predicts the random slopes is not even close to statistically significant. This suggests that the effects of poverty on the random slopes is essentially the same in New York and Pennsylvania. However, we must be cautious in interpreting these results because of the implicit equality restrictions on the error variances across states that are inherent in the dummy variable approach. In addition, the dummy variable tests each coefficient separately rather than simultaneously testing two or more coefficients. The result is that there are a number of separate significance tests with the dummy variable approach, giving rise to multiple testing issues. The separate group analysis to which we now turn allows us to test several hypotheses simultaneously as well as to test hypotheses about variances.

6.1.4 Summary

Thus far we have reviewed primarily the various parameterizations of the conditional LCM first presented in Chapter 5. We did this to explicate exactly what parameters do and do not vary as a function of group membership. For the conditional LCM with a single dichotomous predictor, we were able to assess the shift in the model-implied means of the latent curve factors conditioned on group membership. We could expand this model by including one or more covariates, some of which might interact with the dichotomous variable representing group membership. Despite the many advantages of this single-group conditional LCM, it is critical to appreciate that there are strong assumptions imposed in the estimation of this model.

Most important, the single-group conditional LCM allows for shifts in the conditional means of the latent factors as a function of the covariates, but it assumes that *all other model parameters are equal across groups*. Thus, all factor variances, factor covariances, time-specific error variances, and even the functional form of growth are assumed equal across groups. However, there are both empirical and theoretical situations in which we would like to empirically evaluate the extent to which these equalities hold in a given sample and to allow these parameters to take on different values if needed. It is to these models that we now turn.

6.2 MULTIPLE-GROUP ANALYSIS

Our empirical example assumes that the total crime indices for New York and Pennsylvania differ only in their means of random intercepts, slopes, or in the impact of the covariates on these random variables. However, it is possible that, say, the variances of the random intercepts differ across the states. Or the functional form of the trajectory might even differ within each state. Complications such as these are more easily handled by taking an approach different from the dummy variable analysis. The multiple-group analysis approach to different groups is better suited for these problems.

In multiple-group analysis a separate model and data set are established for each group. These models permit minor or even radical differences in the latent curve model for each group. In this section we explain the multiple-group approach to latent curve models. We start with the unconditional model and then extend it to the conditional model.

6.2.1 Unconditional Latent Curve Model for Multiple Groups

We begin by examining the unconditional latent curve model and investigate whether its parameters are equivalent in two groups. The level 1 model is

$$y_{it}^{(g)} = \alpha_i^{(g)} + \lambda_t^{(g)} \beta_i^{(g)} + \epsilon_{it}^{(g)} \tag{6.14}$$

and the level 2 model is

$$\alpha_i^{(g)} = \mu_\alpha^{(g)} + \zeta_{\alpha i}^{(g)} \tag{6.15}$$

$$\beta_i^{(g)} = \mu_\beta^{(g)} + \zeta_{\beta i}^{(g)} \tag{6.16}$$

where the parameters and assumptions are the same as previously with the exception that a superscript (g) is appended to each variable and parameter to signify the group to which it belongs, where $g = 1, 2, 3, \ldots, G$, and G is the total number of groups. In addition, the variance and covariance parameters of the model are superscripted $[\lambda_t^{(g)}, \psi_{\alpha\alpha}^{(g)}, \psi_{\beta\beta}^{(g)}, \psi_{\alpha\beta}^{(g)}, \text{VAR}(\epsilon_t)^{(g)}]$.

It is interesting to contrast this unconditional multiple group model to the dummy variable model for group differences presented in the last section. Equations (6.1), (6.2), and (6.3) were

$$y_{it} = \alpha_i + \lambda_t \beta_i + \epsilon_{it} \tag{6.17}$$

$$\alpha_i = \mu_\alpha + \gamma_{\alpha D_1} D_{1i} + \zeta_{\alpha i} \tag{6.18}$$

$$\beta_i = \mu_\beta + \gamma_{\beta D_1} D_{1i} + \zeta_{\beta i} \tag{6.19}$$

We can obtain a multiple-group model equivalent to these dummy variable equations by setting $\lambda_t^{(1)} = \lambda_t^{(2)}$, $\psi_{\alpha\alpha}^{(1)} = \psi_{\alpha\alpha}^{(2)}$, $\psi_{\beta\beta}^{(1)} = \psi_{\beta\beta}^{(2)}$, $\psi_{\alpha\beta}^{(1)} = \psi_{\alpha\beta}^{(2)}$, and $\text{VAR}(\epsilon_t)^{(1)} = \text{VAR}(\epsilon_t)^{(2)}$ for each t. The multiple-group formulation allows the means in the level 2 equations to differ ($\mu_\alpha^{(1)} \neq \mu_\alpha^{(2)}$, $\mu_\beta^{(1)} \neq \mu_\beta^{(2)}$). The dummy variable approach allows the conditional means to differ as a function of group membership. In fact, imposing the implicit restrictions of the dummy variable approach to a multiple-group approach leads to the identical empirical results obtained with the crime index in New York and Pennsylvania. That is, the mean intercepts and mean slopes for each state are the same for the multiple-group analysis as reported for the SEM dummy variable approach in the preceding section.

The main difference between the dummy variable and the multiple-group approach is that the multiple-group method enables us easily to test the implicit restrictions in the dummy variable approach. This option raises the question of the order in which to test the restrictions. Many substantive areas lack the specificity to dictate specific hypotheses about the equality of parameters and the order in which they should be tested. Lacking this guidance, the order of testing the equality of parameters is somewhat arbitrary. One possible ordering for two groups is to estimate unconditional models as follows:

1. No constraints in either group except for $\lambda_1^{(g)} = 0$ and $\lambda_2^{(g)} = 1$ for all groups
2. Same as step 1 but set $\lambda_t^{(1)} = \lambda_t^{(2)}$ for $t > 2$
3. Same as step 2 but set $\lambda_t^{(1)} = \lambda_t^{(2)} = t - 1$
4. Choose step 1, 2, or 3 and $\mu_\alpha^{(1)} = \mu_\alpha^{(2)}$ and $\mu_\beta^{(1)} = \mu_\beta^{(2)}$
5. Restrictions in step 4 and set $\psi_{\alpha\alpha}^{(1)} = \psi_{\alpha\alpha}^{(2)}$, $\psi_{\beta\beta}^{(1)} = \psi_{\beta\beta}^{(2)}$, and $\psi_{\alpha\beta}^{(1)} = \psi_{\alpha\beta}^{(2)}$
6. Restrictions in step 5 and set $\text{VAR}(\epsilon_t)^{(1)} = \text{VAR}(\epsilon_t)^{(2)}$

Steps 1 to 3 concentrate on comparing the latent curve trajectory in the two groups. The first step is a minimal constraint. It estimates a freed-loading model (see Chapter 4) for each group without constraining the pattern of the latent curve across groups. If the model of this first step does not fit the data, it does not make sense to go further in this hierarchy. Assuming that step 1 results in an adequate fit, step 2 maintains the flexible nonlinearity of the freed-loading model but imposes the same nonlinearity across groups. The linear trajectory model is imposed on both groups in step 3. The models from steps 1 to 3 are nested so that researchers can compare their fit sequentially to decide which best describes the data. Once the researcher chooses the optimally fitting model of these three, a next step is to test whether the means of the intercepts and slopes are equal across groups as shown in step 4. Steps 5 and 6 are even more restrictive in that they test whether the variances of the random intercepts, random slopes, and the error variances are the same in the different groups.

One potential limitation to this hierarchy is that it is not possible to test explicitly for different numbers of latent factors across groups. For example, consider a situation in which a linear trajectory is optimal for group one, but a quadratic is optimal for group two. The six-step hierarchy above does not provide a mechanism for identifying these two different functional forms when considering the groups simultaneously. One option to circumvent this challenge is to identify the optimal functional form within each group separately and then combine the groups in the multiple-group model to test the equality of parameters of interest. Although a reasonable strategy, this approach does not allow for formal chi-square difference tests between competing models when the models in the different groups are not nested. This should be considered when applying these various strategies in practice.

In sum, we cannot emphasize enough that the steps above are not a rigid sequence in which to test all models. Substantive concerns might dictate a different order. For instance, if a researcher suspects that the means of the intercepts and means of the slopes differ, steps 5 and 6 without the restrictions on the means could be moved ahead of step 4 so that the restrictions of step 4 are imposed last. Or if separate hypotheses govern the means of the intercepts versus those of the slopes, these constraints could be separated into two steps.

We illustrate testing for invariance in the unconditional model for the crime index after we discuss invariance in the conditional model.

6.2.2 Conditional Latent Curve Model for Multiple Groups

Multiple-group methods apply to conditional models as well as unconditional models. To illustrate this, consider Figure 6.2. It contains the conditional latent curve models for the total crime indices for New York and Pennsylvania. The level 1 and level 2 equations for the repeated measures are

$$y_{it}^{(g)} = \alpha_i^{(g)} + \lambda_t^{(g)} \beta_i^{(g)} + \epsilon_{it}^{(g)} \tag{6.20}$$

$$\alpha_i^{(g)} = \mu_\alpha^{(g)} + \sum_{k=1}^{K} \gamma_{\alpha x_k}^{(g)} x_{ki}^{(g)} + \zeta_{\alpha i}^{(g)} \tag{6.21}$$

$$\beta_i^{(g)} = \mu_\beta^{(g)} + \sum_{k=1}^{K} \gamma_{\beta x_k}^{(g)} x_{ki}^{(g)} + \zeta_{\beta i}^{(g)} \tag{6.22}$$

A sufficient condition for model identification is that the model parameters are identified in each group separately. Chapter 5 established the identification of the conditional latent curve model, so this implies that the conditional latent curve model for each group is identified and that therefore so is the separate group analysis.

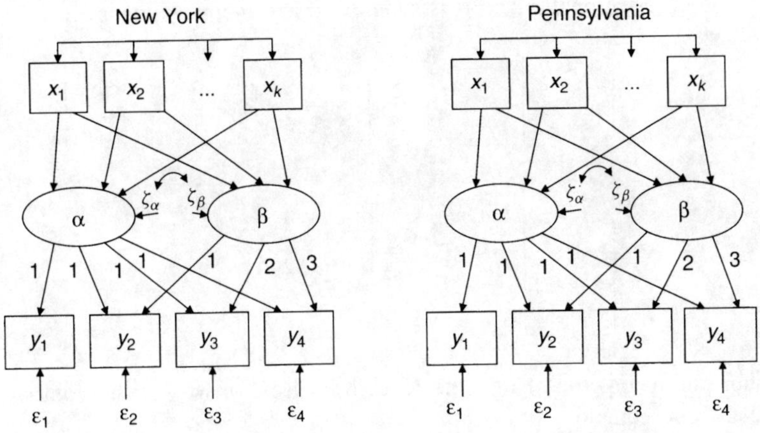

FIGURE 6.2 Multiple-group analysis of conditional latent curve models for crime in New York and Pennsylvania.

SEMs have established procedures for estimating and testing multiple-group analyses (e.g., Bollen, 1989b, pp. 355–369). To present this approach it is useful to express the model in the matrix form of SEM. Write the level 1 model for the gth group as

$$\mathbf{y}^{(g)} = \mathbf{\Lambda}^{(g)}\boldsymbol{\eta}^{(g)} + \boldsymbol{\epsilon}^{(g)} \tag{6.23}$$

where

$$\mathbf{y}^{(g)} = \begin{bmatrix} y_1^{(g)} \\ y_2^{(g)} \\ y_3^{(g)} \\ \vdots \\ y_T^{(g)} \end{bmatrix} \qquad \mathbf{\Lambda}^{(g)} = \begin{bmatrix} 1 & 0 \\ 1 & 1 \\ 1 & \lambda_3^{(g)} \\ \vdots & \vdots \\ 1 & \lambda_T^{(g)} \end{bmatrix}$$

$$\boldsymbol{\eta}^{(g)} = \begin{bmatrix} \alpha_i^{(g)} \\ \beta_i^{(g)} \end{bmatrix} \qquad \boldsymbol{\epsilon}^{(g)} = \begin{bmatrix} \epsilon_{1i}^{(g)} \\ \epsilon_{2i}^{(g)} \\ \epsilon_{3i}^{(g)} \\ \vdots \\ \epsilon_{Ti}^{(g)} \end{bmatrix} \tag{6.24}$$

Write the level 2 model for the gth group as

$$\boldsymbol{\eta}^{(g)} = \boldsymbol{\mu}_\eta^{(g)} + \mathbf{\Gamma}^{(g)}\mathbf{x}^{(g)} + \boldsymbol{\zeta}^{(g)} \tag{6.25}$$

where the observational index, i, is suppressed and the new symbols are

$$\boldsymbol{\mu}_{\eta}^{(g)} = \begin{bmatrix} \mu_{\alpha}^{(g)} \\ \mu_{\beta}^{(g)} \end{bmatrix} \qquad \boldsymbol{\Gamma}^{(g)} = \begin{bmatrix} \gamma_{\alpha x_1}^{(g)} & \gamma_{\alpha x_2}^{(g)} & \cdots & \gamma_{\alpha x_K}^{(g)} \\ \gamma_{\beta x_1}^{(g)} & \gamma_{\beta x_2}^{(g)} & \cdots & \gamma_{\beta x_K}^{(g)} \end{bmatrix} \tag{6.26}$$

$$\mathbf{x}^{(g)} = \begin{bmatrix} x_1^{(g)} \\ x_2^{(g)} \\ \vdots \\ x_K^{(g)} \end{bmatrix} \qquad \boldsymbol{\zeta}^{(g)} = \begin{bmatrix} \zeta_{\alpha}^{(g)} \\ \zeta_{\beta}^{(g)} \end{bmatrix} \tag{6.27}$$

By substituting Eq. (6.25) into the level 1 Eq. (6.23), we get the combined (or reduced-form) equation,

$$\mathbf{y}^{(g)} = \boldsymbol{\Lambda}^{(g)}(\boldsymbol{\mu}_{\eta}^{(g)} + \boldsymbol{\Gamma}^{(g)}\mathbf{x}^{(g)} + \boldsymbol{\zeta}^{(g)}) + \boldsymbol{\epsilon}^{(g)} \tag{6.28}$$

Every SEM implies that the means and covariance matrices of the observed variables are a function of the parameters in the model. These are summarized in the moment structure hypotheses. For the means we have

$$\boldsymbol{\mu}^{(g)} = \boldsymbol{\mu}^{(g)}(\boldsymbol{\theta}) \tag{6.29}$$

where for the gth group $\boldsymbol{\mu}^{(g)}$ is the vector of population means of the observed variables ($\mathbf{y}^{(g)}$, $\mathbf{x}^{(g)}$), $\boldsymbol{\mu}^{(g)}(\boldsymbol{\theta})$ is the implied mean vector, and $\boldsymbol{\theta}$ is the vector of the unknown model parameters in the model. The implied mean vector derives from the expected value of Eq. (6.28) and $\mathbf{x}^{(g)}$ and is

$$\boldsymbol{\mu}^{(g)}(\boldsymbol{\theta}) = \begin{bmatrix} \boldsymbol{\Lambda}^{(g)}(\boldsymbol{\mu}_{\eta}^{(g)} + \boldsymbol{\Gamma}^{(g)}\boldsymbol{\mu}_{\mathbf{x}}^{(g)}) \\ \boldsymbol{\mu}_{\mathbf{x}}^{(g)} \end{bmatrix} \tag{6.30}$$

where $\boldsymbol{\mu}_{\mathbf{x}}^{(g)}$ is the vector of means for the covariates, $\mathbf{x}^{(g)}$, and the other terms are defined as before. The implied covariance structure is

$$\boldsymbol{\Sigma}^{(g)} = \boldsymbol{\Sigma}^{(g)}(\boldsymbol{\theta}) \tag{6.31}$$

where $\boldsymbol{\Sigma}^{(g)}$ is the population covariance matrix of the observed variables and $\boldsymbol{\Sigma}^{(g)}(\boldsymbol{\theta})$ is the implied covariance matrix. The covariance matrix of the observed variables is

$$\begin{aligned} \text{COV}\left(\begin{bmatrix} \mathbf{y}^{(g)} \\ \mathbf{x}^{(g)} \end{bmatrix}, \begin{bmatrix} \mathbf{y}^{(g)\prime} & \mathbf{x}^{(g)\prime} \end{bmatrix}\right) &= E\left(\begin{bmatrix} \mathbf{y}^{(g)} \\ \mathbf{x}^{(g)} \end{bmatrix}\begin{bmatrix} \mathbf{y}^{(g)\prime} & \mathbf{x}^{(g)\prime} \end{bmatrix}\right) \\ &\quad - E\begin{bmatrix} \mathbf{y}^{(g)} \\ \mathbf{x}^{(g)} \end{bmatrix} E\begin{bmatrix} \mathbf{y}^{(g)\prime} & \mathbf{x}^{(g)\prime} \end{bmatrix} \end{aligned} \tag{6.32}$$

If we substitute Eq. (6.28) in for $\mathbf{y}$ and take expected values, we are led to the implied covariance matrix,

$$\boldsymbol{\Sigma}^{(g)}(\boldsymbol{\theta}) = \begin{bmatrix} \boldsymbol{\Lambda}^{(g)}\left(\boldsymbol{\Gamma}^{(g)}\boldsymbol{\Sigma}_{xx}^{(g)}\boldsymbol{\Gamma}^{(g)\prime} + \boldsymbol{\Psi}^{(g)}\right)\boldsymbol{\Lambda}^{(g)\prime} + \boldsymbol{\Sigma}_{\epsilon\epsilon}^{(g)} & \boldsymbol{\Lambda}^{(g)}\boldsymbol{\Gamma}^{(g)}\boldsymbol{\Sigma}_{xx}^{(g)} \\ \boldsymbol{\Sigma}_{xx}^{(g)}\boldsymbol{\Gamma}^{(g)\prime}\boldsymbol{\Lambda}^{(g)\prime} & \boldsymbol{\Sigma}_{xx}^{(g)} \end{bmatrix} \quad (6.33)$$

Equations (6.30) and (6.33) give the model-implied mean vector and covariance matrix for conditional latent curve models for group g. Every group in the analysis has an analogous pair of the implied moment matrices. These same equations yield the model implied mean and implied covariance matrices for unconditional models if $\boldsymbol{\Sigma}_{xx}^{(g)}$ and $\boldsymbol{\Gamma}^{(g)}$ are set to $\mathbf{0}$.

The maximum likelihood estimator for multiple groups is[2]

$$\ln L(\boldsymbol{\theta}) = \sum_{g=1}^{G} \frac{N_g}{N} \ln L_g(\boldsymbol{\theta}) \quad (6.35)$$

where

$$\ln L_g(\boldsymbol{\theta}) = \sum_{i=1}^{N} \ln L_{gi}(\boldsymbol{\theta}) \quad (6.36)$$

and

$$\ln L_{gi}(\boldsymbol{\theta}) = K_i - \tfrac{1}{2}\ln|\boldsymbol{\Sigma}_i^{(g)}(\boldsymbol{\theta})| - \tfrac{1}{2}[\mathbf{z}_i^{(g)} - \boldsymbol{\mu}_i^{(g)}(\boldsymbol{\theta})]'\boldsymbol{\Sigma}_i^{(g)}(\boldsymbol{\theta})[\mathbf{z}_i^{(g)} - \boldsymbol{\mu}_i^{(g)}(\boldsymbol{\theta})] \quad (6.37)$$

The $\mathbf{z}_i^{(g)}$ is the vector of stacked vectors of $\mathbf{y}_i^{(g)}$ and $\mathbf{x}_i^{(g)}$ values, $\boldsymbol{\mu}_i^{(g)}(\boldsymbol{\theta})$ is the model-implied mean vector for the ith case, and $\boldsymbol{\Sigma}_i^{(g)}(\boldsymbol{\theta})$ is the model-implied covariance matrix for the ith case. The N_g is the number of cases in the gth group. The K_i is a constant that does not affect the estimated parameter values. Values of $\boldsymbol{\theta}$ are chosen to maximize the ln likelihood [$\ln L(\boldsymbol{\theta})$] function across all groups. It is a weighted sum where those groups with the most cases receive the greatest weight in the likelihood function. Comparing the ln likelihood value of the fitted model to that of a saturated model where the means and covariances are freely estimated permits a likelihood ratio test. Under the usual assumptions for the ML

[2]The likelihood function below accommodates missing data. When there are no missing data across groups, we could write this as the ML fitting function:

$$F_{g\text{ML}} = \ln|\boldsymbol{\Sigma}^{(g)}(\boldsymbol{\theta})| + \text{tr}\left(\mathbf{S}^{(g)}[\boldsymbol{\Sigma}^{(g)}(\boldsymbol{\theta})]^{-1}\right) - \ln|\mathbf{S}^{(g)}| - (p+q) + [\bar{\mathbf{z}}^{(g)} - \boldsymbol{\mu}^{(g)}(\boldsymbol{\theta})]'[\boldsymbol{\Sigma}^{(g)}(\boldsymbol{\theta})]^{-1}[\bar{\mathbf{z}}^{(g)} - \boldsymbol{\mu}^{(g)}(\boldsymbol{\theta})] \quad (6.34)$$

The $\bar{\mathbf{z}}^{(g)}$ is the vector of sample means of the $\mathbf{y}^{(g)}$ and $\mathbf{x}^{(g)}$. The N_g is the number of cases in the gth group. Values of $\boldsymbol{\theta}$ are chosen to minimize the F_{ML} fitting function across all groups. The likelihood function in the text is more general in that it permits missing data.

estimator, the likelihood ratio test statistic has an asymptotic chi-square distribution with df equal to $\frac{1}{2}G(p+q)(p+q+3) - u$, where u is the number of independent parameters that are estimated across all groups.

Empirical Example Using these methods, tests of the equality of any combination of parameters in LCMs across groups are possible. To illustrate, we return to the unconditional LCM for the crime index in New York and Pennsylvania. Table 6.3 lists a set of hypotheses for testing the degree of invariance across states using this superscript system for two groups ($G = 2$). As stated above, the hierarchy of testing is best dictated by the substantive knowledge of the researcher. Without strong reasons to deviate from the sequence described above, it is that sequence of tests that is followed. The ML estimator provides a straightforward way of performing tests in this hierarchy. Steps up the hierarchy generally lead to increasingly restrictive models. That is, the parameters in one model are a restrictive version of ones that precede it. This creates nested relations. The difference in the chi squares and their degrees of freedom between two nested models provides a significance test of the null hypothesis that the more restrictive model fits the same as the less restrictive model without the restrictions. Failure to reject lends support to the equality restrictions that are introduced. A significant chi-square difference suggests that the restrictions are not accurate and that the groups differ in at least one of the parameters tested.

The unconditional models that correspond to those in Table 6.2 are first. The likelihood ratio test statistics of these stand-alone models are presented in Table 6.4*a*. The likelihood ratio tests that compare these models are presented in

Table 6.4*a* Likelihood Ratio Tests of Multiple-Group Analysis of Total Crime Indexes for New York and Pennsylvania for Unconditional Latent Curve Model ($N = 1157$)

	Model	T_{ML}	df	p-value
1	H_{form} $(\lambda_1 = 0; \lambda_2 = 1)$	3.80	6	0.70
2	$H_{\lambda_3\lambda_4}$: $\lambda_3^{(NY)} = \lambda_3^{(PA)}; \lambda_4^{(NY)} = \lambda_4^{(PA)}$	3.96	8	0.86
3	H_{Λ} : $\lambda_t^{(NY)} = \lambda_t^{(PA)} = t - 1$	8.24	10	0.61
4	$H_{\Lambda\mu} : H_{\Lambda}$ and $\mu_\alpha^{(NY)} = \mu_\alpha^{(PA)}; \mu_\beta^{(NY)} = \mu_\beta^{(PA)}$	63.23	12	0.00
5	$H_{\Lambda\mu\psi} : H_{\Lambda\mu}$ and $\psi_{\alpha\alpha}^{(NY)} = \psi_{\alpha\alpha}^{(PA)}; \psi_{\beta\beta}^{(NY)} = \psi_{\beta\beta}^{(PA)}; \psi_{\alpha\beta}^{(NY)} = \psi_{\alpha\beta}^{(PA)}$ removing constraints on μ	69.23	15	0.00
6	$H_{\Lambda\psi} : H_{\Lambda}$ and $\psi_{\alpha\alpha}^{(NY)} = \psi_{\alpha\alpha}^{(PA)}; \psi_{\beta\beta}^{(NY)} = \psi_{\beta\beta}^{(PA)}; \psi_{\alpha\beta}^{(NY)} = \psi_{\alpha\beta}^{(PA)}$	13.99	13	0.37
7	$H_{\Lambda\psi\Theta_{\epsilon\epsilon}} : H_{\Lambda\psi}$ and $\Theta_{\epsilon\epsilon}^{(NY)} = \Theta_{\epsilon\epsilon}^{(PA)}$	18.75	17	0.34

Table 6.4*b* Likelihood Ratio Difference Tests Between Nested Constraints of Multiple-Group Analysis for Total Crime Indexes

Model Comparison	diff − T_{ML}	df	*p*-Value
2 vs. 1	0.16	2	0.92
3 vs. 2	4.28	2	0.12
4 vs. 3	54.99	2	0.00
5 vs. 4	6.00	3	0.11
6 vs. 3	5.75	3	0.12
7 vs. 6	4.76	4	0.31

Table 6.4*b*. The test of $H_{\text{form}}(\lambda_1 = 0; \lambda_2 = 1)$ suggests that the freed-loading model does an excellent job of fitting the covariance matrix and means for both Pennsylvania and New York ($p = 0.70$). The H_{form} denotes that only the form of the model and the $\lambda_1 = 0$ and $\lambda_2 = 1$ constraints are imposed across the states. The freed loadings, λ_3 and λ_4, provide the means to fit nonlinearity in the trajectories. Since these loadings are unconstrained across the two states, they permit different nonlinear forms in New York and Pennsylvania. In the next model we test whether the freed loadings are equal in the different states by using the hypothesis $H_{\lambda_3\lambda_4} : \lambda_3^{(\text{NY})} = \lambda_3^{(\text{PA})}; \lambda_4^{(\text{NY})} = \lambda_4^{(\text{PA})}$, and $p = 0.86$ for the LRT suggests an excellent fit. Comparing this model to the prior one where the loadings could differ by state, Table 6.4*b* shows a chi-square difference test of 0.16 with 2 degrees of freedom (df). The lack of significance of the comparison suggests that the freed loadings take the same value in both states. In light of these results, we turn to a test of whether the linear trajectory model is adequate by imposing the linear trajectory model for both states ($H_\Lambda : \lambda_t^{(\text{NY})} = \lambda_t^{(\text{PA})} = t - 1$). The fit of this model also is excellent ($p = 0.61$) and there is no significant improvement of fit compared to the less restrictive model that permits the last two loadings to be freed. Thus, at this point we find that the latent *linear* trajectory model has a good fit for both states.

When we examine the equality of the means ($H_{\Lambda\mu} : H_\Lambda$ *and* $\mu_\alpha^{(\text{NY})} = \mu_\alpha^{(\text{PA})}$; $\mu_\beta^{(\text{NY})} = \mu_\beta^{(\text{PA})}$), we find a jump of 55 in the chi-square test statistic with 2 df that is highly statistically significant ($p < 0.01$) compared with the prior linear trajectory model that constrains only the Λ's (H_Λ). This is strong evidence against assuming that the two states have equal means in *both* the initial levels and slopes. Adding the restriction that the variances and covariances of the disturbances of the level 1 model are equal across groups hardly changes the chi-square from the previous model that introduced the restrictions on the means (chi-square difference of 6.00 with 3 df, $p = 0.11$). Removing the constraints on μ and estimating $H_{\Lambda\psi}$ leads to a nonsignificant chi-square ($p = 0.37$), and comparing $H_{\Lambda\psi}$ to H_Λ results in a nonsignificant chi-square difference ($p = 0.12$). Furthermore, adding an equality restriction for the error variances across states ($H_{\Lambda\psi\Theta_{\epsilon\epsilon}}$) results in a nonsignificant chi-square (18.75, df $= 17$, $p = 0.34$) and a nonsignificant chi-square difference with $H_{\Lambda\psi}(p = 0.31)$.

What do we conclude from this multiple-group comparison of the unconditional model? Our analysis provides strong evidence consistent with using the linear trajectory model for the crime index in both New York and Pennsylvania over this eight-month period. Furthermore, we have evidence that the variances and the covariance for the random intercepts and random slopes are equal, as are the error variances. We have strong evidence of a difference in the means of α_i or β_i in New York and Pennsylvania. Thus, the evidence points toward an unconditional model with a linear trajectory but with differences in means of the intercepts or slopes. The significant differences in means of the intercepts or slopes are consistent with the dummy variable results reported in Table 6.1. However, here we are able to test explicitly for group differences in model parameters that are assumed equal in the single-group conditional LCM.

With the unconditional multiple-group model determined, we turn to the conditional model. Recall that earlier we used the dummy variable approach to examine whether the percent of the population living in poverty in New York and Pennsylvania had the same impact on the random intercepts and random slopes for each state. Here we use the multiple-group method to test these relationships. The unconditional model suggests H_Λ as a starting point. As expected, Table 6.5 shows that the model fits quite well ($p = 0.75$). The next constraint is that the regression coefficient for poverty's impact on the random intercepts and random slopes is equal in New York and Pennsylvania while maintaining the linear trajectory model for both ($H_{\Lambda\Gamma}$). Adding the restrictions of making the corresponding γ's equal

Table 6.5*a* Likelihood Ratio Tests for Conditional Latent Curve Model of Total Crime Indexes for New York and Pennsylvania with Percent Population in Poverty as a Covariate ($N = 1157$)

No.	Model	T_{ML}	df	*p*-Value
1	H_Λ : $\lambda_t^{(NY)} = \lambda_t^{(PA)} = t - 1$	10.21	14	0.75
2	$H_{\Lambda\Gamma} : H_\Lambda$ and $\gamma_{\alpha x_k}^{(NY)} = \gamma_{\alpha x_k}^{(PA)}; \gamma_{\beta x_k}^{(NY)} = \gamma_{\beta x_k}^{(PA)}$	15.34	16	0.50
3	$H_{\Lambda\Gamma\mu} : H_{\Lambda\Gamma}$ and $\mu_\alpha^{(NY)} = \mu_\alpha^{(PA)}; \mu_\beta^{(NY)} = \mu_\beta^{(PA)}$	72.23	18	0.00
4	$H_{\Lambda\Gamma\mu\psi} : H_{\Lambda\Gamma\mu}$ and $\psi_{\alpha\alpha}^{(NY)} = \psi_{\alpha\alpha}^{(PA)}; \psi_{\beta\beta}^{(NY)} = \psi_{\beta\beta}^{(PA)}; \psi_{\alpha\beta}^{(NY)} = \psi_{\alpha\beta}^{(PA)}$	77.73	21	0.00
5	$H_{\Lambda\Gamma\mu\psi\Theta_{\epsilon\epsilon}} : H_{\Lambda\Gamma\mu}$ and $\text{VAR}(\epsilon_t)^{(NY)} = \text{VAR}(\epsilon_t)^{(PA)}$ removing constraints on μ	81.86	25	0.00
6	$H_{\Lambda\Gamma\psi} : H_{\Lambda\Gamma}$ and $\psi_{\alpha\alpha}^{(NY)} = \psi_{\alpha\alpha}^{(PA)}; \psi_{\beta\beta}^{(NY)} = \psi_{\beta\beta}^{(PA)}; \psi_{\alpha\beta}^{(NY)} = \psi_{\alpha\beta}^{(PA)}$	21.45	19	0.31
7	$H_{\Lambda\Gamma\psi\epsilon} : H_{\Lambda\Gamma\psi}$ and $\text{VAR}(\epsilon_t)^{(NY)} = \text{VAR}(\epsilon_t)^{(PA)}$	26.13	23	0.29

Table 6.5*b* Likelihood Ratio Difference Tests Between Nested Constraints of Multiple-Group Analysis for Total Crime Indexes

Model Comparison	Diff. − T_{ML}	df	*p*-Value
2 vs. 1	5.13	2	0.08
3 vs. 2	56.89	2	0.00
4 vs. 3	5.50	3	0.14
5 vs. 4	4.13	4	0.39
6 vs. 2	6.11	3	0.11
7 vs. 6	4.68	4	0.32

across the states ($H_{\Lambda\Gamma}$) has an excellent fit ($p = 0.50$). The chi-square difference for $H_{\Lambda\Gamma}$ and H_{Λ} is small and not statistically significant. Hence the evidence suggests that the impact of poverty on the random intercepts is equal across states, as is the impact of poverty on the slopes.

If we add the restriction of equal intercepts across states ($H_{\Lambda\Gamma\mu}$), the chi-square jumps by almost 57 with df $= 2$, resulting in a statistically significant decline in fit. It seems unlikely that both intercepts in the conditional model are equal in New York and Pennsylvania. Adding constraints on the disturbance variances and covariances ($H_{\Lambda\Gamma\mu\psi}$), and further adding equality restrictions for the repeated measures error variances ($H_{\Lambda\Gamma\mu\psi\Theta_{\epsilon\epsilon}}$), do not lead to further significant declines in fit. In fact, if we remove the restrictions on the intercepts and examine $H_{\Lambda\Gamma\psi}$ and $H_{\Lambda\Gamma\psi\Theta_{\epsilon\epsilon}}$, each of these models has a good fit and there is no significant decline in fit by having these equality constraints.

How do these results compare to the dummy variable models in Table 6.2 that we used in the preceding section? Consistent with the dummy variable results without the interaction term, we find a significant difference in the intercepts in the equations for α_i or β_i for the unconditional model. In the conditional model, the dummy variable approach suggested a marginally significant difference in the impact of poverty on α_i in New York versus Pennsylvania, whereas the separate group analysis was consistent with assuming that all poverty impacts were the same across groups. There are at least two plausible explanations for this slight discrepancy. One is that the unequal error variances in the repeated measures across the states might have contributed to a possible difference across methods. As we mentioned in Section 6.1, the method assumes that the variances in the errors and exogenous variables do not differ over group. Our multiple-group analysis suggests that this assumption is not violated, so this is a less likely explanation. Another possibility is that the multiple-group analysis uses a simultaneous test for all regression coefficients at the same time, whereas the dummy variable approach tested each equation separately. So the difference is one between a simultaneous test and multiple tests. Indeed, when we alter the hypothesis to test only for poverty's impact on the intercepts to differ (i.e., $H_{\Lambda\gamma_{\alpha x_k}}$) instead of a difference in poverty's impact on both the intercepts and slopes (i.e., $H_{\Lambda\Gamma}$) as was done above, there is a marginally significant difference in the intercepts for the separate group analysis. This marginally significant difference matches the dummy variable approach, suggesting that the dummy variable and multiple-group analyses lead to similar conclusions. The ability of the multiple-group analysis to perform both simultaneous tests of parameters and tests on variances and covariances gives it more flexibility than the dummy variable approach. Fortunately, in this empirical example the variances and covariances were equal, so that the multiple-group and dummy variable approaches result in similar conclusions. However, there are instances where variances differ, and this can lead to different conclusions. Without testing these assumptions, it is difficult to know whether these approaches will yield similar conclusions for other empirical applications. Thus, it is prudent to test the implicit assumptions of the dummy variable approach with the multiple-group approach.

6.3 UNKNOWN GROUP MEMBERSHIP

With both the dummy variable and multiple-group approaches, the group membership of each case in the sample is known. In some situations we might hypothesize that the data derive from more than one population without knowing from which group an observation comes. Mixture modeling provides a method for estimating models when group membership is unknown. It makes use of mixed modeling techniques and full information estimation. These techniques derive from a literature on SEM mixture modeling that permits the estimation of latent groups along with SEMs (Jedidi et al., 1994, 1997; Yung, 1994, 1997; Arminger and Stein, 1997; Arminger et al., 1999). Recent work has applied these ideas to latent curve modeling (e.g., Muthén and Shedden, 1999; Li et al., 2001; Muthén, 2001; Bauer and Curran, 2003, 2004). The techniques simultaneously cluster cases into groups and estimate the latent curves within each group. To date, these techniques have largely been confined to methodological papers and there is relatively little experience in empirical research. It is an area undergoing rapid development. This section gives an overview of the approach. The equations to represent our level 1 and level 2 models in this case are

$$y_{it} = \sum_{g=1}^{G} p_i^{(g)}[\alpha_i^{(g)} + \lambda_t^{(g)}\beta_i^{(g)} + \epsilon_{it}^{(g)}] \tag{6.38}$$

$$\alpha_i^{(g)} = \mu_\alpha^{(g)} + \sum_{k=1}^{K} \gamma_{\alpha x_k}^{(g)} x_{ik} + \zeta_{\alpha i}^{(g)} \tag{6.39}$$

$$\beta_i^{(g)} = \mu_\beta^{(g)} + \sum_{k=1}^{K} \gamma_{\beta x_k}^{(g)} x_{ik} + \zeta_{\beta i}^{(g)} \tag{6.40}$$

where y_{it} is the value of the dependent variable for the itth case and time and $p_i^{(g)}$ is the probability that the ith case belongs to the gth group with all $p_i^{(g)} \geq 0$ and $\sum_{g=1}^{G} p_i^{(g)} = 1$. In this expression, y_{it} is a function of the trajectory equation governing each group and the probability that the case belongs to that group. For instance, suppose that we have two groups;

$$y_{it} = p_i^{(1)}[\alpha_i^{(1)} + \lambda_t^{(1)}\beta_i^{(1)} + \epsilon_{it}^{(1)}] + p_i^{(2)}[\alpha_i^{(2)} + \lambda_t^{(2)}\beta_i^{(2)} + \epsilon_{it}^{(2)}] \tag{6.41}$$

where $p_i^{(1)} + p_i^{(2)} = 1$. It also is interesting to note that when group membership is known, all the probabilities in Eq. (6.38) are zero except for the group in which the case belongs and that probability is 1. In this situation of known membership, Eq. (6.38) equals (6.20) once we restrict the left-hand side of the equation to $y_{it}^{(g)}$. In this case of known membership, we can employ the multiple- group analysis described in Section 6.2.

However, when membership is unknown, a different procedure is followed. First, we will use the SEM matrix notation to rewrite Eqs. (6.38), (6.39), and (6.40) into the matrix notation that is similar to that used in Section 6.2:

$$\mathbf{y} = \sum_{g=1}^{G} p_i^{(g)} \left[\boldsymbol{\Lambda}^{(g)}\boldsymbol{\eta}^{(g)} + \boldsymbol{\epsilon}^{(g)}\right] \tag{6.42}$$

$$\boldsymbol{\eta}^{(g)} = \boldsymbol{\mu}_\eta^{(g)} + \boldsymbol{\Gamma}^{(g)}\mathbf{x}^{(g)} + \boldsymbol{\zeta}^{(g)} \tag{6.43}$$

Using this notation and given that an observation is in group g, the implied mean vector and implied covariance matrix equal Eqs. (6.30) and (6.33).

Jedidi et al. (1997, pp. 57–58) prove that if the separate group model with known groups is identified, the finite mixture model with unknown groups is identified when for each group, the observed variables come from a multivariate normal distribution. We retain this conditional multivariate normal assumption for both identification and estimation of the model. Define $\mathbf{z}'^{(g)} = \left[\mathbf{y}'^{(g)} \quad \mathbf{x}'^{(g)}\right]$ and assume that within each group the distribution of the observed variables $\mathbf{z}^{(g)}$ is multivariate normal. Under this assumption the unconditional distribution of $\mathbf{z}$ is a finite mixture of these multivariate normal distributions and the likelihood function for a sample of $i = 1, 2, \ldots, N$ values is

$$L = \prod_{i=1}^{N} \sum_{g=1}^{G} p_i^{(g)} (2\pi)^{-(p+q)/2} |\boldsymbol{\Sigma}^{(g)}(\boldsymbol{\theta})|^{-1/2} \times \exp -\frac{1}{2}\left\{[\mathbf{z}_i - \boldsymbol{\mu}^{(g)}(\boldsymbol{\theta})]' \left[\boldsymbol{\Sigma}^{(g)}(\boldsymbol{\theta})\right]^{-1} [\mathbf{z}_i - \boldsymbol{\mu}^{(g)}(\boldsymbol{\theta})]\right\} \tag{6.44}$$

where L is a function of $p_i^{(g)}$ and all of the parameters in $\boldsymbol{\theta}$ (Jedidi et al., 1997, p. 44).

Researchers have taken a variety of approaches to maximize the likelihood function (e.g., Jedidi et al., 1994, 1997; Arminger et al., 1999). Many of these employ the EM algorithm. There are several cautionary notes to emphasize for estimating models with unknown group membership. One is that local optimal solutions appear more likely with these types of models than with models with known memberships. It is recommended that the researcher try several different sets of starting values when examining a model. A second characteristic is that the assumption of multivariate normality is more important for these mixture models than it is for the usual SEM. One reason is that multinormality of the observed variables within each group is often unrealistic. Dummy exogenous variables are a good example where this assumption is violated. Arminger et al. (1999) find evidence that unlike single-group analysis, the mixture model ML estimator does not lead to consistent estimators when this distributional assumption is violated. Another reason for concern comes from the work of Bauer and Curran (2003, 2004). Their papers illustrate the possibility that violation of the multivariate normality assumption can lead to the impression of multiple groups even when the data are generated from one group. In other words, multiple groups might be created simply to do a better job of fitting the nonnormal distribution of the data.

Arminger et al. (1999) propose a conditional mixture model that is somewhat less restrictive in its distributional assumptions. Their model applies in a situation like ours where there are exogenous **x** variables. Under this condition, they make the assumption that the distribution of **y** *conditional* on **x** is multivariate normal. This does not require distributional assumptions for **x** and the assumption of multivariate normality for the conditional distribution is more realistic than for the unconditional distributions. Though this is an attractive alternative assumption for these models, there still is the danger of creating nonexistent groups to capture the nonnormality of the conditional distributions.

6.3.1 Empirical Example

We return to the crime index data from New York and Pennsylvania to illustrate this estimator. Recall that initially we pooled the data and examined conditional mean differences in the trajectory parameters as a function of state membership. We then split the sample and evaluated equality constraints of key model parameters as a function of group membership. Whether a given reporting community was embedded in Pennsylvania or New York was an observed variable, and we could thus unambiguously identify group membership. However, it could be hypothesized that there are two or more important groups within the sample, but the group membership variable was *not* observed. That is, group membership is latent. We can use the growth mixture models to attempt to recover these unobserved groups.

To accomplish this, we return to the pooled sample consisting of data from both states. We fit SEM mixture models and extracted one, two, and three group solutions (Muthén and Muthén, 2001). We must then make some determination as to the optimal fitting solution of the three. Unfortunately, we cannot use a likelihood ratio test to determine the number of groups because this test does not follow a central chi-square distribution in this situation (McLachlan and Basford, 1988, sec. 1.10; Arminger et al., 1999, p. 483). However, we can use the likelihood-based Bayesian information criterion (BIC; Schwarz, 1978; Raftery, 1995). Examination of the BICs indicated that the three-cluster model exhibited the best fit. Table 6.6

Table 6.6 ML Parameter Estimates (Asymptotic Standard Errors) for the Three-Cluster Mixture Model for Total Crime Index Using Pooled Sample from New York and Pennsylvania ($N = 1157$)

	Cluster		
Parameter	One	Two	Three
μ_α	50.97(0.56)	58.29(0.59)	47.28(0.56)
μ_β	1.27(0.10)	1.09(0.07)	0.89(0.19)
$\psi_{\alpha\alpha}$	25.85(3.31)	22.06(3.17)	62.28(7.55)
$\psi_{\beta\beta}$	0.31(0.28)	0.25(0.09)	3.91(0.99)
$\psi_{\alpha\beta}$	−0.81(0.67)	−0.40(0.34)	−5.11(2.02)
VAR(ϵ_1)	7.47(1.58)	1.10(0.37)	25.11(4.21)
VAR(ϵ_2)	6.82(1.31)	1.09(0.22)	24.86(2.91)
VAR(ϵ_3)	5.63(0.94)	0.74(0.23)	20.24(2.68)
VAR(ϵ_4)	11.19(1.81)	2.18(0.49)	28.25(4.56)

reports the ML estimates of the means and variances from this SEM mixture model with three clusters.[3]

These results suggest that different trajectories hold for each of the three clusters. The mean intercepts range from 47 to 58 and the mean slopes vary from 0.89 to 1.27. The relative magnitudes of the estimates of the variances of the random intercepts and random slopes are similar for the first two clusters, but considerably larger for the third. Similarly, the asymptotic standard errors are larger for cluster 3. A similar pattern of larger point estimates and standard errors in cluster 3 compared to the first two clusters holds for the error variances. Furthermore, this example illustrates some of the complications that occur with such mixture models. The three-cluster models exhibited more instability than did the two-cluster model even though the log likelihood and BIC were better in the three-cluster models. In addition, the cluster models had convergence problems when the starting values for the estimates were varied. More specifically, those solutions with more clusters tended to converge less frequently than those with fewer clusters. In fact, the two-cluster model converged more frequently and it converged to the same solution far more frequently than did the three-cluster model. A case could be made that the two-cluster model is better for these data, despite it not having the lowest BIC value. Thus it would be premature to accept the three- or two-cluster solution without replication and substantive evidence that would support this division of reporting units. As this empirical example and the distributional assumptions for these models illustrate, researchers need considerable caution when employing these models to estimate unknown groups (see Hipp and Bauer, 2004).

There are many interesting extensions of these mixture models, including consideration of one or more covariates predicting the latent trajectories within cluster,

[3]Mplus 2 (Muthén and Muthén, 2001) was used to estimate these models. The ML estimator used is documented at http://www.statmodel.com/mplus/techappen.pdf.

one or more covariates predicting the probability of cluster membership, and cluster membership predicting one or more distal outcomes (e.g., Arminger et al., 1999; Muthén and Shedden, 1999; Muthén and Muthén, 2001). There is also a class of highly related techniques commonly referred to as semiparametric group-based trajectory analysis (e.g., Nagin, 1999; Nagin and Tremblay, 2001). In this approach, random components are not typically allowed at the level of the latent factors (i.e., $\Psi = 0$), and two or more mixtures are extracted to serve as "points of support" for the population distribution of trajectory parameters. We do not explore these models here, but see Nagin (1999; 2005), Jones et al., (2001), and Nagin and Tremblay (2001) for further details.

6.4 CONCLUSIONS

It is not uncommon for the cases in a sample to derive from more than one population. In this chapter we presented techniques with which to analyze and compare two or more groups. The dummy variable approach led to a conditional LCM where dummy variables code group membership and permit these dummy variables to directly influence the random intercepts and random slopes. This was a simple way to assess whether groups differed in their mean initial levels or mean slopes. However, implicit in the dummy variable approach was the assumption that the groups followed the same functional form in their trajectories and that the groups shared the same variances of random intercepts, random slopes, and error as well as the covariance between the random intercepts and random slopes. When these assumptions were questionable, another way to analyze the data was with a multiple-group analysis. The SEM approach provided maximum flexibility in that it permitted an easy examination of the possible differences in all of these model parameters. Finally, we also mentioned the case where groups seemed likely but case membership was not known. In these models we estimated both the latent curve model and the group membership at the same time. Taken together, these three methods comprised a flexible set of tools for analyzing different groups.

APPENDIX 6A: Case-by-Case Approach to Analysis of Various Groups

Nearly all of this chapter treated the SEM approach to group analysis. In this appendix we describe briefly the case-by-case approach to group analysis. A complication ignored in this appendix is the treatment of missing data in the case-by-case approach. In this approach, we estimate the latent curve level 1 model separately for each case no matter what group it is in. For the crime index data, the case-by-case approach would estimate the trajectory regression for each police reporting unit in New York and Pennsylvania. Recall that this model permits the error variances to differ by case but assumes that within each case the error variance is homoscedastic over time. This leads to an intercept and slope estimate for each case in both samples, and researchers can plot and analyze these estimates to gain insights into the trajectories for each state.

6A.1 Dummy Variable Method

A test of whether the mean intercept or mean slope differ by state requires regressions of the estimates of random intercepts and the random slopes on the dummy variable (D_1) for state ($D_1 = 1$ for Pennsylvania; 0 for New York). Within rounding, we obtain the same results as the SEM approach for the differences in coefficients for the dummy variable method.

6A.2 Multiple-Group Analysis

The case-by-case approach also permits a multiple-group analysis analogous to the SEM approach. The method of estimation of the level 1 model proceeds as before: For each case in the sample, the repeated measure is regressed on the appropriately coded time trend variable using OLS regression. Next, the $\widehat{\alpha}_i^{(g)}$'s and $\widehat{\beta}_i^{(g)}$'s are separated into the groups to which they belong. Then within each group, the $\widehat{\alpha}_i^{(g)}$'s and $\widehat{\beta}_i^{(g)}$'s are regressed on the covariates, $x_{ik}^{(g)}$'s. The latter step provides estimates of the intercepts and the regression coefficients of Eqs. (6.21) and (6.22).

Testing whether the intercepts or covariate slopes differ across groups is straightforward. Consider the test for equal slopes in two groups for a covariate predicting the random slopes ($\widehat{\beta}_i^{(g)}$),

$$z = \frac{\widehat{\gamma}_{\beta x_k}^{(1)} - \widehat{\gamma}_{\beta x_k}^{(2)}}{\text{s.e.}(\widehat{\gamma}_{\beta x_k}^{(1)} - \widehat{\gamma}_{\beta x_k}^{(2)})} \tag{6A.1}$$

where

$$\text{s.e.}(\widehat{\gamma}_{\beta x_k}^{(1)} - \widehat{\gamma}_{\beta x_k}^{(2)}) = \sqrt{\text{var}(\widehat{\gamma}_{\beta x_k}^{(1)}) + \text{var}(\widehat{\gamma}_{\beta x_k}^{(2)})} \tag{6A.2}$$

The $\text{var}(\widehat{\gamma}_{\beta x_k}^{(g)})$ is the estimated variance of $\widehat{\gamma}_{\beta x_k}^{(g)}$ from the OLS regression. For two independent groups the covariance of the coefficients is zero and no covariance term is required in this expression. In large samples, the test statistic, z, in Eq. (6A.1) follows a standardized normal distribution. The null hypothesis of the test is H_o : $\gamma_{\beta x_k}^{(1)} = \gamma_{\beta x_k}^{(2)}$. With this, a researcher can test whether two regression coefficients are equal across the groups. A similar test is possible for testing the equality of the intercepts of these equations. Although the simplicity of these tests is an advantage, there are some complications that emerge when there are greater than two groups, when a researcher wants to test several coefficients simultaneously, or when the object of testing are the variances of the errors or disturbances of the equations. Although we can devise methods to handle these complications,[4] it is more straightforward to use the SEM separate group testing described earlier in the chapter.

[4]For example, we could test the equality of several coefficients and the intercepts in two groups by using software that permits the estimation of seemingly unrelated regressions with equality constraints on the corresponding coefficients across equations (e.g., SAS).

6A.3 Unknown Group Membership

The case-by-case approach to unknown group membership begins by estimating the level 1 model, $y_{it} = \alpha_i + \lambda_t\beta_i + \epsilon_{it}$, as usual. That is, perform a separate regression of the y variable on the trend variable for each case and save the $\widehat{\alpha}_i$s and $\widehat{\beta}_i$s for all cases. Interestingly, the case-by-case estimation of the level 1 model is unaltered no matter what the number of groups. The next step is to perform a cluster analysis treating the $\widehat{\alpha}_i$'s and $\widehat{\beta}_i$'s as the variables. Cluster analysis is a method of estimating group membership when group membership is unknown. There are numerous clustering algorithms to choose from, but they typically form preliminary clusters and then measure the distance of each observation from the cluster classifying an observation in the cluster to which it is closest.

To illustrate this we look at the crime index data from New York and Pennsylvania. We run the case-by-case OLS regression for each. Even though we know which $\widehat{\alpha}_i$'s and $\widehat{\beta}_i$'s belong to which state, we merge these to illustrate our clustering procedure. We use the Proc Fastclus from SAS (*SAS/STAT User's Guide*, 2000). It employs an iterative algorithm for minimizing the sum of squared Euclidean distances from the cluster means. It begins by selecting observations to serve as cluster seeds, and then observations are assigned to the cluster with the nearest seed. The algorithm updates this process until changes in the cluster seeds become small or zero. This clustering algorithm, like others, is sensitive to the variance of the variables such that those variables with larger variances exert more influence on the cluster than those with smaller variances. Standardizing the $\widehat{\alpha}_i$'s and $\widehat{\beta}_i$'s to have means of zero and variances of 1 can avoid this problem. Given that the pooled data derive from two states, we ran the model setting the number of clusters to two. Table 6A.1 cross-classifies the actual state from which an observation derives with the cluster into which it was classified.

If the two clusters corresponded to New York and Pennsylvania, we would expect all the cases to fall in either the main diagonal [1,1 and 2,2 cells] or the opposite diagonal [1,2 and 2,1]. The dispersion of cases across the cells suggests either that there is considerable misclassification in the cluster analysis or that the two clusters do not correspond to New York and Pennsylvania. Instead, it might be picking up some unobserved categories that separate the observations. When we

Table 6A.1 Cross-Classification of State of Case with Cluster into Which It Was Placed for the $\widehat{\alpha}_i$'s and $\widehat{\beta}_i$'s from the Total Crime Index Data ($N = 1002$)

	Cluster	
State	One	Two
New York	237	149
Pennsylvania	378	238

forced a three-cluster solution, we found that one cluster had only five cases in it and the other two clusters shifted somewhat.

6A.4 Appendix Summary

In the appendix we gave a brief overview of how researchers could use the case-by-case approach to address issues of multiple-group analysis. The approach remains appealing in its simplicity, but it does have limitations in terms of easily treating missing data, testing equality of variances and covariances, and providing some of the other additional information that is available with the SEM approach.

CHAPTER SEVEN

Multivariate Latent Curve Models

Thus far, we have considered only *univariate* latent curve models (LCMs). That is, although there are multiple repeated measures on a dependent variable (making it multivariate in the sense of the repeated measures), some consider this a univariate LCM given that repeated measures are for just one outcome (e.g., just reading achievement or just rates of crime). In Chapters 2, 3, and 4, we explored a variety of issues associated with univariate LCMs without considering exogenous predictors of the latent factors. Chapters 5 and 6 extended these models to incorporate one or more exogenous predictors of the growth factors, leading to conditional univariate LCMs. A key premise of this conditional model was that the exogenous variables were either *time-invariant*—that is, exogenous covariates that stay the same for an individual—or if time-varying, they were measured at or before the first wave of the repeated measure. However, there are clearly many situations in applied research in which we are interested not only in repeated measures of a given variable over time, but in one or more predictor variables that vary over the same or similar range of time as the repeated measure. The focus of this chapter is the analysis of repeated measures of two or more variables.

The chapter opens with a brief review of the univariate LCM without predictors, and then we review the inclusion of one or more time-invariant covariates. Next we expand this model to allow for repeated measures on the covariates but continue to model random growth only in the primary outcome. The next modification to this model allows for simultaneous growth processes in both the primary outcome and the time-varying covariates. An expansion of this to the multivariate LCM with one or more time-invariant covariates follows. This, in turn, leads to a general multivariate modeling framework referred to as the *autoregressive latent trajectory* (ALT) *model*. We conclude with a general notational framework that encompasses all variations of the multivariate trajectory model.

7.1 TIME-INVARIANT COVARIATES

We begin by reviewing briefly the inclusion of time-invariant covariates (TICs) in the standard univariate LCM. Recall that in Chapter 2 the level 1 equation of the

Latent Curve Models: A Structural Equation Perspective, by Kenneth A. Bollen and Patrick J. Curran

unconditional linear LCM was

$$y_{it} = \alpha_i + \beta_i \lambda_t + \epsilon_{it} \tag{7.1}$$

and the level 2 equations were

$$\alpha_i = \mu_\alpha + \zeta_{\alpha_i} \tag{7.2}$$

$$\beta_i = \mu_\beta + \zeta_{\beta_i} \tag{7.3}$$

Equations (7.2) and (7.3) highlight that there were no predictors of the latent growth factors because the right-hand terms consist only of an overall mean and individual deviation from this mean. Results from Chapter 2 demonstrated that these level 1 and level 2 equations could be expressed in terms of structural equation matrices such that Eq. (7.1) was equivalent to[1]

$$\mathbf{y}_i = \mathbf{\Lambda}_{\mathbf{y}\eta} \boldsymbol{\eta}_i + \boldsymbol{\epsilon}_i \tag{7.4}$$

and Eq. (7.2) and (7.3) were equivalent to

$$\boldsymbol{\eta}_i = \boldsymbol{\mu}_\eta + \boldsymbol{\zeta}_i \tag{7.5}$$

Further, the reduced-form expression is created by substituting Eq. (7.5) into Eq. (7.4), resulting in

$$\mathbf{y}_i = \mathbf{\Lambda}_{\mathbf{y}\eta} \boldsymbol{\mu}_\eta + \mathbf{\Lambda}_{\mathbf{y}\eta} \boldsymbol{\zeta}_i + \boldsymbol{\epsilon}_i \tag{7.6}$$

Finally, the mean and covariance structure implied by Eq. (7.6) expressed as a function of the model parameters in vector $\boldsymbol{\theta}$ is

$$\boldsymbol{\mu}(\boldsymbol{\theta}) = \mathbf{\Lambda}_{\mathbf{y}\eta} \boldsymbol{\mu}_\eta \tag{7.7}$$

and

$$\mathbf{\Sigma}(\boldsymbol{\theta}) = \mathbf{\Lambda}_{\mathbf{y}\eta} \mathbf{\Psi} \mathbf{\Lambda}'_{\mathbf{y}\eta} + \boldsymbol{\theta}_{\epsilon\epsilon} \tag{7.8}$$

respectively.

In Chapter 5 we extended the level 2 equations [e.g., Eqs. (7.2) and (7.3)] to include exogenous predictor variables. The matrix expression for the level 1 model remains as before [e.g., Eq. (7.4)], but the level 2 model defined in Eq. (7.5) is expanded to include the exogenous covariates such that

$$\boldsymbol{\eta}_i = \boldsymbol{\mu}_\eta + \mathbf{\Gamma}_{\eta\mathbf{x}} \mathbf{x}_i + \boldsymbol{\zeta}_i \tag{7.9}$$

where $\mathbf{\Gamma}_{\eta\mathbf{x}}$ is a matrix of fixed coefficients of the regression of $\boldsymbol{\eta}_i$ on $\mathbf{x}_i$, where $\mathbf{x}_i$ is a $Q \times 1$ vector of observed time-invariant exogenous variables. The i subscript

[1]The $\mathbf{\Lambda}_{\mathbf{y}\eta}$ has subscripts here because later in the chapter we introduce a second repeated measure and we want to distinguish the two series of repeated measures. Using only $\mathbf{\Lambda}$ without subscripts could lead to confusion about the two series.

signifies that the values of $\mathbf{x}$ can vary over cases, but the lack of a t subscript means that $\mathbf{x}$ does not change over time.[2] We can substitute Eq.(7.9) into Eq. (7.4) to create the reduced form expression such that

$$\mathbf{y}_i = \mathbf{\Lambda}_{\mathbf{y}\eta}\boldsymbol{\mu}_\eta + \mathbf{\Lambda}_{\mathbf{y}\eta}\mathbf{\Gamma}_{\eta\mathbf{x}}\mathbf{x}_i + \mathbf{\Lambda}_{\mathbf{y}\eta}\boldsymbol{\zeta}_i + \boldsymbol{\epsilon}_i \tag{7.10}$$

with the addition of the exogenous covariates in $\mathbf{x}_i$ the model implied mean structure is now

$$\boldsymbol{\mu}(\boldsymbol{\theta}) = \begin{bmatrix} \boldsymbol{\mu}_{\mathbf{y}}(\boldsymbol{\theta}) \\ \boldsymbol{\mu}_{\mathbf{x}}(\boldsymbol{\theta}) \end{bmatrix} = \begin{bmatrix} \mathbf{\Lambda}_{\mathbf{y}\eta}\boldsymbol{\mu}_\eta + \mathbf{\Lambda}_{\mathbf{y}\eta}\mathbf{\Gamma}_{\eta\mathbf{x}}\boldsymbol{\mu}_{\mathbf{x}} \\ \boldsymbol{\mu}_{\mathbf{x}} \end{bmatrix} \tag{7.11}$$

where $\boldsymbol{\mu}(\boldsymbol{\theta})$ is a $(T + Q) \times 1$ vector with the means for y ordered first and the means for x ordered second. The corresponding covariance structure is

$$\mathbf{\Sigma}(\boldsymbol{\theta}) = \begin{bmatrix} \mathbf{\Sigma}_{yy}(\boldsymbol{\theta}) & \mathbf{\Sigma}_{yx}(\boldsymbol{\theta}) \\ \mathbf{\Sigma}_{xy}(\boldsymbol{\theta}) & \mathbf{\Sigma}_{xx}(\boldsymbol{\theta}) \end{bmatrix} \tag{7.12}$$

where $\mathbf{\Sigma}(\boldsymbol{\theta})$ is a $(T + Q) \times (T + Q)$ symmetric covariance matrix with y ordered first and x ordered second, with elements

$$\mathbf{\Sigma}_{yy}(\boldsymbol{\theta}) = \mathbf{\Lambda}_{\mathbf{y}\eta}(\mathbf{\Gamma}_{\eta\mathbf{x}}\mathbf{\Sigma}_{xx}\mathbf{\Gamma}_{\eta\mathbf{x}}' + \mathbf{\Psi})\mathbf{\Lambda}_{\mathbf{y}\eta}' + \mathbf{\Theta}_{\epsilon\epsilon} \tag{7.13}$$

$$\mathbf{\Sigma}_{xy}(\boldsymbol{\theta}) = \mathbf{\Sigma}_{xx}\mathbf{\Gamma}_{\eta\mathbf{x}}'\mathbf{\Lambda}_{\mathbf{y}\eta}' \tag{7.14}$$

$$\mathbf{\Sigma}_{yx}(\boldsymbol{\theta}) = \mathbf{\Sigma}_{xy}'(\boldsymbol{\theta}) \tag{7.15}$$

$$\mathbf{\Sigma}_{xx}(\boldsymbol{\theta}) = \mathbf{\Sigma}_{xx} \tag{7.16}$$

Note that the exogenous variables x_{iq} are *not* subscripted by t, reflecting that the observed values do not vary as a function of time. These are *time-invariant covariates* (TICs). The definition of what constitutes a TIC often varies as a function of substantive theory or characteristics of the empirical data. For example, some TICs unambiguously reflect a variable that does not vary with time. Examples might include year of birth, gender, ethnicity, or country of origin. In this case only a single measure is needed given that the variable does not change.

However, other types of TICs are less clear. For example, we might consider a variable that is theoretically believed not to vary as a function of time (e.g., stable personality traits), but repeated assessments might lead to changing observed values (possibly due to measurement error or context effects). In this situation, we might conceptualize a TIC latent variable that is imperfectly captured by the observed variable and we would need to take account of the measurement error to prevent it from confounding our analysis. Alternatively, other measures might vary as a function of time, yet only the initial time assessment is of interest. For example, a

[2]In those conditional models where a time-varying variable is the covariate, a t subscript could be added to represent time. But the logical consistency of the model demands that the time-varying variable be observed no later than the first repeated measure.

time-specific level of antisocial behavior on entry into kindergarten might influence developmental trajectories of reading ability through grade 4. Although antisocial behavior is expected to develop over time in young children, the initial measure is treated like an exogenous TIC predictor of the latent growth factors, assuming that later antisocial behavior has no impact.

Treating the initial measure of a variable like a TIC is sometimes done in latent curve applications. However, if the explanatory variable is time varying and later as well as earlier values influence the other repeated measure, treating this as a time-invariant predictor creates a misspecified LCM and probable bias. There are several powerful alternative ways that repeated assessments of a predictor variable can enter into the LCM.

First we consider including time-varying covariates as direct predictors of the repeated dependent variable over time. These influences can be contemporaneous or lagged effects. We then extend this model to allow for the estimation of random growth processes for both variables over time. Finally, we explore the simultaneous incorporation of autoregression and latent curve structures to model a set of two different repeated measures.

7.1.1 Empirical Example: Conditional Model of Reading and Math Achievement

In Chapter 4 we presented a variety of nonlinear trajectory models to characterize fixed and random components of the development of math ability in a sample of $N = 1767$ children. We found that there was significant variability in childhood math ability in kindergarten, and there was significant variability in individual rates of increase in math ability throughout elementary school.

However, in addition to the repeated measures of math achievement, there are repeated measures of reading achievement. Suppose that math ability depends partially on the child's reading skills. This might occur given the need for a child to accurately comprehend a written mathematics problem prior to solving it. Thus, it could be useful to explore the relation between developmental trajectories of math and of reading simultaneously. It is to these models that we now turn.

The earlier univariate unconditional curve models of math ability in Chapter 4 found that a freed factor loading model fit optimally. This final nonlinear LCM had a good fit according to the RMSEA, but a less favorable fit according to the other fit measures [$T_{\text{ML}}(18) = 78.1$, $p < 0.001$, IFI $= 0.89$, TLI $= 0.87$, RMSEA $= 0.04(0.03, 0.05)$].[3] Further, the magnitude of the estimated factor loadings reflected that of the total change in math ability observed from kindergarten to grade 6, 23% occurred between kindergarten and first grade, 23% between first and second; 17% between second and third; 17% between third and fourth; 9% between fourth and fifth; and 11% between fifth and sixth. A starting point from which to expand this LCM of math ability is to include the initial assessment of reading ability.

[3]Although the likelihood ratio test statistic is statistically significant, this is partially due to the greater statistical power that accompanies a large sample size ($N = 1767$). We are thus augmenting our evaluation of model fit with several other indices along with consideration of covariance and mean residuals.

Table 7.1 Coefficient Estimates for Random Intercepts and Random Slopes Regressed on Gender, Minority Status, and Initial Reading Ability

Predictor Variable	Math Intercept Factor	Math Slope Factor
Child minority status	−2.871 (0.325)	−2.330 (0.727)
Child gender	−0.843 (0.323)	−1.026 (0.721)
Initial reading ability	0.461 (0.028)	0.003 (0.075)

We next regressed the latent intercept and slope factor of math ability on three exogenous covariates: child gender, child ethnicity, and child reading ability at the initial assessment. This model fit was similar to the unconditional model [$T_{\mathrm{ML}}(33) = 128.2$, $p < 0.001$, RMSEA $= 0.04$ (0.03, 0.05), IFI $= 0.91$, TLI $=$ 0.87]. The regressions of the latent curve factors on the three covariates reflected that compared to majority children, minority children reported significantly lower initial levels of math ability and significant negative impact on the rate of change in math ability over time. See Table 7.1 for resulting parameter estimates and standard errors. Further, compared to males, females reported significantly lower initial levels of math ability but did not report any significant differences in rate of change in math over time. Finally, children reporting higher levels of reading ability at the initial time period also reported higher math ability at the initial period; importantly, initial reading ability was not predictive in rates of change in math.

One limitation of this model is that developmental theory would predict that reading ability is developing as a function of time. Thus, treating reading ability as a time-invariant predictor of the latent curve factors of math ability may not be appropriate to test our theoretical model. Indeed, this may be one source of model misfit. A reasonable extension of this LCM of math ability is to explicitly incorporate the time-varying influences of reading ability. In the next sections we explore several ways to incorporate time-varying predictors.

7.2 TIME-VARYING COVARIATES

The first method for incorporating repeated measures of covariates is to regress the repeated measures of the dependent variable directly on the covariates. To accomplish this, we modify the level 1 equation to include these *time-varying covariates* (TVCs). Modification of the level 1 equation is a departure from prior chapters in that it is the first time that the repeated measure is predicted directly by something other than a function of time.

Denote the new repeated measure by w_{it}, where the subscripts show that w varies for individual i and time t. Using w differentiates them from the TICs denoted x. The repeated dependent variable is still y_{it}, indicating that its value is for individual i at time t.

One model regresses the dependent measure y_{it} directly on w_{it} by including the influence of the time-specific measures of w in the level 1 equation. This results in

$$y_{it} = \alpha_i + \beta_i \lambda_t + \gamma_t w_{it} + \epsilon_{it} \tag{7.17}$$

indicating that variable y for individual i at time point t is an additive combination of an individually varying intercept (α_i), an individually varying slope multiplied by time ($\beta_i\lambda_t$), a time-specific influence of the covariate w ($\gamma_t w_{it}$), and the usual individual and time-specific disturbance (ϵ_{it}). All of the same assumptions hold as were explicated in Chapter 2 for the unconditional model (see Table 2.1). However, given the inclusion of the TVC, we have the added assumptions that COV $(\epsilon_{it}, w_{it+s}) = 0$ for all i, t, and s. Finally, Eq. (7.17) defines the regression of the repeated measure of y on the TVC w within the same time t; however, lagged effects between y and w (e.g., we may regress y at time t onto w at time $t-1$ to examine the prospective relation between w and y; see, e.g., Curran et al., 1998) are possible.[4]

Next consider the form of the level 2 equations given the inclusion of the TVCs in Eq.(7.17). When there are no direct predictors of the latent curve factors, the *unconditional TVC model* results. The level 2 equations are thus the same as for the unconditional LCM, namely

$$\alpha_i = \mu_\alpha + \zeta_{\alpha_i} \tag{7.18}$$

$$\beta_i = \mu_\beta + \zeta_{\beta_i} \tag{7.19}$$

The matrix expression for Eq.(7.17) is

$$\mathbf{y}_i = \mathbf{\Lambda}_{\mathbf{y}\eta}\boldsymbol{\eta}_i + \mathbf{\Gamma}_{\mathbf{yw}}\mathbf{w}_i + \boldsymbol{\epsilon}_i \tag{7.20}$$

where $\mathbf{y}_i$ is a $T \times 1$ vector of repeated measures for individual i, $\mathbf{\Lambda}_{\mathbf{y}\eta}$ is a $T \times m$ matrix of factor loadings for T time points and the m random coefficients of the latent curve factors ($m = 2$ for a linear model, $m = 3$ for a quadratic model, etc.), $\boldsymbol{\eta}_i$ is a $m \times 1$ vector of random coefficients of the latent curve variables, $\mathbf{\Gamma}_{\mathbf{yw}}$ is a $T \times T$ matrix of regression coefficients of the repeated measures on the TVCs (assuming T assessments were taken on both y and w), $\mathbf{w}_i$ is a $T \times 1$ vector of TVCs, and $\boldsymbol{\epsilon}_i$ is a $T \times 1$ vector of time-specific disturbances. The covariance structure of $\boldsymbol{\epsilon}_i$ is denoted $\mathbf{\Theta}_{\epsilon\epsilon}$ and is a $T \times T$ symmetric covariance matrix. For example, for $T = 4$,

$$\mathbf{\Theta}_{\epsilon\epsilon} = \begin{pmatrix} \text{VAR}(\epsilon_1) & 0 & 0 & 0 \\ 0 & \text{VAR}(\epsilon_2) & 0 & 0 \\ 0 & 0 & \text{VAR}(\epsilon_3) & 0 \\ 0 & 0 & 0 & \text{VAR}(\epsilon_4) \end{pmatrix} \tag{7.21}$$

where VAR(ϵ_1) is the time-specific disturbance for the $t = 1$ measure of y_{it}, and so on.

A matrix expression for Eqs.(7.18) and (7.19) is

$$\boldsymbol{\eta}_i = \boldsymbol{\mu}_\eta + \boldsymbol{\zeta}_i \tag{7.22}$$

[4]Note that although the value of the regression parameter γ_t can vary over t (e.g., the fixed effect of the TVC can vary across time), at the time of writing it is not possible to estimate a random component for this effect within the SEM (as is possible in the general multilevel modeling framework).

where η_i is defined as above, μ_η is a $m \times 1$ vector of latent factor means, and ζ_i is a $m \times 1$ vector of disturbance variables. The covariance structure of ζ_i is denoted Ψ and is a $m \times m$ symmetric covariance matrix. For example, for $m = 2$ (e.g., a linear LCM),

$$\mathbf{\Psi} = \begin{pmatrix} \psi_{\alpha_y\alpha_y} & \psi_{\beta_y\alpha_y} \\ \psi_{\alpha_y\beta_y} & \psi_{\beta_y\beta_y} \end{pmatrix} \tag{7.23}$$

where the ψ elements represent the variances and covariances of the latent curve random coefficients.

As before, substituting Eq. (7.22) into Eq. (7.20) results in the reduced form for the unconditional TVC model

$$\mathbf{y}_i = \left(\mathbf{\Lambda}_{\mathbf{y}\eta}\boldsymbol{\mu}_\eta + \mathbf{\Gamma}_{\mathbf{yw}}\mathbf{w}_i\right) + \left(\mathbf{\Lambda}_{\mathbf{y}\eta}\boldsymbol{\zeta}_i + \boldsymbol{\epsilon}_i\right) \tag{7.24}$$

where the first parenthetical term represents the fixed effects (including the time-specific contribution of the TVC) and the second parenthetical term represents the random effects.[5] The path diagram for the unconditional TVC model is shown in Figure 7.1.[6] The TVCs are exogenous and covary with the random latent curve factors as shown in Figure 7.1.

7.2.1 Interpretation of Effects

Inclusion of the TVCs in the level 1 equation modifies the interpretations of parameter estimates relative to the models from earlier chapters. Specifically, the repeated measures of y are now a *joint* function of the random coefficients plus the time-specific influences associated with the TVCs. Given this simultaneous estimation, we can think of the contributions of the growth process and the TVCs in two equivalent ways.

First, conceptualize the TVCs as covariates where we estimate the time-specific influence of the covariate and then examine the underlying growth process of the repeated measure on y net the effects of w. That is, the influence of w_{it} on y_{it} is removed from y_{it}, and the growth process is estimated on the "adjusted" measure of y. However, the joint contribution of the growth process and the TVCs is symmetric, so that it is equivalent to think of the time-specific regressions reflecting the influence of the TVCs on the repeated measures above and beyond the influence of the random growth process. This interpretation may be less natural in many research settings, but it is appropriate to consider the effects of the TVCs as the

[5]We could easily express the model implied mean and covariance structure stemming from Eq. (7.24) as we did for the univariate conditional LCM above [e.g., Eqs. (7.11) and (7.12)]. However, for the TVC model and all multivariate models to follow, we defer the presentation of these model-implied moment structures until later because we will develop a general matrix expression that can be used to define all of the models we discuss here.

[6]Here and in other figures in this chapter, we omit the labels for the disturbance terms, although we do show their paths. We do this to avoid overcrowding the figures.

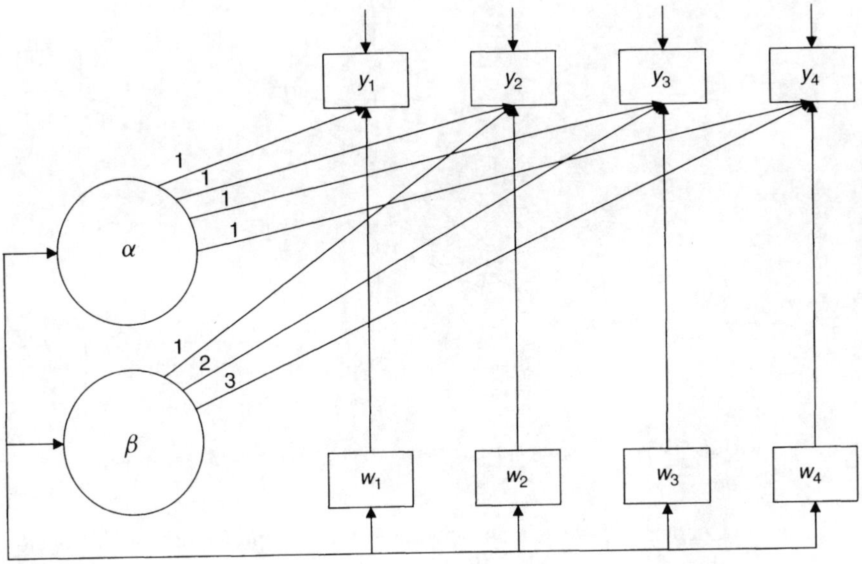

FIGURE 7.1 Path diagram of unconditional time-varying covariate model.

time-specific prediction of the repeated measure after controlling the influence of the underlying growth process.

For example, Curran et al. (1998) provide an illustration in which a TVC model was used to empirically evaluate the time-specific impact of becoming married on heavy alcohol use, above and beyond the normative trajectory of alcohol use. The motivating question was: Does becoming married for the first time exert a significant negative impact on heavy alcohol use above and beyond what changes would be expected from normative development of alcohol use in early adulthood? The TVC model allowed for an evaluation of this effect.

To reiterate, the joint influence of the random growth process and the time-specific contributions of the TVCs are estimated simultaneously, and each effect is net the other. However, the specific interpretation of these effects can vary depending on the theoretical question. We illustrate this with an empirical example after we discuss extending this model by adding exogenous variables to create a conditional model.

7.3 SIMULTANEOUS INCLUSION OF TIME-INVARIANT AND TIME-VARYING COVARIATES

The unconditional TVC model described above allowed for the joint contribution of the latent curve process and the time-specific measures of the TVCs. Further, we considered this an unconditional model given that we did not regress the latent trajectory factors on any time-invariant covariates. However, just as we were able to incorporate exogenous time-invariant covariates into the standard univariate LCM

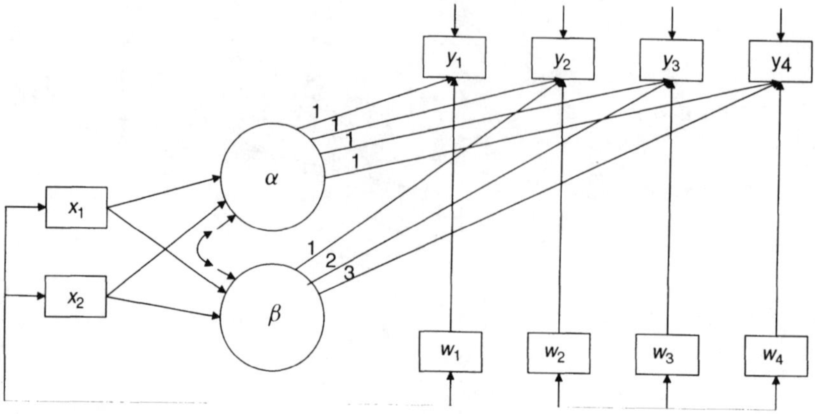

FIGURE 7.2 Path diagram of a time-varying and time-invariant covariate model.

(e.g., as in Chapter 5), we can also include TICs as predictors of the random intercepts and coefficients in the TVC latent curve model.

As a starting point, consider a hypothetical situation in which we are interested in four repeated measures of y (say, math ability). We would like to include four repeated measures of a time-varying covariate w (say, reading ability) as well as two time-invariant covariates (say, child gender and ethnicity). Thus, our goal is to define a model that includes the joint effects of the TVCs and the random latent growth process on the repeated measures of y while modeling individual variability in the latent growth factors as a function of one or more TICs. The path diagram for this model is shown in Figure 7.2.

To incorporate the time-varying covariates (TVCs) and the time-invariant covariates (TICs) simultaneously, we will expand both the level 1 and level 2 equations. For the TVCs, our level 1 equation remains as above, namely,

$$y_{it} = \alpha_i + \beta_i \lambda_t + \gamma_t w_{it} + \epsilon_{it} \tag{7.25}$$

where all terms are defined as above. In matrix form, our level 1 equation (which was given in scalar terms in Eq. (7.17)] is defined as

$$\mathbf{y}_i = \mathbf{\Lambda}_{\mathbf{y}\eta} \boldsymbol{\eta}_i + \mathbf{\Gamma}_{\mathbf{yw}} \mathbf{w}_i + \boldsymbol{\epsilon}_i \tag{7.26}$$

where all matrices are defined as before [i.e., as for Eq. (7.20)].

However, whereas in the unconditional TVC model the level 2 equations were expressed only as a function of a mean and deviation term [i.e., Eq. (7.22)], the level 2 equations now include Q-exogenous predictor variables:

$$\alpha_i = \mu_\alpha + \sum_{q=1}^{Q} \gamma_{\alpha q} x_{iq} + \zeta_{\alpha_i} \tag{7.27}$$

$$\beta_i = \mu_\beta + \sum_{q=1}^{Q} \gamma_{\beta q} x_{iq} + \zeta_{\beta_i} \tag{7.28}$$

Definitions of the notation are the same as above. The matrix expression of Eqs. (7.27) and (7.28) is

$$\boldsymbol{\eta}_i = \boldsymbol{\mu}_\eta + \boldsymbol{\Gamma}_{\eta \mathbf{x}} \mathbf{x}_i + \boldsymbol{\zeta}_i \tag{7.29}$$

where all else is as defined above.

Finally, the reduced form of the combined TIC and TVC model are derived by substituting Eq. (7.29) into Eq. (7.26) and expanding all terms. However, this expression becomes quite unwieldy, and we do not present this in scalar form here. Instead, we can consider the matrix expression that expands the unconditional TVC matrices above such that

$$\mathbf{y}_i = \left(\boldsymbol{\Lambda}_{\mathbf{y}\eta}\boldsymbol{\mu}_\eta + \boldsymbol{\Lambda}_{\mathbf{y}\eta}\boldsymbol{\Gamma}_{\eta\mathbf{x}}\mathbf{x}_i + \boldsymbol{\Gamma}_{\mathbf{yw}}\mathbf{w}_i\right) + \left(\boldsymbol{\Lambda}_{\mathbf{y}\eta}\boldsymbol{\zeta}_i + \boldsymbol{\epsilon}_i\right) \tag{7.30}$$

We again defer the presentation of the model-implied moment structures stemming from Eq. (7.30) until later in this chapter.

7.3.1 Interpretation of Conditional TVC Latent Curve Model

Care must be taken in the interpretation of the model parameters, especially those related to the regressions of the latent growth factors on the TICs in the presence of the TVCs. The reason is that the fixed and random effects of the latent growth factors reflect stability and change in the repeated measures of y *net the effects of the TVCs.* Thus, whereas in Chapter 4 we could unambiguously interpret the relation between a TIC and the latent curves in terms of change in y, we now must consider this interpretation in terms of change in y net the effects of the time-specific measure of w. Of course, this is not a limitation of the model; indeed, this is the very interpretation in which we are interested.

7.3.2 Empirical Example: Conditional TVC Model of Reading and Math Achievement

The reading and math data help us to illustrate this model. From the conditional TIC model, we will incorporate the two exogenous TICs (gender and minority status). This model also incorporates the repeated measures of reading achievement in the contemporaneous prediction of math achievement (while retaining the equality constraints on the TVC regressions). The resulting model corresponds to Eq. (7.30).

This model fits the data well [$T_{\text{ML}}(76) = 203.2$, $p < 0.001$, RMSEA $= 0.03$ (0.03, 0.04), IFI $= 0.95$, TLI $= 0.93$]. As with the unconditional TVC model, there was a significant and positive prediction of math ability from the contemporaneous influence of reading ability ($\hat{\gamma} = 0.41$, $p < 0.0001$). Further, girls reported significantly lower initial math scores than those of boys ($\hat{\gamma} = -0.70$, $p < 0.05$), as did minority children compared to nonminority children ($\hat{\gamma} = -2.88$, $p < 0.001$). However, there were no significant differences in rates of change in math over time as a function of either gender or minority status.

Although the current model allows for incorporation of the repeated measures of both reading and math achievement, what this model has not yet considered is the possibility that a LCM underlies the repeated measures of reading achievement. Developmental theory would predict that there would be random and fixed components underlying the developmental process of *both* reading and math ability. In the next section we explore a model that permits these dual development processes.

7.4 MULTIVARIATE LATENT CURVE MODELS

The time-varying covariate LCM is a powerful approach for incorporating repeated measures of a covariate w as a direct predictor of the repeated measures y in the presence of the growth process for y. An important aspect of this model is that although the TVCs are allowed to covary with one another, these repeated measures are not structured as a function of time. In other words, we have not estimated a latent curve process for the covariates themselves. Although not all repeated measures follow a LCM, there may be reasons why it might be of interest to estimate a latent curve process for both the repeated measures on y and on w. This is called the *multivariate latent curve model*.[7]

The parameterization of the multivariate LCM is a straightforward extension of the models that we have described. Indeed, we simply expand our expressions so that there are random intercepts and slopes for our repeated measures on y and separate random curve components for our repeated measures on w. We are functionally estimating latent growth factors for each repeated measure and then will relate these two variables to one another via the random intercepts and slopes for each repeated measure.

Specifically, the level 1 and level 2 equations for y are

$$y_{it} = \alpha_{iy} + \beta_{iy}\lambda_t + \epsilon_{yit} \tag{7.31}$$

and

$$\alpha_{iy} = \mu_{\alpha y} + \zeta_{\alpha y_i} \tag{7.32}$$

$$\beta_{iy} = \mu_{\beta y} + \zeta_{\beta y_i} \tag{7.33}$$

respectively. However, there also are level 1 and level 2 equations for w, which are

$$w_{it} = \alpha_{iw} + \beta_{iw}\lambda_t + \epsilon_{wit} \tag{7.34}$$

[7]The following models might be more accurately termed *bivariate* latent curve models given that we are considering only repeated measures for outcomes y and w. However, we retain the use of the term *multivariate* here given the generality of these methods to expand to two or more outcome measures of interest.

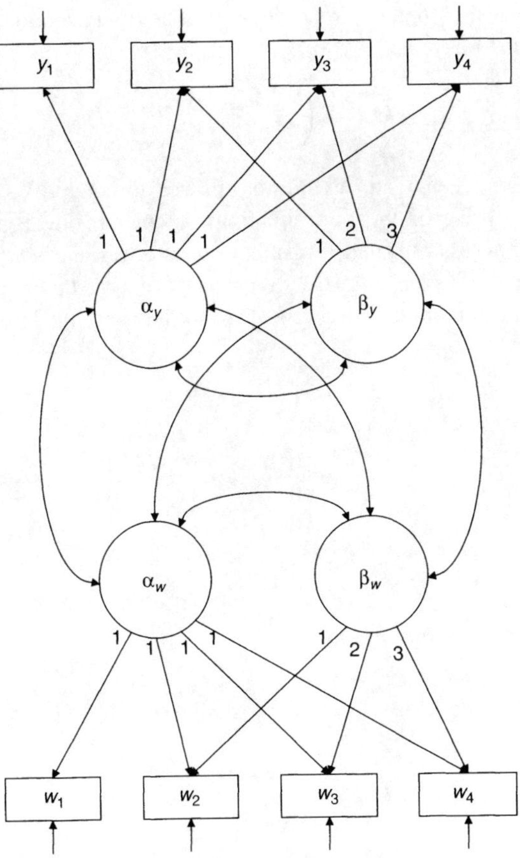

FIGURE 7.3 Path diagram of a bivariate latent curve model.

and

$$\alpha_{iw} = \mu_{\alpha w} + \zeta_{\alpha w_i} \tag{7.35}$$

$$\beta_{iw} = \mu_{\beta w} + \zeta_{\beta w_i} \tag{7.36}$$

respectively. Compared to the standard univariate LCM, the only significant difference in these equations is that they are further subscripted with y and w to differentiate the repeated measure of interest. The notation is naturally becoming more complex given that we now must track parameters relating to y and w. A path diagram example to represent this model is shown in Figure 7.3.

As before, these equations are expressed compactly in reduced matrix form. Consider a single data vector that is ordered with the T-repeated measures on y followed by the T-repeated measures on w. We thus have a stacked vector where

$$\begin{pmatrix} \mathbf{y}_i \\ \mathbf{w}_i \end{pmatrix} = \begin{pmatrix} \mathbf{\Lambda}_{\mathbf{y}\eta}\boldsymbol{\mu}_{\eta_\mathbf{y}} + \mathbf{\Lambda}_{\mathbf{y}\eta}\boldsymbol{\zeta}_{\mathbf{y}_i} + \boldsymbol{\epsilon}_{\mathbf{y}_i} \\ \mathbf{\Lambda}_{\mathbf{w}\eta}\boldsymbol{\mu}_{\eta_\mathbf{w}} + \mathbf{\Lambda}_{\mathbf{w}\eta}\boldsymbol{\zeta}_{\mathbf{w}_i} + \boldsymbol{\epsilon}_{\mathbf{w}_i} \end{pmatrix} \tag{7.37}$$

where all is defined as before. Importantly, the covariance structure of the level 1 residuals is defined as

$$\boldsymbol{\Theta}_{\epsilon\epsilon} = \begin{pmatrix} \boldsymbol{\Theta}_{\boldsymbol{\epsilon_y \epsilon_y}} & \boldsymbol{\Theta}_{\boldsymbol{\epsilon_y \epsilon_w}} \\ \boldsymbol{\Theta}_{\boldsymbol{\epsilon_w \epsilon_y}} & \boldsymbol{\Theta}_{\boldsymbol{\epsilon_w \epsilon_w}} \end{pmatrix} \tag{7.38}$$

If the residuals between y and w are not allowed to covary, Eq. (7.38) is a block diagonal matrix. However, there are situations in which residuals might covary (e.g., within time and across repeated measures), these covariances would be defined in $\boldsymbol{\Theta}_{\boldsymbol{\epsilon_w \epsilon_y}}$. For example, for $T = 4$ assessments on y and w, the (symmetric) structure of $\boldsymbol{\Theta}_{\epsilon\epsilon}$ for within time and across repeated measures residual covariation is given as

$$\boldsymbol{\Theta}_{\epsilon\epsilon} = \begin{pmatrix} \mathrm{VAR}(\epsilon_{y1}) \\ 0 & \mathrm{VAR}(\epsilon_{y2}) \\ 0 & 0 & \mathrm{VAR}(\epsilon_{y3}) \\ 0 & 0 & 0 & \mathrm{VAR}(\epsilon_{y4}) \\ \mathrm{COV}(\epsilon_{y1}, \epsilon_{w1}) & 0 & 0 & 0 & \mathrm{VAR}(\epsilon_{w1}) \\ 0 & \mathrm{COV}(\epsilon_{y2}, \epsilon_{w2}) & 0 & 0 & 0 & \mathrm{VAR}(\epsilon_{w2}) \\ 0 & 0 & \mathrm{COV}(\epsilon_{y3}, \epsilon_{w3}) & 0 & 0 & 0 & \mathrm{VAR}(\epsilon_{w3}) \\ 0 & 0 & 0 & \mathrm{COV}(\epsilon_{y4}, \epsilon_{w4}) & 0 & 0 & 0 & \mathrm{VAR}(\epsilon_{w4}) \end{pmatrix} \tag{7.39}$$

where $\mathrm{VAR}(\epsilon_{y1})$ is the residual variance of ϵ_{y1} at $t = 1$, $\mathrm{COV}(\epsilon_{y1}, \epsilon_{w1})$ is the covariance between the residuals of y and w at $t = 1$, and so on.

Similarly, the covariance structure of the level 2 deviations is defined as

$$\boldsymbol{\Psi} = \begin{pmatrix} \boldsymbol{\Psi}_{\boldsymbol{\eta_y \eta_y}} & \boldsymbol{\Psi}_{\boldsymbol{\eta_y \eta_w}} \\ \boldsymbol{\Psi}_{\boldsymbol{\eta_w \eta_y}} & \boldsymbol{\Psi}_{\boldsymbol{\eta_w \eta_w}} \end{pmatrix} \tag{7.40}$$

where $\boldsymbol{\Psi}$ is the $(m_y + m_w) \times (m_y + m_w)$ symmetric covariance matrix for m_y-factors for y and m_w-factors for w. It is common to freely estimate all elements in $\boldsymbol{\Psi}$, reflecting that all latent curve factors covary with all other latent curve factors. For example, for a linear LCM estimate for both y and w, the elements of $\boldsymbol{\Psi}$ are

$$\boldsymbol{\Psi} = \begin{pmatrix} \psi_{\alpha_y \alpha_y} \\ \psi_{\alpha_y \beta_y} & \psi_{\beta_y \beta_y} \\ \psi_{\alpha_y \alpha_w} & \psi_{\alpha_w \beta_y} & \psi_{\alpha_w \alpha_w} \\ \psi_{\alpha_y \beta_w} & \psi_{\beta_w \beta_y} & \psi_{\beta_w \alpha_w} & \psi_{\beta_w \beta_w} \end{pmatrix} \tag{7.41}$$

where $\psi_{\alpha_y\alpha_y}$ is the variance of the intercept factor for y, $\psi_{\alpha_y\beta_y}$ is the covariance between the intercept and slope factors for y, and so on.

These covariances are often transformed into correlations to allow for interpretation on a standardized metric. In general, this is given as

$$\boldsymbol{\rho}_{\zeta\zeta} = \mathbf{D}_{\zeta}^{-1}\boldsymbol{\Psi}\mathbf{D}_{\zeta}^{-1} \tag{7.42}$$

where $\mathbf{D}_{\zeta}^{-1}$ is the inverse of a diagonal matrix of the square roots of the diagonal elements of $\boldsymbol{\Psi}$ and zeros elsewhere.

In general, the interpretations of the elements of $\boldsymbol{\rho}_{\zeta\zeta}$ are made in precisely the same way as any given correlation. For example, a positive correlation between the intercept factors for y and w (i.e., ρ_{α_y,α_w}) reflects that, on average, individual intercept values that are above the mean intercept on y tend to be above the mean intercept on w, and vice versa. Similarly, a negative correlation between the two linear slope factors for y and w, ρ_{β_y,β_w}, reflects that, on average, individual slope values that are above the mean slope on y tend to be below the mean slope on w, and vice versa. Sometimes the correlation between the slope factors is interpreted as the extent to which two variables "travel together" through time (e.g., McArdle, 1989). This can be a useful conceptualization in many research applications.[8]

There are several additional important characteristics of the multivariate LCM. One is that although we are hypothesizing that the repeated measures of two variables are related over time, this is strictly at the level of the random trajectory components. That is, the only direct relation between variables y and w over time is the covariance structure among the random intercepts and slopes. This model parameterization does not incorporate time-specific relations among the repeated measures (e.g., the time t measure of y predicting the time $t+1$ measure of w). This may be maximally consistent with the theoretical question of interest and thus not pose any problem (e.g., Curran and Willoughby, 2003). However, if time-specific lagged influences are of interest, we must expand this model to allow for these effects. Later we present one method for incorporating these expansions.

A second characteristic of this model is that there are no predictors of any of the random trajectory factors. The random trajectory factors are allowed to covary freely with one another but are not expressed as a function of any other predictor. Next we consider the inclusion of two types of predictors in the multivariate model. We first include TICs much the same as was done in the univariate case. We then extend this model to include one or more of the latent growth factors as predictors themselves.

7.4.1 Multivariate Latent Curve Model with TICs

The unconditional multivariate LCM allows for direct tests of the relation between two growth processes at the level of the random coefficients. That is, we can

[8]Note, however, that the values of these correlations are influenced by the values used to code time. See Chapter 3 for further details.

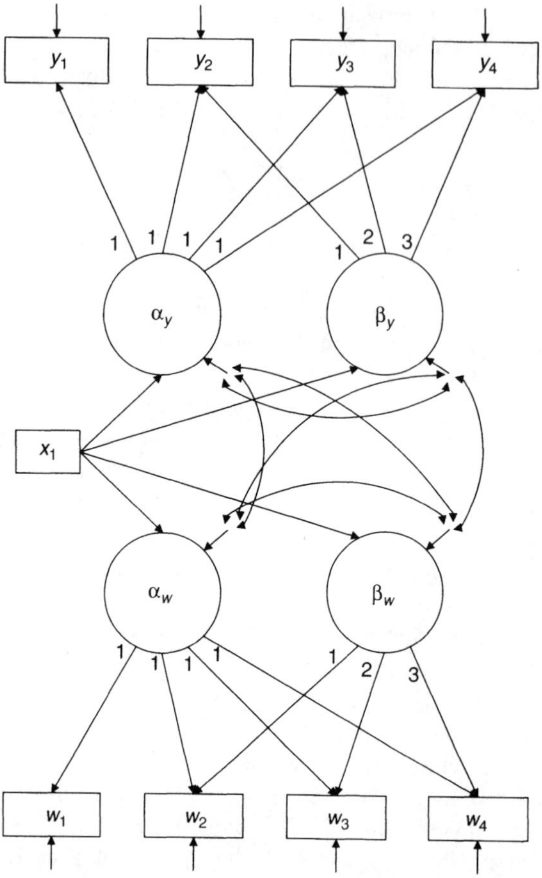

FIGURE 7.4 Path diagram of a latent curve model for two repeated measures and one exogenous covariate.

estimate fixed and random effects for each repeated measure while examining the covariation of the random effects across repeated measures. However, because this model is unconditional, it does not incorporate any exogenous predictors. It is a straightforward extension of the unconditional multivariate LCM to include one or more correlated exogenous predictors. We refer to this as the *conditional multivariate LCM*.

Consider a situation of four repeated observations taken on each of two repeated measures and a time-invariant covariate. Theory may dictate LCMs for both repeated measures with the random intercepts and random slopes of each series related. Furthermore, a single covariate x_1 might predict one or more of the random coefficients determining the trajectory of each repeated measures. One possible parameterization of this model is presented in Figure 7.4.

There are two key differences in the conditional multivariate model compared to the unconditional counterpart. First, this model includes a single time-invariant exogenous predictor of the latent growth factors, although this could easily be expanded to include two or more covariates. Second, each of the latent factors includes a single arrow denoting the presence of a disturbance; this reflects that some portion of the variance of each latent factor is unexplained by the exogenous predictor. Importantly, the covariance structure among the latent factors is now at the level of this disturbance variance. In other words, the factor covariances reflect the linear association between the unexplained variances associated with each growth factor.

To formalize this model, we need to incorporate structural coefficients into the parameter matrices. Because we are only considering predictors of the latent growth factors, the level 1 equations remain as before:

$$y_{it} = \alpha_{iy} + \beta_{iy}\lambda_t + \epsilon_{yit} \tag{7.43}$$

$$w_{it} = \alpha_{iw} + \beta_{iw}\lambda_t + \epsilon_{wit} \tag{7.44}$$

However, the level 2 equations are expanded to include the presence of the exogenous predictors. In general, for Q-predictors, the level 2 equations for y and w are

$$\alpha_{iy} = \mu_{\alpha y} + \sum_{q=1}^{Q} \gamma_{q\alpha_y} x_{iq} + \zeta_{\alpha y_i} \tag{7.45}$$

$$\beta_{iy} = \mu_{\beta y} + \sum_{q=1}^{Q} \gamma_{q\beta_y} x_{iq} + \zeta_{\beta y_i} \tag{7.46}$$

and

$$\alpha_{iw} = \mu_{\alpha w} + \sum_{q=1}^{Q} \gamma_{q\alpha_w} x_{iq} + \zeta_{\alpha w_i} \tag{7.47}$$

$$\beta_{iw} = \mu_{\beta w} + \sum_{q=1}^{Q} \gamma_{q\beta_w} x_{iq} + \zeta_{\beta w_i} \tag{7.48}$$

respectively. Although the level 2 equations reflect the estimation of all possible regressions of each latent factor to all predictors, this need not be the case.

In matrix terms, the level 1 and level 2 equations are expressed as a stacked vector where

$$\begin{pmatrix} \mathbf{y}_i \\ \mathbf{w}_i \end{pmatrix} = \begin{pmatrix} \mathbf{\Lambda}_{\mathbf{y}\eta}\boldsymbol{\eta}_{yi} + \boldsymbol{\epsilon}_{yi} \\ \mathbf{\Lambda}_{\mathbf{w}\eta}\boldsymbol{\eta}_{wi} + \boldsymbol{\epsilon}_{wi} \end{pmatrix} \tag{7.49}$$

and

$$\begin{pmatrix} \boldsymbol{\eta}_{yi} \\ \boldsymbol{\eta}_{wi} \end{pmatrix} = \begin{pmatrix} \boldsymbol{\mu}_{\eta_\mathbf{y}} + \mathbf{\Gamma}_{\eta_\mathbf{y}}\mathbf{x}\mathbf{x}_i + \boldsymbol{\zeta}_{yi} \\ \boldsymbol{\mu}_{\eta_\mathbf{w}} + \mathbf{\Gamma}_{\eta_\mathbf{w}}\mathbf{x}\mathbf{x}_i + \boldsymbol{\zeta}_{wi} \end{pmatrix} \tag{7.50}$$

respectively. As before, we can substitute the level 2 equations into the level 1 equations to express the reduced-form equation as

$$\begin{pmatrix} \mathbf{y}_i \\ \mathbf{w}_i \end{pmatrix} = \begin{pmatrix} \boldsymbol{\Lambda}_{\mathbf{y}\eta_\mathbf{y}}\boldsymbol{\mu}_{\eta_\mathbf{y}} + \boldsymbol{\Lambda}_{\mathbf{y}\eta_y}\boldsymbol{\Gamma}_{\eta_\mathbf{y}\mathbf{x}}\mathbf{x}_i + \boldsymbol{\Lambda}_{\mathbf{y}\eta_y}\boldsymbol{\zeta}_{\mathbf{yi}} + \boldsymbol{\epsilon}_{yi} \\ \boldsymbol{\Lambda}_{\mathbf{w}\eta_w}\boldsymbol{\mu}_{\eta_\mathbf{w}} + \boldsymbol{\Lambda}_{\mathbf{y}\eta_\mathbf{w}}\boldsymbol{\Gamma}_{\eta_w\mathbf{x}}\mathbf{x}_i + \boldsymbol{\Lambda}_{\mathbf{w}\eta_w}\boldsymbol{\zeta}_{\mathbf{wi}} + \boldsymbol{\epsilon}_{wi} \end{pmatrix} \tag{7.51}$$

This stacked vector of T assessments of y and T assessments of w is precisely as was used in the unconditional multivariate LCM described above [i.e., Eq. (7.37)]. The key difference here is that we are incorporating fixed regressions of the latent curve factors ($\boldsymbol{\eta}_{yi}$ and $\boldsymbol{\eta}_{wi}$) on our set of covariates ($\mathbf{x}_i$). We defer presentation of the model-implied moment structures to the end of this chapter.

7.4.2 Multivariate Latent Curve Model with TICs and Factor Regressions

Recall that the conditional multivariate LCM described above regressed the latent factors onto the exogenous covariates, but no structure was estimated among the latent factors themselves. This parameterization allowed for the residuals of the latent factors to covary both within and across repeated measures but did not regress one latent curve factor on another latent curve factor. Given the flexibility of the general structural equation model, this is a straightforward extension (e.g., McArdle, 1989).

Consider the same hypothetical example above with four repeated measures on two variables (denoted y and w) and a single exogenous predictor (denoted x_1). However, say that the underlying theoretical model posited that the random intercept governing one repeated measure predicts the random slope that underlies the other repeated measure, and vice versa. In substantive terms, this might reflect that the model implied level of antisocial behavior at the initial period was predictive of rates of change in substance use over the four assessments. In statistical terms, we are simply regressing the latent slope factor of one repeated measure on the latent intercept factor of the other. Figure 7.5 presents an illustration.

Note that the key difference in this model is that the latent intercept factor for repeated measure y is a predictor of the latent slope factor for repeated measure w, and vice versa. Thus, instead of estimating a covariance between these factors, this is parameterized as a regression coefficient. Further, this now implies that the intercept factors for y and w are *intervening variables* or *mediators* of the exogenous variable x_1 in the prediction of the latent slope factors for each repeated measure. Because of the similarities between this model and the one described in detail above, we do not present the matrix expression and model-implied moment structure here. See McArdle (1989), MacCallum et al. (1997), and Muthén and Curran (1997) for further details.

7.4.3 Empirical Example: Multivariate Conditional Latent Curve Model of Reading and Math Achievement

Prior to estimating a multivariate LCM for the simultaneous inclusion of reading and math, it is often beneficial to identify the optimal univariate LCM for each

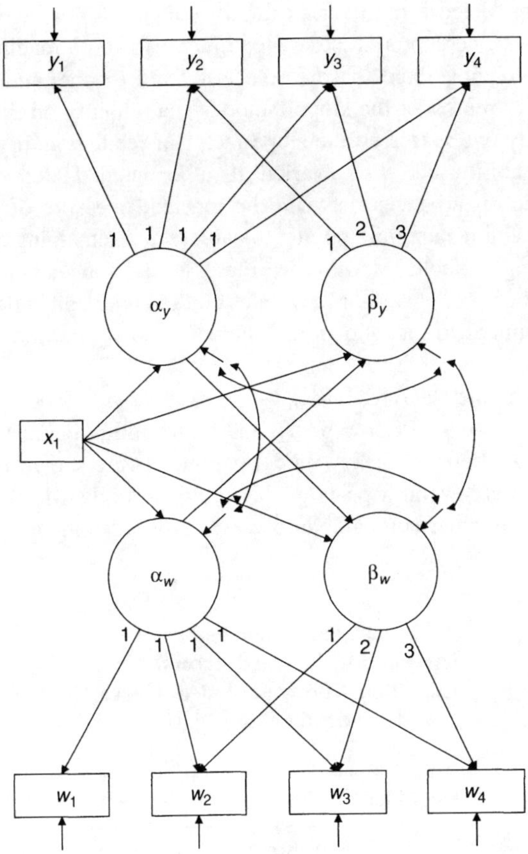

FIGURE 7.5 Path diagram of a latent curve model for two repeated measures and one exogenous covariate with random intercepts influencing random slopes.

process separately and then combine these two processes within a single model. The freed-loading model best characterizes the repeated assessments of math ability. We next considered the univariate LCM of reading ability.

After fitting several alternative unconditional models to the repeated measures of reading ability (linear, polynomial, and freed loading), we concluded that the freed-factor loading model best represented the characteristics of the observed data for the sample of $N = 1767$ children. The final model fit the data moderately well [$T_{\text{ML}}(18) = 129.8$, $p < 0.001$, RMSEA $= 0.06$ (0.05, 0.07), IFI $= 0.89$, TLI $=$ 0.87]. Further, the magnitude of the estimated factor loadings reflected that of the total change in reading ability observed from kindergarten to grade 6, 23% occurred between kindergarten and first grade, 23% between first and second; 15% between second and third; 15% between third and fourth; 13% between fourth and fifth; and 11% between fifth and sixth. As was found with math ability, these results reflect that there is a generally increasing trajectory in reading ability throughout elementary school, but the rate of increase is slowing as time progresses.

We next constructed a multivariate LCM with the following structure: (1) we estimated a latent intercept and latent slope factor for both reading and math with the first factor loading fixed to zero, the last fixed to one, and all others freely estimated; (2) we regressed the slope factor of math ability on the intercept factor of reading ability, we regressed the slope factor of reading ability on the intercept factor of math ability, and we covaried all other latent factors; (3) we covaried the time-specific disturbances between the repeated measure of math ability and reading ability within each time point t to capture any remaining association of the math and reading variables; and (4) we regressed all four latent curve factors on the two exogenous time-invariant covariates child gender and child minority status. This model is similar to that shown in Figure 7.5, so we do not present this again here.

This model fit the data well [$T_{\text{ML}}(94) = 295.3$, $p < 0.001$, RMSEA $= 0.03$ (0.03, 0.04), IFI $= 0.93$, TLI $= 0.91$], and the resulting parameter estimates and asymptotic standard errors for key model parameters are shown in Table 7.2.

We can interpret several aspects of this multivariate LCM. First, the estimated factor loadings are rather comparable in value across reading and math, suggesting

Table 7.2 Parameter Estimates (Asymptotic Standard Errors) for Conditional Bivariate Latent Curve Model for Reading and Math (N = 1767)

Parameter	Reading	Math
μ_α	20.15(0.318)	19.13(.311)
μ_β	9.63(17.59)	27.36(6.82)
$\lambda_{1\beta}$	0	0
$\lambda_{2\beta}$	0.23(0.008)	0.23(0.008)
$\lambda_{3\beta}$	0.46(0.009)	0.46(0.010)
$\lambda_{4\beta}$	0.61(0.010)	0.63(0.011)
$\lambda_{5\beta}$	0.76(0.011)	0.80(0.011)
$\lambda_{6\beta}$	0.89(0.013)	0.89(0.01)
$\lambda_{7\beta}$	1	1
$\psi_{\alpha\alpha}$	18.13(3.68)	10.82(3.46)
$\psi_{\beta\beta}$	97.12(33.91)	12.88(13.29)
$\psi_{\alpha\beta}$	−1.10(7.60)	8.95(1.75)

Predictor	α_{reading}	β_{reading}	α_{math}	β_{math}
Gender	0.504(0.364)	1.10(1.33)	−0.503(0.353)	−1.51(0.83)
Minority	−1.48(0.366)	2.45(3.43)	−3.48(0.354)	−0.93(0.95)
α_{reading}	–	–	–	0.882(0.336)
α_{math}	–	2.103(0.918)	–	–

that these two variables are developing at similar rates over time. We could impose equality constraints on these loadings to test this, but we do not do this here. Second, minority children are reporting significantly lower levels of both initial reading and math ability, but do not report significant differences in rates of change in either repeated measure. Third, there are no statistically significant differences as a function of gender in either starting point or rate of change for reading or math. Finally, there is a statistically significant and positive prediction of each random slope as a function of the cross-domain intercept factor. That is, initial reading ability significantly and positively predicts subsequent changes in math ability, and initial math ability significantly and positively predicts subsequent changes in reading ability. Unlike the previous TIC and TVC models discussed above, the multivariate LCM allows for the estimation of these cross-domain prospective predictions.

7.4.4 Summary

The multivariate LCM provides a method to model two repeated measures whereby the random intercepts and random slopes of each series are related.

The relationship can be nondirectional covariances, or the random intercept of one series might affect the random slope of another. Furthermore, covariates might affect the random intercepts and slopes governing the repeated measures. However, there are no *direct* relations estimated between the time-specific measures within or across the series, although we did consider covariances of the errors of the repeated measures across the different series.

There may be theoretical or empirical situations in which time-specific effects relate directly to the repeated observations themselves. In fact, it might even be hypothesized that these two types of change occur simultaneously. Interestingly, we have discussed two separate modeling approaches that allow for each of these relations, but in isolation of the other. The TVC model allowed for the time-specific prediction of y from w, but this approach did not estimate latent curve components for both y and w. In contrast, the multivariate LCM did explicitly estimate latent curve components for both y and w, but this approach did not allow for time-specific influences of one measure on the other. We will now turn to combining features of the TVC and multivariate LCM to allow for a more comprehensive model for longitudinal data than either the TVC or multivariate LCM provides alone.

7.5 AUTOREGRESSIVE LATENT TRAJECTORY MODEL

We have described two approaches to incorporating repeated measures for two variables. The first fitted a LCM to one set of variables denoted y and simultaneously regressed the repeated measures of y on a second set of variables denoted w. This is the TVC model, and this allowed us to covary out the time-specific measures of w when estimating the latent curves associated with y. A latent curve structure was *not* fitted to the covariates w directly (see, e.g., Figure 7.1). The next extension fitted a latent curve structure to the repeated measures on both y and w, resulting in the multivariate LCM. In this model, the repeated measures on y and w were related to one another solely at the level of the latent curve factors, but there were no time-specific structural relations among the repeated measures (see, e.g., Figure 7.3).[9] We are thus left with the need to choose between relating two sets of repeated measures either at the level of the time-specific influences or at the level of the latent curves, but not both.

There have been several analytic approaches designed to model time-specific and random curve components simultaneously. Important examples include McArdle's latent difference score (LDS) model (McArdle, 2001; McArdle and Hamagami, 2001) and Kenny and Zautra's state–trait–error model (1995). We have proposed a model that combines elements of the TVC and multivariate LCM that we have termed the *autoregressive latent trajectory* (ALT) *model* (Curran and Bollen, 2001; Bollen and Curran, 2004). It is to this that we now turn.

The analytic developments that support the ALT model become somewhat complex, and we do not present a comprehensive review here. See Curran and Bollen (2001) and Bollen and Curran (2004) for more complete discussions. Also, we defer the presentation of the model-implied moment structures associated with the ALT model until the concluding section, in which we present a general notational framework for all of the models discussed here.

7.5.1 Autoregressive (Simplex) Model

The ALT model is based on the synthesis of two existing modeling traditions. Prior to exploring the ALT model in greater detail, it is helpful to begin with a brief review of the classic *autoregressive* (or *simplex*) *model*, given the importance that autoregression plays in the parameterization of the ALT model. Anderson (1960), Humphreys (1960), Heise (1969), Wiley and Wiley (1970), Jöreskog (1970, 1979), and Werts et al. (1971) developed a modeling framework in which a variable is expressed as an additive function of its immediately preceding value plus a random disturbance. In some cases the autoregressive models included latent variables and measurement error in the observed variables. Here we ignore measurement error and build autoregressive structures in the observed variables.[10] We begin with the univariate autoregressive model and subsequently, extend this to the bivariate case.

Univariate Autoregressive Model Because we are not estimating a latent variable structure for the *univariate autoregressive* (AR) *model*, there is not a level 1 and level 2 distinction in equations. The single equation for the univariate AR model is thus

$$y_{it} = \alpha_{yt} + \rho_{y_t,y_{t-1}} y_{i,t-1} + \epsilon_{yit} \tag{7.52}$$

[9] Although earlier we demonstrated that it is possible to covary the time-specific residuals between y and w, the multivariate LCM does not include regressions directly among the repeated measures.

[10] All of our developments could be extended and applied to models that explicitly incorporate measurement error structures.

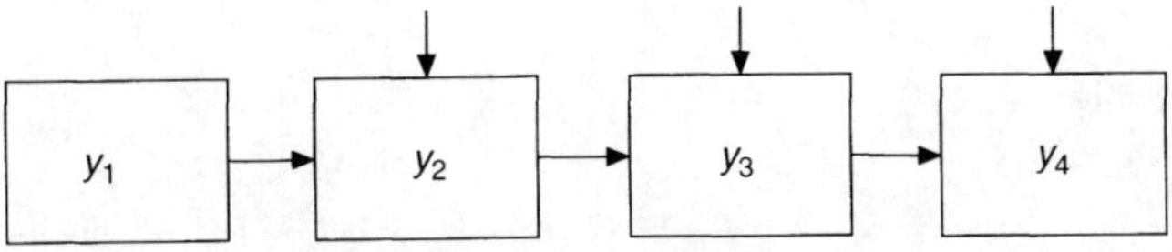

FIGURE 7.6 Path diagram for autoregressive model.

where $E(\epsilon_{it}) = 0$ for all i and t, $\text{COV}(\epsilon_{it}, y_{i,t-1}) = 0$ for all i and $t = 2, 3, \ldots, T$, $E(\epsilon_{it}, \epsilon_{jt}) = 0$ for all t and $i \neq j$, $E(\epsilon_{it}, \epsilon_{jt}) = \text{VAR}(\epsilon_{yt})$ for all t and $i = j$, and $E(\epsilon_{it}, \epsilon_{jt+s}) = 0$ for all s and $i \neq j$. Although it is possible to permit autoregressive disturbances, to keep the model simple we follow the predominant practice and assume nonautocorrelated disturbances $[E(\epsilon_{it}, \epsilon_{it+s}) = 0$ for $s \neq 0]$. The α_{yt} is the (fixed) intercept for the equation for time t. The constant $\rho_{y_t, y_{t-1}}$ is the autoregressive parameter. It gives the impact of the prior value of y on the current one. We treat y_{it} as predetermined for $t = 1$ such that

$$y_{i1} = \alpha_{y1} + \epsilon_{yi1} \tag{7.53}$$

A path diagram for a univariate AR model with $T = 4$ is presented in Figure 7.6.

Like the LCM, the autoregressive model expresses the repeated measures of y for individual i at time point t as a function of a set of parameters. However, the autoregressive model does this in a very different way from that of the LCM. Importantly, the autoregressive model implies that later observations are a direct function of earlier observations plus some time-specific error. The strength of association of this earlier influence on the later measure is reflected in the $\rho_{y_t, y_{t-1}}$ parameter in Eq. (7.52). As we have discussed before, the LCM instead approaches the modeling of the repeated measures as a function of an unobserved random growth process that gave rise to the time-specific observed measures. We combine these two processes in a moment.

Bivariate Autoregressive Cross-Lagged Model The univariate AR model can be expanded to consider two sets of repeated measures over time, and this is sometimes referred to as the *autoregressive cross-lagged* (ARCL) *model*. Although several parameterizations of the ARCL model are possible, we focus here on the standard expression in which autoregressive parameters are estimated within each measure, and cross-lagged parameters are estimated between each measure. Specifically,

$$y_{it} = \alpha_{yt} + \rho_{y_t y_{t-1}} y_{i,t-1} + \rho_{y_t w_{t-1}} w_{i,t-1} + \epsilon_{yit} \tag{7.54}$$

$$w_{it} = \alpha_{wt} + \rho_{w_t y_{t-1}} y_{i,t-1} + \rho_{w_t w_{t-1}} w_{i,t-1} + \epsilon_{wit} \tag{7.55}$$

As before, we assume that the error terms (ϵ's) have means of zero, are not autocorrelated, and are uncorrelated with the right-hand-side variables, although ϵ_{yit} might correlate with ϵ_{wit}. Note that for this model we again treat the y_{i1} and w_{i1}

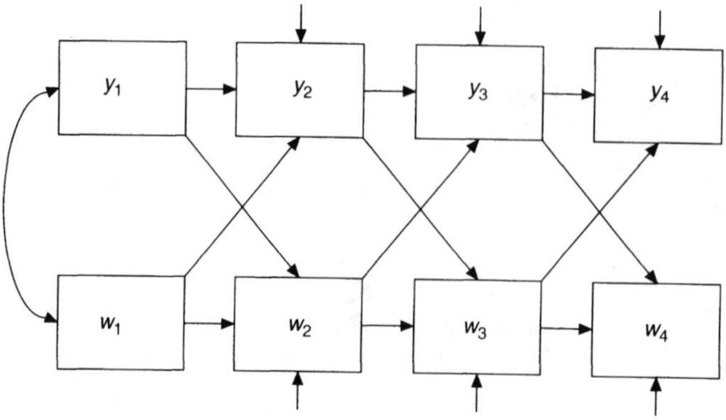

FIGURE 7.7 Path diagram of autoregressive cross-lagged model.

variables as predetermined such that

$$y_{i1} = \alpha_{y1} + \epsilon_{yi1} \tag{7.56}$$

$$w_{i1} = \alpha_{w1} + \epsilon_{wi1} \tag{7.57}$$

A path diagram of the bivariate ARCL model is presented in Figure 7.7.

As is evident in both Eqs. (7.54) and (7.55) as well as in Figure 7.7, each downstream measure of variable y and w is expressed as a combined function of the prior value of each measure. That is, variable y at time t is an additive combination of y at time $t-1$, w at time $t-1$, and a time-specific residual. If a significant cross-lagged effect is found, the interpretation is that the earlier measure of w prospectively predicts the later measure of y above and beyond the influence of the earlier measure of y. This is why this model is sometimes called a *residualized change model*, in that the later measure of y is "residualized" as a function of the earlier measure of y while testing the regression of the later y on the earlier measure of w (and vice versa). Finally, it is also evident how the multivariate TVC model described earlier shares certain similarities with the bivariate autoregressive model. Clearly, there are significant differences, but the lagged effects of y on earlier measures of w are much like the TVC contributions described above. It is this similarity on which we will now capitalize.

7.5.2 Unconditional Bivariate ALT Model

We now turn to the combination of elements of the time-varying covariate LCM, the multivariate LCM, and the autoregressive models to result in the ALT model.[11] Although the initial temptation might simply be to estimate autoregressive and cross-lagged parameters within the standard multivariate LCM, this is problematic.

[11] Although we begin with a discussion of the bivariate ALT model, there are several interesting developments associated with the univariate ALT model as well. See Bollen and Curran (2004) and Curran and Bollen (2001) for further details.

Note that in the standard univariate and bivariate autoregressive models, we treated the initial assessment ($t = 1$) as predetermined [i.e., Eqs. (7.53), (7.56), and (7.57)]. This allows for the circumvention of the potential for bias stemming from the problem of an "infinite regress" (Bollen and Curran, 2004). The source of this bias stems from the fact that although we observed our measures of y and w at $t = 1$, we do *not* observe these variables at prior time points. Omitting these influences may lead to bias throughout our system of equations.

This potential bias is avoided in the autoregressive models because we treat the initial measures as predetermined, and thus all omitted prior influences are "absorbed" into the means, variances, and covariances of the initial (exogenous) measures. However, this potential bias would not be avoided if we were simply to introduce autoregressive structures among the repeated measures in the standard multivariate LCM. The reason is that the initial measures of each series are *not* exogenous to the system given that these serve as indicators on the latent curve factors. We must modify the model parameterization to allow for the proper modeling of the potential bias from the infinite regress when simultaneously estimating autoregressive and latent curve structures.

The necessary parameterization is straightforward when there are $T = 4$ or more observed repeated assessments. (Later we describe methods for estimating these models based on $T = 3$ assessments.) When $T \geq 4$, we will treat the initial measure on each measure y and w as predetermined (as we defined earlier), and we will covary these measures with the latent curve factors. It is then possible to estimate autoregressive and cross-lagged parameters without introducing bias stemming from the omitted prior measures of each repeated measure.

Recall that the level 1 and level 2 equations for the unconditional multivariate LCM for y and w are

$$y_{it} = \alpha_{iy} + \beta_{iy}\lambda_t + \epsilon_{yit} \tag{7.58}$$

$$w_{it} = \alpha_{iw} + \beta_{iw}\lambda_t + \epsilon_{wit} \tag{7.59}$$

respectively, and the level 2 equations for the linear latent curve factors for y and w are

$$\alpha_{iy} = \mu_{\alpha y} + \zeta_{\alpha y_i} \tag{7.60}$$

$$\beta_{iy} = \mu_{\beta y} + \zeta_{\beta y_i} \tag{7.61}$$

and

$$\alpha_{iw} = \mu_{\alpha w} + \zeta_{\alpha w_i} \tag{7.62}$$

$$\beta_{iw} = \mu_{\beta w} + \zeta_{\beta w_i} \tag{7.63}$$

respectively. However, we will make two important changes to the level 1 equations. First, we define the initial assessment of each variable to be predetermined such that

$$y_{i1} = \mu_{y1} + \epsilon_{yi1} \tag{7.64}$$

$$w_{i1} = \mu_{w1} + \epsilon_{wi1} \tag{7.65}$$

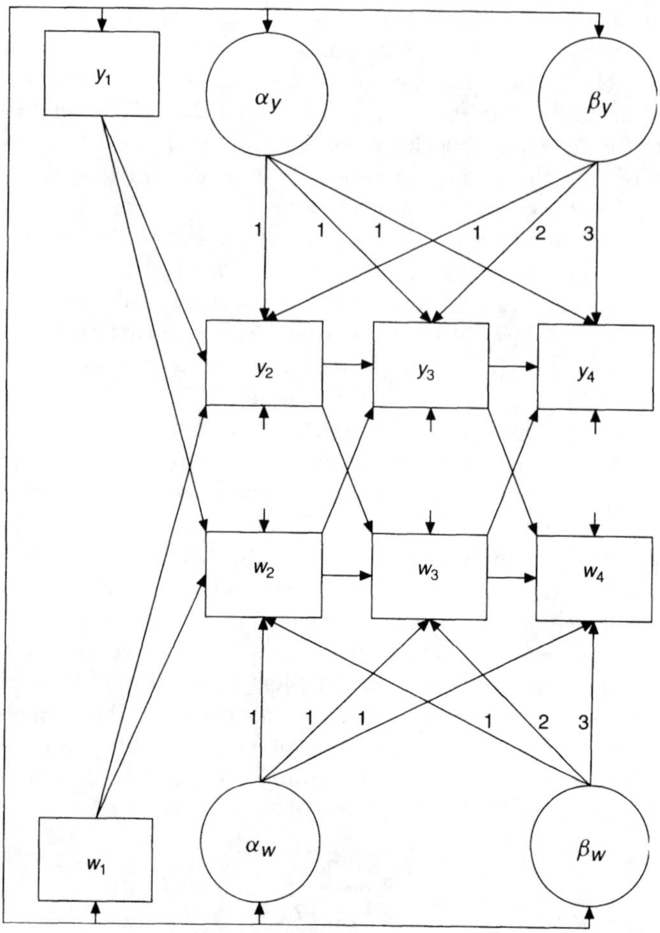

FIGURE 7.8 Path diagram of bivariate autoregressive latent trajectory model with predetermined wave 1 variables.

Second, we introduce both time-adjacent autoregressive parameters within each repeated measure as well as cross-lagged parameters between each repeated measure.[12] The level 1 equations for $t > 1$ are now

$$y_{it} = \alpha_{yi} + \Lambda_{yt2}\beta_{yi} + \rho_{y_t y_{t-1}} y_{i,t-1} + \rho_{y_t w_{t-1}} w_{i,t-1} + \epsilon_{yit} \tag{7.66}$$

$$w_{it} = \alpha_{wi} + \Lambda_{wt2}\beta_{wi} + \rho_{w_t y_{t-1}} y_{i,t-1} + \rho_{w_t w_{t-1}} w_{i,t-1} + \epsilon_{wit} \tag{7.67}$$

As before, we assume that the error terms (ϵ's) have means of zero, are not autocorrelated, and are uncorrelated with the right-hand-side variables and random

[12]In a moment we will describe various constraints on ρ that are required to identify the ALT model for $T = 3$ and $T = 4$.

coefficients, although ϵ_{yit} might correlate with ϵ_{wit}. Finally, the level 2 equations remain as before [i.e., Eqs. (7.60) through (7.63)]. Figure 7.8 is the path diagram of this model for $T = 4$ waves of data.

An examination of Eqs. (7.66) and (7.67) reveals that these are a synthesis of the bivariate cross-lagged model and the bivariate latent trajectory model. In addition to permitting a LCM with separate random intercepts and slopes for each variable series, the bivariate ALT model allows a lagged autoregressive effect and cross-lagged effect. This hybrid model leads to considerable flexibility not available in either the latent curve or autoregressive modeling framework separately. Through a series of restrictions described in Bollen and Curran (2004), we can estimate a variety of alternative structures associated with latent curve and autoregressive parts of the model. Indeed, one parameterization leads to the standard multivariate LCM parts, and one parameterization leads to the autoregressive parts, thus further highlighting the hybrid nature of this modeling approach.

Unconditional Univariate ALT Model for $T = 3$ We noted earlier that when there are at least four repeated assessments (i.e., $T \geq 4$), we can treat the initial measures of each repeated measure as predetermined to avoid potential bias from the omission of influences from measures occurring prior to our initial assessment. When $T = 4$, the model is identified if an equality constraint is placed on ρ, the autoregressive parameter (e.g., $\rho_{y_t, y_{t-1}} = \rho_y$ for all t); when $T > 4$, the model is identified even without this equality constraint on ρ within each series and over time. However, when $T = 3$, the predetermined ALT model is not identified even with the equality constraints on ρ. There is a way in which this model can be identified through the use of specific nonlinear constraints on the factor loadings. We describe this approach here, but see Bollen and Curran (2004) for complete technical developments.

To demonstrate that it is possible to identify the ALT model with only three waves of data, we treat y_{i1} as *endogenous* rather than as *predetermined*. The general structure is thus similar to a standard linear unconditional LCM with $T = 3$ repeated assessments. However, at the first time period, we add the equation

$$y_{i1} = \alpha_i + \rho_{1,0} y_{i,0} + \epsilon_{i0} \tag{7.68}$$

where $\rho_{1,0}$ is the autoregressive coefficient for the impact of $y_{i,0}$ on $y_{i,1}$. The $y_{i,0}$ occurs prior to the first wave of data that we observe; $y_{i,0}$ is thus an omitted variable. If we expand this equation by repeated substitution for the y's, this becomes an infinite series. If the autoregressive parameters, $\rho_{t,t-1}$, are equal to ρ with $|\rho| < 1$, the series converges and the values to which the factor loadings for the first observed measure converge are

$$\Lambda_{11} = (1 - \rho)^{-1} \tag{7.69}$$

$$\Lambda_{12} = -\rho(1 - \rho)^{-2} \tag{7.70}$$

Given this convergence, we can impose constraints on the factor loadings to account for the variables that were omitted prior to our initial assessment.

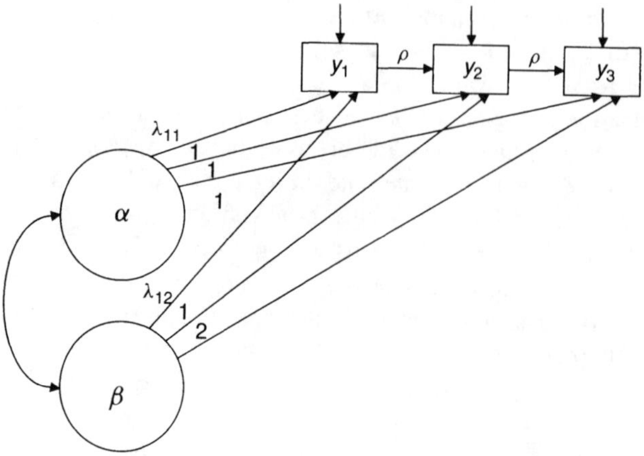

FIGURE 7.9 Path diagram of ALT model for three waves of data.

A linear unconditional univariate ALT model for $T = 3$ is presented in Figure 7.9. Note that this is similar in form to the unconditional linear LCM discussed in Chapter 2 except for two critical differences. First, there is an autoregressive parameter ρ estimated between the time-adjacent observed measures. Second, the factor loadings relating the initial measure of y are estimated and nonlinearly constrained as a function of ρ as defined in Eqs. (7.69) and (7.70). This model parameterization allows for the joint contribution of the random growth process and the autoregressive component (Bollen and Curran, 2004).

Importantly, note that as ρ goes to zero, Λ_{11} goes to 1 and Λ_{12} goes to 0, which correspond directly to the factor loadings fixed for the initial $t = 1$ measure for the standard LCM without autoregressions. The univariate ALT model for $T = 3$ thus simplifies to the standard linear LCM when $\rho = 0$. Further, the $T = 3$ univariate ALT extends naturally to the bivariate ALT. See Bollen and Curran (2004) for further analytic details and for methods to extend this to the bivariate case.

7.5.3 Conditional Bivariate ALT Model

We can incorporate one or more exogenous predictors in the bivariate ALT model as well. This is again accomplished by the extension of the equations for the random trajectories. Specifically, we modify Eqs. (7.60) through (7.63) to include time-invariant covariates x_{i1} and x_{i2} such that

$$\alpha_{yi} = \mu_{y\alpha} + \gamma_{\alpha y1}x_{i1} + \gamma_{\alpha y2}x_{i2} + \zeta_{y\alpha i} \tag{7.71}$$

$$\beta_{yi} = \mu_{y\beta} + \gamma_{\beta y1}x_{i1} + \gamma_{\beta y2}x_{i2} + \zeta_{y\beta i} \tag{7.72}$$

and

$$\alpha_{wi} = \mu_{w\alpha} + \gamma_{\alpha w1}x_{i1} + \gamma_{\alpha w2}x_{i2} + \zeta_{w\alpha i} \tag{7.73}$$

$$\beta_{wi} = \mu_{w\beta} + \gamma_{\beta w1}x_{i1} + \gamma_{\beta w2}x_{i2} + \zeta_{w\beta i} \tag{7.74}$$

As before, the set of γ's represent the fixed regressions of the random trajectory components on the two correlated exogenous variables. Although we do not show it here, it is possible to have the random intercepts as explanatory variables in the equations above.

In the models described thus far, we have treated the initial repeated measures as predetermined, so we must regress these initial measures y_{i1} and w_{i1} on the set of exogenous measures. Thus, the conditional equations for the $t = 1$ assessments for y_{i1} and w_{i1} are

$$y_{i1} = \mu_{y1} + \gamma_{y1}x_{i1} + \gamma_{y2}x_{i2} + \epsilon_{yi1} \tag{7.75}$$

$$w_{i1} = \mu_{w1} + \gamma_{w1}x_{i1} + \gamma_{w2}x_{i2} + \epsilon_{wi1} \tag{7.76}$$

where all notation is defined as before. These equations give the impact of the covariates on the first wave of data for each of the repeated measures. Figure 7.10 is a path diagram example of the model for four waves of data.

7.5.4 Empirical Example: Income Data from the National Longitudinal Survey of Youth

Our empirical example comes from Bollen and Curran (2004). It utilizes data drawn from the National Longitudinal Survey of Youth (NLSY) of Labor Market

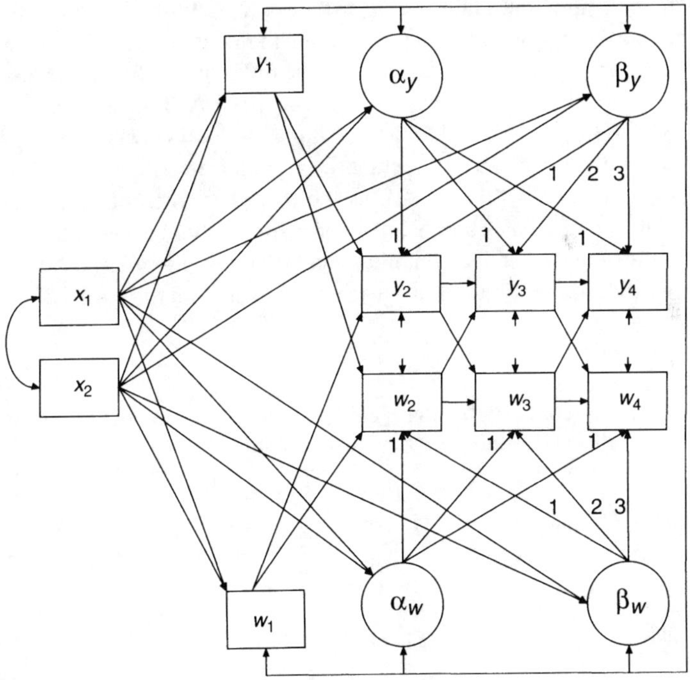

FIGURE 7.10 Path diagram of conditional bivariate ALT model with two exogenous covariates.

Experience, a study initiated in 1979 by the U.S. Department of Labor to examine the transition of young people into the labor force. We extracted a subsample of data consisting of $N = 3912$ individuals assessed at two-year intervals from 1986 to 1994. The outcome variable is the respondent's total net family income for the prior calendar year. We used a square-root transformation of the net income data to reduce the kurtosis and skewness of the original data. Either the autoregressive or the latent trajectory model are plausible structures for these data. We fit an unconditional ALT model to test the presence of autoregressive and latent curve components.

Table 7.3 contains the overall fit statistics for the autoregressive (AR) model, the latent curve model (LCM), and the ALT model for these data using the maximum likelihood fitting function. Given the large sample ($N = 3912$), it is not surprising that the likelihood ratio test statistic (denoted T_{ML} in the table) is statistically significant for all models. However, all of the measures indicate that the fit of either of the ALT models, including the ρ parameters, is superior to the standard autoregressive or latent curve alternatives. For instance, the likelihood ratio test statistics are substantially lower than the corresponding df for the ALT model compared to the AR model or the LCM. Similarly, the RMSEAs increased from a marginal 0.15 and 0.10 fit for the AR and LC models, respectively, to a substantially better 0.04 value for the ALT model. As described in Bollen and Curran (2004), we cannot perform a nested likelihood ratio test for ALT vs. autoregressive models or the latent curve vs. autoregressive model; however, the IFI and RMSEA rank the ALT model as the optimally fitting model relative to the AR and LC models.

We can, however, compute a nested test of the joint contribution of the autoregressive parameters within the ALT model. The difference between the test statistics for the ALT model with ρ freely estimated and the ALT model with ρ fixed to zero is $T_{\Delta} = 203.28 - 26.2 = 177.1$ with $\text{df}_{\Delta} = 7 - 3 = 4$, which is highly significant, indicating the necessity to include the ρ parameters. Choosing between the two ALT models, one with the ρ's free and the other where they are set equal is more difficult. The RMSEA and IFI fit indices differ little between the two models, whereas the difference between the likelihood ratio test statistics is statistically significant and the information-based measures are superior for the ALT model with

Table 7.3 Overall Fit of Autoregressive, Latent Curve, and Autoregressive Latent Trajectory Models for Net Income, 1986–1994 ($N = 3912$)

Overall Fit	Autoregressive	Latent Curve	ALT $\rho_{t,t-1}$ Free	ALT $\rho_{t,t-1}$ Equal	ALT $\rho_{t,t-1} = 0$
T_{ML}	534.1	412.8	26.2	62.4	203.28
df	6	10	3	6	7
p-value	<0.001	<0.001	<0.001	<0.001	<0.001
IFI	0.95	0.96	0.99	0.99	0.98
RMSEA	0.15	0.10	0.04	0.05	0.08

Table 7.4 ML Estimates of the Parameters in the ALT Model with Free ρ's for the NLSY Data on Net Income, 1986–1994 ($N = 3912$)

	Income				
Parameter	1986	1988	1990	1992	1994
α	1.00 (–)	1.00 (–)	1.00 (–)	1.00 (–)	1.00 (–)
β	0 (–)	1.00 (–)	2.00 (–)	3.00 (–)	4.00 (–)
ρ	—	0.15 (0.028)	0.27 (0.028)	0.33 (0.042)	0.40 (0.060)
VAR(ϵ)	—	1.45 (0.087)	1.36 (0.055)	1.42 (0.081)	1.45 (0.122)
R^2	—	0.57	0.63	0.64	0.66

unconstrained ρ's. The evidence suggests a tendency to favor the ALT with free ρ's, and this is the model that we feel optimally characterizes the observed data.

The overall model-fit measures are just one part of the model assessment. It also is necessary to examine the model parameter estimates and their properties when assessing a model. Table 7.4 reports the estimates, asymptotic standard errors, and R^2 values for the income variables in each year. The magnitude of the autoregressive parameter tends to increase with time such that the impact of 1986 $\sqrt{\text{income}}$ on 1988 $\sqrt{\text{income}}$ is 0.15, and the autoregressive parameter for 1992 $\sqrt{\text{income}}$ on 1994 $\sqrt{\text{income}}$ grows to 0.40. This suggests an increasing ability for earlier income to predict later income as a person ages. The autoregressive relation must be interpreted in conjunction with the latent curve process so that, for example, for each unit change in the 1992 $\sqrt{\text{income}}$, we expect a 0.40 difference in the 1994 $\sqrt{\text{income}}$ *net* of the latent curve of an individual's income. In addition, the R^2 values are moderately high, ranging from 0.57 to 0.66. The $\widehat{\text{VAR}}(\alpha)$ and $\widehat{\text{VAR}}(\beta)$ (and asymptotic s.e.'s) are 1.79 (0.221) and 0.033 (0.019), respectively. Both indicate statistically significant (one-tailed, $p < 0.05$) individual variability in the initial income level and rate of change, although the variability in the slope has lower statistical significance than the intercepts. It is interesting to note that in the latent curve model without autoregression, the $\widehat{\text{VAR}}(\alpha)$ and $\widehat{\text{VAR}}(\beta)$ are 1.933 and 0.122, both values much larger than their respective counterparts from the ALT model. The implication is that had we mistakenly assumed that the latent curve model was the one of choice, we would have estimated far more individual variability in income than is probably true.

7.5.5 Summary

We have explored a variety of flexible and powerful LCMs that allow for the testing of many different types of relations between two variables over time. The time-varying covariate model estimates a latent curve structure for the repeated measures on y while regressing the measures of y on a set of repeatedly measured covariates w; however, there is no latent curve structure estimated for the time-varying covariates. In contrast, the multivariate LCM estimates latent curve structures for the set of measures on both y and w, but there are no time-specific structures estimated between y and w; all relations are modeled at the level of the latent curve factors. Finally, we described a hybrid autoregressive latent trajectory (ALT) model that allows for the estimation of relations both at the level of the latent curve factors and between time-specific measures of each repeated measure. Taken together, these three approaches may correspond closely with many theoretical hypotheses that arise in the behavioral sciences.

We conclude this chapter with the presentation of a general matrix expression that provides a single framework for defining all of the models we have discussed thus far. This not only demonstrates the close relations among these various models but also allows for a compact and efficient method for expressing model-implied moment structures.

7.6 GENERAL EQUATION FOR ALL MODELS

In this section we present a general matrix expression that allows for the representation of a broad class of longitudinal models, including the autoregressive, latent curve, TVC, and ALT models within a single expression. Such a general expression allows for a unified framework to approach model identification, model estimation, and model fit for a variety of model parameterizations (full details are given in Bollen and Curran, 2004).

The starting point is a general equation from which we can impose restrictions in order to derive all of the models that we have so far discussed. The model we use has two equations,

$$\boldsymbol{\eta}_i = \boldsymbol{\mu} + \mathbf{B}\boldsymbol{\eta}_i + \boldsymbol{\zeta}_i \tag{7.77}$$

$$\mathbf{o}_i = \mathbf{P}\boldsymbol{\eta}_i \tag{7.78}$$

where Eq. (7.77) provides the structural relations between variables such that $\boldsymbol{\eta}_i$ is a vector that contains both the repeated measures and the random intercepts and random slopes, $\boldsymbol{\mu}$ is a vector of means or intercepts, $\mathbf{B}$ is a coefficient matrix that gives the coefficients for the relationships of $\boldsymbol{\eta}_i$'s on each other, and $\boldsymbol{\zeta}_i$ is the disturbance vector for the variables in $\boldsymbol{\eta}_i$. We assume that $E(\boldsymbol{\zeta}_i) = \mathbf{0}$. The nature of the covariances of $\boldsymbol{\zeta}_i$ with $\boldsymbol{\eta}_i$ will vary depending on the model, but for identification purposes at least some of these covariances will be zero or known values. Equation (7.78) functions to pick out the observed variables, $\mathbf{o}_i$, from the latent variables of Eq. (7.77).

More specifically,

$$\boldsymbol{\eta}_i = \begin{bmatrix} \mathbf{y}_i \\ \mathbf{w}_i \\ \mathbf{x}_i \\ \boldsymbol{\alpha}_i \\ \boldsymbol{\beta}_i \end{bmatrix} \tag{7.79}$$

where $\mathbf{y}_i$ and $\mathbf{w}_i$ are two variables measured repeatedly for T time periods, $\mathbf{x}_i$ is a $q \times 1$ vector of exogenous determinants of the latent curve parameters of the repeated measures, $\boldsymbol{\alpha}_i$ is the 2×1 vector of α_{yi} and α_{wi}, the random intercepts for the two sets of repeated measures, and $\boldsymbol{\beta}_i$ is the 2×1 vector of β_{yi} and β_{wi} the random slopes for the two repeated measures.[13] The $\boldsymbol{\mu}$ vector is

$$\boldsymbol{\mu} = \begin{bmatrix} \boldsymbol{\mu}_y \\ \boldsymbol{\mu}_w \\ \boldsymbol{\mu}_x \\ \boldsymbol{\mu}_\alpha \\ \boldsymbol{\mu}_\beta \end{bmatrix} \tag{7.80}$$

where $\boldsymbol{\mu}_y$ and $\boldsymbol{\mu}_w$ are vectors of means/intercepts for the $\mathbf{y}_i$ and $\mathbf{w}_i$ repeated measures observed, $\boldsymbol{\mu}_x$ is the vector of means for the exogenous covariates in the model, $\boldsymbol{\mu}_\alpha$ is a vector of means/intercepts for random intercepts α_{yi} and α_{wi}, and $\boldsymbol{\mu}_\beta$ is a vector of the means/intercepts of β_{yi} and β_{wi}.

The $\mathbf{B}$ matrix is

$$\mathbf{B} = \begin{bmatrix} \mathbf{B_{yy}} & \mathbf{B_{yw}} & \mathbf{B_{yx}} & \mathbf{B}_{\mathbf{y}\alpha} & \mathbf{B}_{\mathbf{y}\beta} \\ \mathbf{B_{wy}} & \mathbf{B_{ww}} & \mathbf{B_{wx}} & \mathbf{B}_{\mathbf{w}\alpha} & \mathbf{B}_{\mathbf{w}\beta} \\ \mathbf{B_{xy}} & \mathbf{B_{xw}} & \mathbf{B_{xx}} & \mathbf{B}_{\mathbf{x}\alpha} & \mathbf{B}_{\mathbf{x}\beta} \\ \mathbf{B}_{\alpha\mathbf{y}} & \mathbf{B}_{\alpha\mathbf{w}} & \mathbf{B}_{\alpha\mathbf{x}} & \mathbf{B}_{\alpha\alpha} & \mathbf{B}_{\alpha\beta} \\ \mathbf{B}_{\beta\mathbf{y}} & \mathbf{B}_{\beta\mathbf{w}} & \mathbf{B}_{\beta\mathbf{x}} & \mathbf{B}_{\beta\alpha} & \mathbf{B}_{\beta\beta} \end{bmatrix} \tag{7.81}$$

where the double-subscript notation in the partition matrix indicates that the submatrix contains those coefficients related to effects among the subscripted variables. For instance, $\mathbf{B_{yy}}$ contains the effects of the repeated y variables on each other, and $\mathbf{B}_{\beta\mathbf{x}}$ contains the impact of the exogenous $\mathbf{x}_i$ on the random slopes, β_{yi} and β_{wi}, for the y's and w's. In all of our models, $\mathbf{x}_i$ consists of exogenous variables, so that there are no other observed or latent variables that influence them. In addition, in none of our models do the repeated measures affect the random intercepts and random slopes. So for our models, the $\mathbf{B}$ matrix simplifies to

$$\mathbf{B} = \begin{bmatrix} \mathbf{B_{yy}} & \mathbf{B_{yw}} & \mathbf{B_{yx}} & \mathbf{B}_{\mathbf{y}\alpha} & \mathbf{B}_{\mathbf{y}\beta} \\ \mathbf{B_{wy}} & \mathbf{B_{ww}} & \mathbf{B_{wx}} & \mathbf{B}_{\mathbf{w}\alpha} & \mathbf{B}_{\mathbf{w}\beta} \\ \mathbf{0} & \mathbf{0} & \mathbf{0} & \mathbf{0} & \mathbf{0} \\ \mathbf{0} & \mathbf{0} & \mathbf{B}_{\alpha\mathbf{x}} & \mathbf{B}_{\alpha\alpha} & \mathbf{B}_{\alpha\beta} \\ \mathbf{0} & \mathbf{0} & \mathbf{B}_{\beta\mathbf{x}} & \mathbf{B}_{\beta\alpha} & \mathbf{B}_{\beta\beta} \end{bmatrix} \tag{7.82}$$

The disturbance vector for Eq. (7.77) is

$$\boldsymbol{\zeta}_i = \begin{bmatrix} \boldsymbol{\epsilon}_{yi} \\ \boldsymbol{\epsilon}_{wi} \\ \boldsymbol{\epsilon}_{xi} \\ \boldsymbol{\zeta}_{\alpha i} \\ \boldsymbol{\zeta}_{\beta i} \end{bmatrix} \tag{7.83}$$

[13]In the case of quadratic LCMs, the $\boldsymbol{\beta}_i$ vector would be enlarged to include the additional latent factors for the random coefficients of the quadratic term.

with covariance matrix $\boldsymbol{\Psi}$. The $\mathbf{P}$ matrix is

$$\mathbf{P} = \begin{bmatrix} \mathbf{I}_T & \mathbf{0} & \mathbf{0} & \mathbf{0} & \mathbf{0} \\ \mathbf{0} & \mathbf{I}_T & \mathbf{0} & \mathbf{0} & \mathbf{0} \\ \mathbf{0} & \mathbf{0} & \mathbf{I}_Q & \mathbf{0} & \mathbf{0} \end{bmatrix} \tag{7.84}$$

where $\mathbf{I}_T$ is a $T \times T$ identity matrix with dimensions that depend on the number of repeated measures and $\mathbf{I}_Q$ is a $Q \times Q$ identity matrix with Q exogenous variables. The matrix picks out the observed variables in a given model, where $\mathbf{o}_i$ is

$$\mathbf{o}_i = \begin{bmatrix} \mathbf{y}_i \\ \mathbf{w}_i \\ \mathbf{x}_i \end{bmatrix} \tag{7.85}$$

We can place restrictions on this general model expression that leads directly to some of the more standard models that we have described earlier. Given space constraints, we do not detail these restrictions here, but see Bollen and Curran (2004) for further details.

7.6.1 Examples of General Notation

To demonstrate the generality of these matrix expressions, we demonstrate how specific restrictions lead to several well-known longitudinal models.

Unconditional Univariate Latent Curve Model In the case of a unconditional latent curve model for $\mathbf{y}_i$, we would modify the matrices so that

$$\boldsymbol{\eta}_i = \begin{bmatrix} \mathbf{y}_i \\ \alpha_i \\ \beta_i \end{bmatrix} \tag{7.86}$$

where $\mathbf{y}_i$ denotes column vector of repeated measures and α_i and β_i are the scalar values of the intercept and linear slope.[14] The mean structure is given as

$$\boldsymbol{\mu} = \begin{bmatrix} \mathbf{0} \\ \mu_\alpha \\ \mu_\beta \end{bmatrix} \tag{7.87}$$

where the $T \times 1$ vector $\mathbf{0}$ represents the zero fixed intercepts for the repeated measures in a latent trajectory model, and μ_α and μ_β are the scalar means of the linear latent curve factors. The $\mathbf{B}$ matrix is

$$\mathbf{B} = \begin{bmatrix} \mathbf{0} & \mathbf{B}_{\mathbf{y}\alpha} & \mathbf{B}_{\mathbf{y}\beta} \\ \mathbf{0} & \mathbf{0} & \mathbf{0} \\ \mathbf{0} & \mathbf{0} & \mathbf{0} \end{bmatrix} \tag{7.88}$$

[14]These can be expressed as scalars because we are only considering a linear LCM of a single repeated measure. These would expand to vectors for higher-order polynomials or when moving to the multivariate model.

where

$$\mathbf{B}_{\mathbf{y}\alpha} = \begin{bmatrix} 1 \\ 1 \\ \vdots \\ 1 \end{bmatrix} \tag{7.89}$$

and

$$\mathbf{B}_{\mathbf{y}\beta} = \begin{bmatrix} 0 \\ 1 \\ \vdots \\ T-1 \end{bmatrix} \tag{7.90}$$

define the linear latent curve. Additional columns in **B** would be included to define more complex latent curve functions (e.g., quadratic, cubic, etc.).

The remaining matrices are

$$\boldsymbol{\zeta}_i = \begin{bmatrix} \boldsymbol{\epsilon}_i \\ \zeta_{\alpha i} \\ \zeta_{\beta i} \end{bmatrix} \tag{7.91}$$

$$\mathbf{o}_i = \begin{bmatrix} \mathbf{y}_i \end{bmatrix} \tag{7.92}$$

and

$$\mathbf{P} = \begin{bmatrix} \mathbf{I} & 0 & 0 \end{bmatrix} \tag{7.93}$$

Thus, the standard unconditional LCM can thus be fully parameterized by placing specific restrictions on the general matrix expression.

Conditional Univariate Latent Curve Model For the conditional univariate LCM, we will extend the equations above to incorporate the regression of the random latent curve factors on the joint influences of the set of exogenous covariates in the $Q \times 1$ vector **x**. First, we expand $\boldsymbol{\eta}_i$ such that

$$\boldsymbol{\eta}_i = \begin{bmatrix} \mathbf{y}_i \\ \mathbf{x}_i \\ \alpha_i \\ \beta_i \end{bmatrix} \tag{7.94}$$

where $\mathbf{y}_i$ is a $T \times 1$ vector of repeated measures, $\mathbf{x}_i$ is a $Q \times 1$ vector of exogenous covariates, and α_i and β_i are scalar values for the random intercept and random slope. The intercept/mean vector is thus

$$\boldsymbol{\mu} = \begin{bmatrix} \mathbf{0} \\ \boldsymbol{\mu}_x \\ \mu_\alpha \\ \mu_\beta \end{bmatrix} \tag{7.95}$$

The **B** matrix is now

$$\mathbf{B} = \begin{bmatrix} \mathbf{0} & \mathbf{0} & \mathbf{B}_{\mathbf{y}\alpha} & \mathbf{B}_{\mathbf{y}\beta} \\ \mathbf{0} & \mathbf{0} & \mathbf{0} & \mathbf{0} \\ \mathbf{0} & \mathbf{B}_{\alpha\mathbf{x}} & 0 & 0 \\ \mathbf{0} & \mathbf{B}_{\beta\mathbf{x}} & 0 & 0 \end{bmatrix} \tag{7.96}$$

with disturbance matrix

$$\boldsymbol{\zeta}_i = \begin{bmatrix} \boldsymbol{\epsilon}_{yi} \\ \boldsymbol{\epsilon}_{xi} \\ \zeta_{\alpha i} \\ \zeta_{\beta i} \end{bmatrix} \tag{7.97}$$

The final two matrices are

$$\mathbf{P} = \begin{bmatrix} \mathbf{I}_T & \mathbf{0} & \mathbf{0} \\ \mathbf{0} & \mathbf{I}_Q & \mathbf{0} \end{bmatrix} \tag{7.98}$$

and

$$\mathbf{o}_i = \begin{bmatrix} \mathbf{y}_i \\ \mathbf{x}_i \end{bmatrix} \tag{7.99}$$

Note that all of these expressions are parallel to those given for the unconditional univariate LCM above, but we have simply expanded these to incorporate the influences of $\mathbf{x}_i$.

Time-Varying Covariate Model Next, we consider the TVC latent curve model defined in Eq. (7.20). The component matrices for a linear TVC model are

$$\boldsymbol{\eta}_i = \begin{bmatrix} \mathbf{y}_i \\ \mathbf{w}_i \\ \alpha_i \\ \beta_i \end{bmatrix} \tag{7.100}$$

$$\boldsymbol{\mu} = \begin{bmatrix} \mathbf{0} \\ \boldsymbol{\mu}_w \\ \mu_\alpha \\ \mu_\beta \end{bmatrix} \tag{7.101}$$

$$\mathbf{B} = \begin{bmatrix} \mathbf{0} & \mathbf{B}_{\mathbf{yw}} & \mathbf{B}_{\mathbf{y}\alpha} & \mathbf{B}_{\mathbf{y}\beta} \\ \mathbf{0} & \mathbf{0} & \mathbf{0} & \mathbf{0} \\ \mathbf{0} & \mathbf{0} & 0 & 0 \\ \mathbf{0} & \mathbf{0} & 0 & 0 \end{bmatrix} \tag{7.102}$$

$$\boldsymbol{\zeta}_i = \begin{bmatrix} \boldsymbol{\epsilon}_{yi} \\ \boldsymbol{\epsilon}_{wi} \\ \zeta_{\alpha i} \\ \zeta_{\beta i} \end{bmatrix} \tag{7.103}$$

$$\mathbf{P} = \begin{bmatrix} \mathbf{I}_T & \mathbf{0} & \mathbf{0} & \mathbf{0} \\ \mathbf{0} & \mathbf{I}_T & \mathbf{0} & \mathbf{0} \end{bmatrix} \tag{7.104}$$

$$\mathbf{o}_i = \begin{bmatrix} \mathbf{y}_i \\ \mathbf{w}_i \end{bmatrix} \tag{7.105}$$

Unconditional Multivariate Latent Curve Model We next expand the TVC latent curve model so that there are latent factors for the repeated observations on both y and on w as defined in Eq. (7.37). The component matrices for the unconditional multivariate LCM are

$$\boldsymbol{\eta}_i = \begin{bmatrix} \mathbf{y}_i \\ \mathbf{w}_i \\ \boldsymbol{\alpha}_i \\ \boldsymbol{\beta}_i \end{bmatrix} \tag{7.106}$$

$$\boldsymbol{\mu} = \begin{bmatrix} \mathbf{0} \\ \mathbf{0} \\ \boldsymbol{\mu}_\alpha \\ \boldsymbol{\mu}_\beta \end{bmatrix} \tag{7.107}$$

$$\mathbf{B} = \begin{bmatrix} \mathbf{0} & \mathbf{0} & \mathbf{B}_{\mathbf{y}\alpha} & \mathbf{B}_{\mathbf{y}\beta} \\ \mathbf{0} & \mathbf{0} & \mathbf{B}_{\mathbf{w}\alpha} & \mathbf{B}_{\mathbf{w}\beta} \\ \mathbf{0} & \mathbf{0} & \mathbf{0} & \mathbf{0} \\ \mathbf{0} & \mathbf{0} & \mathbf{0} & \mathbf{0} \end{bmatrix} \tag{7.108}$$

$$\boldsymbol{\zeta}_i = \begin{bmatrix} \boldsymbol{\epsilon}_{yi} \\ \boldsymbol{\epsilon}_{wi} \\ \boldsymbol{\zeta}_{\alpha i} \\ \boldsymbol{\zeta}_{\beta i} \end{bmatrix} \tag{7.109}$$

$$\mathbf{P} = \begin{bmatrix} \mathbf{I}_T & \mathbf{0} & \mathbf{0} & \mathbf{0} \\ \mathbf{0} & \mathbf{I}_T & \mathbf{0} & \mathbf{0} \end{bmatrix} \tag{7.110}$$

$$\mathbf{o}_i = \begin{bmatrix} \mathbf{y}_i \\ \mathbf{w}_i \end{bmatrix} \tag{7.111}$$

We could easily expand these matrices to include the effects of exogenous covariates, but we do not demonstrate these here.

Unconditional Univariate ALT Model The unconditional ALT model for a single repeated measure in this notation defines **B** as

$$\mathbf{B} = \begin{bmatrix} \mathbf{B}_{\mathbf{yy}} & \mathbf{B}_{\mathbf{y}\alpha} & \mathbf{B}_{\mathbf{y}\beta} \\ \mathbf{0} & \mathbf{0} & \mathbf{0} \\ \mathbf{0} & \mathbf{0} & \mathbf{0} \end{bmatrix} \tag{7.112}$$

where $\mathbf{B_{yy}}$ contains the autoregressive parameters among the repeated measures y, and

$$\mathbf{B_{y\alpha}} = \begin{bmatrix} 0 \\ 1 \\ \vdots \\ 1 \end{bmatrix} \qquad \mathbf{B_{y\beta}} = \begin{bmatrix} 0 \\ 1 \\ \vdots \\ T-1 \end{bmatrix} \tag{7.113}$$

for a model where y_{1i} is predetermined. In addition, we have

$$\boldsymbol{\mu} = \begin{bmatrix} \boldsymbol{\mu}_y \\ \mu_\alpha \\ \mu_\beta \end{bmatrix} \tag{7.114}$$

with

$$\boldsymbol{\mu}_y = \begin{bmatrix} \mu_{y1} \\ 0 \\ \vdots \\ 0 \end{bmatrix} \tag{7.115}$$

and

$$\boldsymbol{\zeta}_i = \begin{bmatrix} \boldsymbol{\epsilon}_i \\ \zeta_{\alpha i} \\ \zeta_{\beta i} \end{bmatrix} \tag{7.116}$$

The variances of ϵ_{1i}, $\zeta_{\alpha i}$, and $\zeta_{\beta i}$ are equal to the variances of the predetermined variables, y_{1i}, α_i, and β_i, respectively. We would need only to incorporate the components associated with vector $\mathbf{x}$ to expand these expressions for the conditional univariate ALT model.

Conditional Multivariate ALT Model As an example of the general notation for a more complicated model, consider the conditional, ALT model for two repeated measures. The $\boldsymbol{\eta}_i$, $\boldsymbol{\mu}$, $\boldsymbol{\zeta}_i$ and $\mathbf{P}$ matrices are the same as in the general model in Eqs. (7.79), (7.80), (7.83), and (7.84), respectively. The $\mathbf{B}$ matrix is

$$\mathbf{B} = \begin{bmatrix} \mathbf{B_{yy}} & \mathbf{B_{yw}} & \mathbf{B_{yx}} & \mathbf{B_{y\alpha}} & \mathbf{B_{y\beta}} \\ \mathbf{B_{wy}} & \mathbf{B_{ww}} & \mathbf{B_{wx}} & \mathbf{B_{w\alpha}} & \mathbf{B_{w\beta}} \\ \mathbf{0} & \mathbf{0} & \mathbf{0} & \mathbf{0} & \mathbf{0} \\ \mathbf{0} & \mathbf{0} & \mathbf{B_{\alpha x}} & \mathbf{0} & \mathbf{0} \\ \mathbf{0} & \mathbf{0} & \mathbf{B_{\beta x}} & \mathbf{0} & \mathbf{0} \end{bmatrix} \tag{7.117}$$

where the covariates $\mathbf{x}$ directly affect the random intercepts and random slopes and the random intercepts and slopes directly affect the repeated measures.

Summary To summarize thus far, we provided a general matrix expression that allows for an explicit representation of the proposed ALT model. An added advantage is that a variety of well-known models are included in this general expression through the use of specific restrictions on one or more of the parameter matrices. This not only highlights the logical relations among many of these alternative model parameterizations, but also allows for a unified framework to consider the implied moments, model identification, estimation, and fit. It is to these that we now turn.

7.7 IMPLIED MOMENT MATRICES

We noted earlier that each model has an implied covariance matrix $[\boldsymbol{\Sigma}(\boldsymbol{\theta})]$ and implied mean vector $[\boldsymbol{\mu}(\boldsymbol{\theta})]$ that provide functions of the model parameters that exactly predict the population covariance matrix and the mean vector for the observed variables. To find these implied moment matrices, we make use of the reduced form of Eq. (7.77):

$$\boldsymbol{\eta}_i = (\mathbf{I} - \mathbf{B})^{-1}(\boldsymbol{\mu} + \boldsymbol{\zeta}_i) \tag{7.118}$$

We then substitute this equation for $\boldsymbol{\eta}_i$ in Eq. (7.78), and this leads to

$$\mathbf{o}_i = \mathbf{P}(\mathbf{I} - \mathbf{B})^{-1}(\boldsymbol{\mu} + \boldsymbol{\zeta}_i) \tag{7.119}$$

From this it follows that the implied mean vector of the observed variables is

$$\boldsymbol{\mu}(\boldsymbol{\theta}) = E(\mathbf{o}_i) = \mathbf{P}(\mathbf{I} - \mathbf{B})^{-1}\boldsymbol{\mu} \tag{7.120}$$

and the implied covariance matrix of observed variables is

$$\begin{aligned} \boldsymbol{\Sigma}(\boldsymbol{\theta}) &= E(\mathbf{o}_i\mathbf{o}_i') - E(\mathbf{o}_i)E(\mathbf{o}_i') \\ &= \mathbf{P}(\mathbf{I} - \mathbf{B})^{-1}\boldsymbol{\Psi}(\mathbf{I} - \mathbf{B})^{-1\prime}\mathbf{P}' \end{aligned} \tag{7.121}$$

By substituting the values of the matrices and vectors that correspond to the particular model of interest, we can use Eqs. (7.120) and (7.121) to find the implied moments for any of these models.

These implied moment matrices are helpful in determining the identification of the model parameters, in testing the model fit, and in parameter estimation. Consider identification. Identification concerns whether it is possible to find unique values for the model parameters in $\boldsymbol{\theta}$ when we have the population moments ($\boldsymbol{\mu}$ and $\boldsymbol{\Sigma}$) of the observed variables. If so, the model is identified. If not, the model is underidentified. For instance, if we impose the constraints on Eq. (7.77) that lead us to the AR model, we can see that this model will be identified with only two waves of data. This is not surprising since the univariate two-wave AR model is essentially a simple regression model. Having greater than two waves of data leads to an overidentified model.

Turning to the univariate LCM and substituting into the implied moment matrices in Eqs. (7.120) and (7.121), we find that the model will be identified with three or more waves of data. Identification of the univariate ALT model is more demanding. Where the y_{i1} variable is predetermined, we need five waves of data to identify the ALT model without further restrictions on the model parameters. Five waves of data also are sufficient to identify the conditional ALT model as well.

Each wave of data is extremely costly in many areas of research. Therefore, it is valuable to examine the conditions under which a univariate ALT model is identified with fewer than five waves of data. We restrict ourselves to the unconditional ALT model, but the conditions we describe are sufficient to identify the conditional ALT model. We now turn to this issue.

ALT Model with $T = 4$ The unconditional univariate, four-wave ALT model is not identified unless we add an additional assumption. One common assumption in AR models is that the autoregressive coefficient is equal over time ($\rho = \rho_{y_2 y_1} = \rho_{y_3 y_2} = \rho_{y_4 y_3}$). When we have four indicators, y_{i1}, y_{i2}, y_{i3}, and y_{i4}, we can identify the model. The four-wave model with a linear trajectory is

$$y_{it} = \alpha_i + \lambda_t \beta_i + \rho y_{i,t-1} + \epsilon_{it} \tag{7.122}$$

where $\lambda_t = t - 1$ with $t = 2, 3, 4$ (keeping in mind that the $t = 1$ measure of y is treated as predetermined). Taking the means of the endogenous y_{i2}, y_{i3}, and y_{i4}, we can rewrite these equations in a matrix expression to get

$$\begin{bmatrix} \mu_{y_2} \\ \mu_{y_3} \\ \mu_{y_4} \end{bmatrix} = \begin{bmatrix} 1 & 1 & \mu_{y_1} \\ 1 & 2 & \mu_{y_2} \\ 1 & 3 & \mu_{y_3} \end{bmatrix} \begin{bmatrix} \mu_\alpha \\ \mu_\beta \\ \rho \end{bmatrix} \tag{7.123}$$

Assuming that the 3×3 matrix to the right of the equal sign is nonsingular, and manipulating these equations algebraically, we find that

$$\begin{bmatrix} \mu_\alpha \\ \mu_\beta \\ \rho \end{bmatrix} = \begin{bmatrix} 1 & 1 & \mu_{y_1} \\ 1 & 2 & \mu_{y_2} \\ 1 & 3 & \mu_{y_3} \end{bmatrix}^{-1} \begin{bmatrix} \mu_{y_2} \\ \mu_{y_3} \\ \mu_{y_4} \end{bmatrix} \tag{7.124}$$

Since the means of the observed variables (μ_{y_t}) are known to be identified, the equation demonstrates the identification of μ_α, μ_β, and ρ provided that the inverse on the right-hand side exists.

Substitution into Eq. (7.121) of the matrices that correspond to this ALT model gives the implied covariance matrix for this model. Manipulation of the equations for the variances and covariances of the y_{it} variables with each other establishes the identification of $\text{VAR}(\alpha_i)$, $\text{VAR}(\beta_i)$, $\text{VAR}(y_{1i})$, $\text{COV}(\alpha_i, \beta_i)$, $\text{COV}(\alpha_i, y_{1i})$, $\text{COV}(\beta_i, y_{1i})$, and $\text{VAR}(\epsilon_{1i})$ with $\text{VAR}(\epsilon_{4i})$. In fact, the model is overidentified with one degree of freedom. The linear ALT model is overidentified with $T \geq 5$ even with the ρ parameters freely estimated.

ALT Model with $T = 3$ As we described earlier, there is not sufficient information to uniquely identify the ALT model using the predetermined initial assessment approach that we use for $T \geq 4$. However, this model can still be estimated through the use of two nonlinear constraints on the factor loadings for the initial measures. We demonstrate this through the expression of the implied moments for this model by using Eqs. (7.120) and (7.121) and substituting in the appropriate matrix values for the ALT model with an endogenous y_{i1}. By manipulating these equations, we can establish the identification of the ALT model for three waves of data.

For instance, the results above show that

$$\Lambda_{11} = (1 - \rho)^{-1} \tag{7.125}$$

and

$$\Lambda_{12} = -\rho(1 - \rho)^{-2} \tag{7.126}$$

under the assumption of an equal autoregressive parameter. For three waves of data, the implied means are

$$\mu_{y_1} = \Lambda_{11}\mu_\alpha + \Lambda_{12}\mu_\beta \tag{7.127}$$

$$\mu_{y_2} = \mu_\alpha + \mu_\beta + \rho\mu_{y_1} \tag{7.128}$$

$$\mu_{y_3} = \mu_\alpha + 2\mu_\beta + \rho\mu_{y_2} \tag{7.129}$$

Manipulating these equations algebraically, we find that

$$\rho = \frac{2\mu_{y_2} - \mu_{y_1} - \mu_{y_3}}{2\mu_{y_1}} \tag{7.130}$$

With ρ identified, so are Λ_{11} and Λ_{12}. We then have

$$\mu_\alpha = \mu_{y_1} - 2\rho\mu_{y_1} + \rho\mu_{y_2} \tag{7.131}$$

and

$$\mu_\beta = \frac{\mu_{y_1} - \Lambda_{11}\mu_\alpha}{\Lambda_{12}} \tag{7.132}$$

More complicated expressions show that the other parameters of the ALT model with three waves of data are identified.[15]

[15]One complication in estimating the ALT model with y_{i1} endogenous is that it requires nonlinear constraints for the values of $\Lambda_{11}$and Λ_{12}. Among the structural-equation-modeling software that estimates the models we describe in this paper, CALIS in SAS (SAS Institute, 2000) and LISREL (Jöreskog, et al., 1999) are two widely available software packages that currently allow nonlinear constraints. This is true even when $|\rho| < 1$. We note that the linear trajectory assumption and the constant ρ value might be reasonable for short periods of time, but less realistic for longer time periods. If this is the case, the results above will at best be only approximately correct.

7.8 CONCLUSIONS

We have explored a wide variety of models in this chapter. We began with a review of the unconditional and conditional univariate LCM. The term *univariate* may seem like a misnomer given that we are considering multiple repeated measures of our dependent variable; however, we use this term to refer more specifically to the fact that we are considering change over time in a single set of repeated measures of interest—thus the term *univariate*. We then argued that in some situations it may be empirically or theoretically important to consider repeated measures of predictors that are used as time-invariant covariates but may themselves change over time. We extended the conditional TIC trajectory model to incorporate time-varying covariates, and this resulted in an important expansion of the standard conditional model. A further extension of the model occurred by permitting separate latent curve models for two time-varying repeated measures, and we discussed this in a section on multivariate LCM. Finally, we noted that a critical characteristic of the multivariate LCM was that any relations thought to exist between two repeated measures is estimated solely at the level of the trajectories. Any deviation of a specific time point from this trajectory is treated as random error. To overcome this, we concluded with the ALT model, which allowed for the simultaneous estimation of relations between two repeated measures at the level of both the trajectories and the repeated measures.

Taken together, we have reviewed a large set of models that are available for examining the relation between two or more variables over time. In practice, it is sometimes difficult to choose between two or even more than two possible modeling strategies. In general, it is often best to turn to theory as a guide. However, in many cases a given theory may not be developed sufficiently to distinguish among some of the nuances of these various modeling frameworks. Thus, theory should also be accompanied by a thoughtful strategy of testing between competing models so that an empirically informed decision is made in choosing models.

CHAPTER EIGHT

Extensions of Latent Curve Models

The preceding chapters provide a firm grounding in latent curve models (LCMs). We can extend these models in a number of directions. This chapter gives an overview of two topics that are useful additions to those already covered. We start by discussing dichotomous and ordinal repeated observed variables in latent curve models. The section that follows illustrates how we can take fuller advantage of the latent variable capabilities for which SEMs are known. That is, we treat multiple indicators and measurement error in latent curve analysis with repeated latent variables and latent covariates. We conclude with a brief discussion of other extensions of LCMs.

8.1 DICHOTOMOUS AND ORDINAL REPEATED MEASURES

Up to this point we have confined our attention to repeated measures that were either continuous or reasonable approximations to continuous variables. In practice, dichotomous, ordinal, or censored variables are common in social science research. Job satisfaction, abortion attitudes, psychological or medical diagnoses, and ratings of company performance are just a few examples of the thousands of ordinal variables that appear in social science inquiries. Two questions arise when we have such variables: (1) What assumptions are violated by treating them as continuous, and (2) how can we include such repeated measures in latent curve modeling? The literature on the analysis of categorical observed variables is vast (see, e.g., Lazarsfeld and Henry, 1968; Bishop et al. 1975; Agresti, 1990; Long, 1997), and we touch on only a small part of it. We restrict ourselves to dichotomous and ordinal repeated measures in which the underlying repeated variable is continuous, and we focus on answering the two preceding questions.[1]

[1] It is important to remember that special attention is needed when the *endogenous* repeated measures are dichotomous or ordinal. No special problems are encountered when one or more *exogenous* covariates are discrete variables.

Latent Curve Models: A Structural Equation Perspective, by Kenneth A. Bollen and Patrick J. Curran

8.1.1 Assumptions Violated[2]

A useful starting point is to recognize the difference between an ideal theoretical measure and the observed measure that is collected. Happiness, for example, might be a subjective variable that well approximates a continuous variable. That is, if we had the appropriate measurement technology, we could have many gradations on which individuals would report their happiness. Yet a survey might force a respondent to choose "yes" or "no" in response to a question about whether they are happy. Other surveys might give respondents three, four, or 10 ordered categories for a self-report on happiness. Even though happiness might be a continuous variable, the response format given will force the variable into dichotomous or ordinal categories.

What about a variable such as being married or being unemployed? Although behavioral variables are more concrete than subjective measures, it is possible to view these and many other behavioral outcomes as tied to an underlying continuous variable. Consider the dichotomous variable of married or not married. Individuals are likely to differ in their propensity to be married. Some individuals have a very low propensity to be married, while others have a far greater propensity. Variables such as age of the individual, availability of potential spouses, and personality could influence the propensity to be married. Although we could view married or not as dichotomous, we also can view this as a dichotomous outcome of a continuous propensity.[3]

We will assume that we are dealing with cases where there is an underlying continuum even though our observed measure is dichotomous or ordinal. We continue to refer to the *observed* repeated measures in a vector **y**. For each **y** we have a corresponding continuous *underlying* variable vector that we will refer to as $\mathbf{y}^*$. Up until this section, we have assumed that $\mathbf{y} = \mathbf{y}^*$; that is, the repeated measures are continuous. When we have dichotomous or ordinal **y**, this equality will not hold. The reason is that the ordinal version of a variable is not equal to its continuous counterpart. In this situation our LCMs will hold for $\mathbf{y}^*$. For example, consider the unconditional LCM for repeated measures:

$$y_{it}^* = \alpha_i + \lambda_t \beta_i + \epsilon_{it} \tag{8.1}$$

where the variables, parameters, and assumptions are the same as before except that the repeated variable is the underlying continuous variable, y_{it}^*, rather than y_{it}. With $y_{it} \neq y_{it}^*$, we have

$$y_{it} \neq \alpha_i + \lambda_t \beta_i + \epsilon_{it} \tag{8.2}$$

[2]The assumption violation approach that we take in this section is based on the approach taken in Bollen (1989b, pp. 433–446).

[3]The use of propensities underlying categorical variables differs by disciplines. Some of the models, such as the dichotomous logistic model, are easily developed without assuming an underlying propensity. However, use of an underlying propensity or underlying continuous variable provides a unifying framework that functions well under a variety of conditions and that corresponds well to substantive areas where the categorical measure is due to crude measurement. For additional discussion, see Long (1997, pp. 40–41).

The first assumption violated by having dichotomous or ordinal repeated measures is that the level 1 equation no longer holds for these repeated measures even if it does hold for the underlying continuous variables.

The second assumption violated by using y_{it} in place of y^*_{it} concerns the distributional assumptions typically made. As we discussed in previous chapters, if the y^*_{it}s come from a distribution with no excess multivariate kurtosis, the maximum likelihood estimator applies and retains its desirable properties (e.g., consistency, asymptotic efficiency). However, the y_{it} values are collapsed versions of the corresponding y^*_{it} values and it is unlikely that the y_{it}'s satisfy the same distributional assumptions as do the original y^*_{it}'s. Hence, the second assumption of no excess multivariate kurtosis is likely to be violated by using y_{it}'s in place of y^*_{it}'s. If this were the only assumption violated, the significance tests and efficiency properties of the ML estimator would probably be affected. Alternative estimators, corrected standard errors and test statistics, and bootstrapping procedures are available to analyze observed variables from distributions with excess kurtosis (e.g., Browne, 1984; Satorra and Bentler, 1988; Bollen and Stine, 1990, 1992). So this assumption violation alone is not critical.

The third assumption that is violated by treating dichotomous or ordinal measures as if they were continuous has more serious consequences. To better understand this, we consider the moment structure hypotheses that characterize SEMs. Using the asterisk symbol for the underlying continuous variables, we can write the moment structure hypotheses as

$$\boldsymbol{\Sigma}^* = \boldsymbol{\Sigma}(\boldsymbol{\theta}) \tag{8.3}$$

$$\boldsymbol{\mu}^* = \boldsymbol{\mu}(\boldsymbol{\theta}) \tag{8.4}$$

where $\boldsymbol{\Sigma}^*$ is the population covariance matrix of $\mathbf{y}^*$, $\boldsymbol{\mu}^*$ is the vector of means of $\mathbf{y}^*$, $\boldsymbol{\Sigma}(\boldsymbol{\theta})$ is the implied covariance matrix, $\boldsymbol{\mu}(\boldsymbol{\theta})$ is the implied mean vector, and $\boldsymbol{\theta}$ is the vector of model parameters. Given that the observed variables $\mathbf{y}$ are collapsed versions of $\mathbf{y}^*$, in nearly all cases $\boldsymbol{\Sigma} \neq \boldsymbol{\Sigma}^*$ and $\boldsymbol{\mu} \neq \boldsymbol{\mu}^*$, so that

$$\boldsymbol{\Sigma} \neq \boldsymbol{\Sigma}(\boldsymbol{\theta}) \tag{8.5}$$

$$\boldsymbol{\mu} \neq \boldsymbol{\mu}(\boldsymbol{\theta}) \tag{8.6}$$

This means that the moment structure hypotheses will typically not hold for the repeated measures observed, even when they do hold for the underlying continuous repeated variables. Thus, the moment structure hypotheses are violated by using dichotomous or ordinal repeated measures. This assumption violation is the most serious violation since its failure means that the consistency of the estimators will generally not hold. Having our estimator converge on the true parameter value as the sample goes to infinity (i.e., consistency) is a basic requirement that we expect for nearly all estimators. So the violation of this hypothesis creates problems, if ignored.

In summary, when we treat dichotomous or ordinal repeated measures as if they are continuous, we violate three assumptions: (1) the level 1 equations no longer hold even if they did hold for the underlying continuous variables; (2) the distributional assumptions for the repeated observed variables are unlikely to hold even if they held for the continuous variables; and (3) the moment structure hypotheses are likely to be violated even if true for the underlying continuous repeated measures. The robustness of the estimators to ordinal repeated measures depends on several factors (Bollen, 1989b, pp. 434–439). Important among them is the number of categories and the distribution of cases among those categories. In general, ordinal variables with many categories and a more symmetric, bell-shaped distribution of cases among categories are less damaging than ordinal variables with few categories, although other factors must be considered (e.g., distribution of cases in categories, size of sample). In most situations it would be prudent to take account of the ordinal nature of the repeated measures rather than treating them as continuous. Next we present a way to take account of ordinality.

8.1.2 Corrective Procedures

Auxiliary Threshold Model Any corrective procedures that analysts employ should address the three assumption violations highlighted in Section 8.1.1. The first assumption violation was that the level 1 model no longer held for the ordinal variable even if it held for the underlying variable. More formally, assume that

$$y_{it}^* = \alpha_i + \lambda_t \beta_i + \epsilon_{it} \tag{8.7}$$

holds for y_{it}^* but not for y_{it}. Part of the corrective procedure is to link y_{it} to y_{it}^*. An auxiliary threshold model is a natural way to show the nonlinear relation between these categorical and continuous variables:

$$y_{it} = c_t \qquad \text{when} \qquad \tau_{c_t - 1} < y_{it}^* \leq \tau_{c_t} \tag{8.8}$$

where $c_t = 1, 2, 3, ldots, C_t$, the total number of ordered categories; $\tau_{c_t - 1}$ and τ_{c_t} are the lower and upper thresholds for category c_t with $\tau_{0_t} = -\infty$ and $\tau_{C_t} = +\infty$; and the thresholds are ordered from lowest to highest, with $C_t - 1$ threshold values. Equations (8.7) and (8.8) connect the ordinal variable to its underlying continuous counterpart such that as the continuous variable lies between two thresholds, the ordinal variable will register in the category that falls between those thresholds. Figure 8.1 illustrates this relation for a five-category variable. All observations on y_{it}^* that are less than τ_{1_t} will appear in category 1 for y_t, those that fall between τ_{1_t} and τ_{2_t} belong in category 2, and so on. See Mehta et al. (2004) for a comprehensive summary of these issues.

The auxiliary threshold equation (8.8) provides the connection between the continuous underlying y_{it}^* variable and y_{it}, but we are still left with a number of unknown parameters. The mean ($\mu_{y_t^*}$) and variance ($\sigma_{y_t^* y_t^*}$) of y_{it}^* are unknown, as are the thresholds from τ_{1_t} to $\tau_{C_t - 1}$. This leads to a total of $2 + C_t - 1$ parameters that need to be identified for each repeated ordinal measure. Furthermore, we have said nothing about the distribution of y_{it}^*. We need to know this distribution to help determine the unknown parameters.

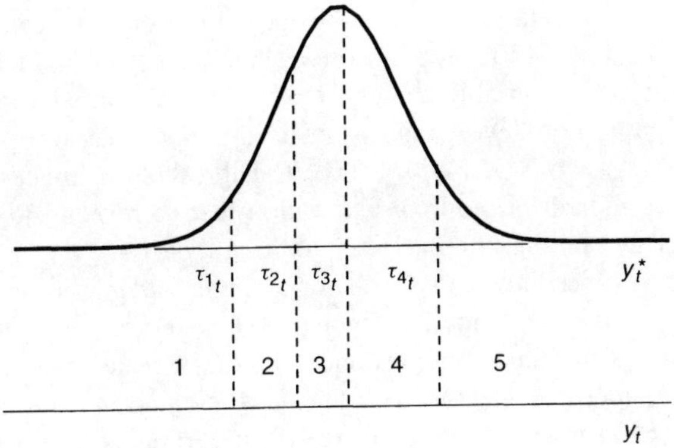

FIGURE 8.1 Underlying continuous variable (y_t^*) mapped into five-category ordinal variable (y_t).

Parameters that we know to be identified are the proportions of cases that fall into each category of the observed ordinal variable. For the tth repeated measure, we refer to these as $\pi_{1t}, \pi_{2t}, ldots, \pi_{Ct}$, where π_{ct} is the population proportion of cases that fall into the cth category of the ordinal variable. Since these proportions must sum to 1, we have only $C_t - 1$ independent proportions with which to work.

To take this process further, we need to make assumptions. One assumption is that of the distribution of the y_{it}^* variable. In SEMs, the most common distributional assumption is that y_{it}^* comes from a normal distribution. Even with this assumption, we still have $2 + C_t - 1$ parameters to identify and just $C_t - 1$ independent proportions. We need two more restrictions to aid in identifying the parameters in the threshold model. Typically, SEM researchers remove this indeterminacy by assuming that y_{it}^* is a standard normal variable with a mean of zero ($\mu_{y_t^*} = 0$) and a variance of 1 ($\sigma_{y_t^* y_t^*} = 1$). The assumption of a standard normal variable, combined with knowledge of the proportion of cases falling into each category, is sufficient to identify the values of the thresholds.[4] For instance,

$$\tau_{j_t} = \Phi^{-1}\left(\sum_{c=1}^{j} \pi_{ct}\right) \tag{8.9}$$

where $j = 1, 2, ldots, C - 1$ and $\Phi^{-1}(\cdot)$ is the inverse of the cumulative standard normal distribution evaluated at the value inside the parentheses. In practice the thresholds would be estimated by substituting the sample proportion in each category for the population proportions.

To illustrate this procedure, we examined data drawn from a sample of $N = 825$ African-American women assessed as part of the 1992 National Longitudinal Survey of Youth (NLSY). Respondents were asked to rate their satisfaction with their community as a place to raise children using a five-category Likert scale ranging

[4]Simulation work by Quiroga (1992) and Flora and Curran (2004) suggest that the calculation of polychoric correlations is robust to modest violations of the assumption of bivariate normality.

from 1 (*poor*) to 5 (*excellent*).[5] The sample proportions in each category are 0.108, 0.274, 0.241, 0.230, 0.147 from poor to excellent, respectively. Assuming that the underlying satisfaction variable comes from a standard normal variable and using these sample proportions, we would estimate the thresholds to be $\tau_{1_1} = -1.238$, $\tau_{2_1} = -0.301$, $\tau_{3_1} = 0.313$, $\tau_{4_1} = 1.051$. The threshold estimates are the cutoff values from a standardized normal distribution that correspond to the cumulative proportion of cases falling into each category and below.

Although this procedure works well for a single variable, problems arise with this assumption given our interests in repeated measures. Key to our interests is the trend in mean values and variances of the repeated measures. Yet if we assume that the repeated measure is a standard variable in each time period, we are forcing it to keep the same mean (zero) and variance (one) no matter how many periods of data we have. This is clearly an inappropriate assumption for most repeated measures. But if we remove the assumption of standard variables, we underidentify the model. We need an alternative set of two restrictions that would be more realistic to assume for repeated measures and would permit us to study trends in their means and variances.

Jöreskog (2001a,b) recommends alternative assumptions for the auxiliary threshold model for ordinal variables that we endorse. He suggests that the mean and variance of y_{it}^* be freely estimated, but that the τ_{1t} and τ_{2t} thresholds be set to 0 and 1, respectively. Furthermore, he suggests that the same assumption be made for each repeated measure. This alternative assumption restricts two thresholds to be equal across time for the repeated measure and estimates the mean and variance for each time period. This alternative assumption of stable thresholds seems more realistic than the assumptions of equal means and variances. However, additional research is required to determine the plausibility of this scaling convention compared to other alternatives. This also requires ordinal variables with at least three categories.[6]

Fortunately, it is relatively easy to transform from the earlier values resulting from assuming a standard variable to the new values that have the first two thresholds at zero and 1 and estimate the mean and variance of the underlying variable (Jöreskog, 2001a,b; Mehta et al., 2004). Label the previous thresholds under the assumption of standardized variables as $\widetilde{\tau}_{1t}, \widetilde{\tau}_{2t}, \ldots, \widetilde{\tau}_{Ct}$. Jöreskog (2001a, p. 20) shows that the new thresholds, means, and standard deviations resulting when $\tau_{1_t} = 0$ and $\tau_{2_t} = 1$ are

$$\mu_{y_t^*} = \frac{-\widetilde{\tau}_{1_t}}{\widetilde{\tau}_{2_t} - \widetilde{\tau}_{1_t}} \tag{8.10}$$

$$\sqrt{\sigma_{y_t^* y_t^*}} = \frac{1}{\widetilde{\tau}_{2_t} - \widetilde{\tau}_{1_t}} \tag{8.11}$$

$$\tau_{c_t} = \frac{\widetilde{\tau}_{c_t} - \widetilde{\tau}_{1_t}}{\widetilde{\tau}_{2_t} - \widetilde{\tau}_{1_t}} \qquad \text{where} \quad c_t = 3, 4, \ldots, C_t - 1 \tag{8.12}$$

[5]In the original data set, the ordinal variable response categories ranged from excellent to poor. We reverse-coded the variable in our analysis, to range from poor to excellent.

[6]We discuss dichotomous variables below.

Table 8.1 Means, Variances, and Thresholds for Coded as Standardized and Coded with 0, 1 Thresholds for Satisfaction with Neighborhood Variable in 1992

	$\mu_{y_t^*}$	$\sqrt{\sigma_{y_t^* y_t^*}}$	τ_{1_1}	τ_{2_1}	τ_{3_1}	τ_{4_1}
Standard	0	1	−1.238	−0.301	0.313	1.051
0, 1 thresholds	1.321	1.067	0	1	1.655	2.442

We apply this transformation to our neighborhood satisfaction ordinal variable. Table 8.1 lists the resulting values under the standard normal distribution assumption ($\mu_{y_t^*} = 0$, $\sigma_{y_t^* y_t^*} = 1$) and under the alternative parameterization that sets the first two thresholds to 0 and 1. As the table shows, setting the first two thresholds to 0 and 1 permits us to estimate the mean of the underlying variable to be 1.321 and its standard deviation to be 1.067.

The preceding material refers to the auxiliary threshold model for a single ordinal variable. We follow a similar procedure for repeated ordinal variables. Setting the first two thresholds to 0 and 1 permits us to estimate the means and standard deviations of each underlying variable. Of course, we are assuming that the first two thresholds are the same over time, but this is a more realistic assumption for most repeated measures than is assuming that the means and standard deviations are unchanged over time. That is, shifts in the thresholds that determine the category in which a case is placed seems less likely to vary over time than do shifts in either the means or variances of a variable. If we expect that all thresholds are stable, then in addition to setting the first two thresholds to zero and 1 for all repeated measures, we can estimate but constrain each additional threshold to be equal over time (e.g., $\tau_{c_1} = \tau_{c_2} = \cdots = \tau_{c_T}$ for all c).

We illustrate this with the neighborhood satisfaction data. Table 8.2 lists the estimates of the means, variances, and thresholds for the five-category ordinal variable of neighborhood satisfaction for five years (1992 to 2000 at two-year intervals), where the same threshold is constrained to be equal over time.[7] The metric of these repeated underlying measures is set by the threshold values. Keeping this in mind and examining the means and standard deviations as we might do with other continuous variables enables us to interpret the changes in the underlying variables over this time period. The mean level of satisfaction is stable in 1992 and 1994, then increases slightly in 1996. A bigger jump occurs from 1996 to 1998, with a smaller increase for the last year, 2000. The standard deviations have no obvious trend but hover in the range 1.07 to 1.15. With standard deviations of roughly 1 and a maximum change of 0.2 over this 18-year period, the underlying satisfaction variables exhibit a modest positive change.

[7]We used PRELIS 2.53 to estimate this model with constrained thresholds (Jöreskog et al., 1999). Using the PRELIS option to test the equality of pairs of thresholds, we found that the thresholds for 1996 were the only ones that indicated a statistically significant departure from equality. Given the multiple testing problem, the size of the sample, and the lack of a clear rationale for expecting a difference, we estimated the thresholds with the equality constraint.

Table 8.2 Estimation of Means, Variances, and Equal Thresholds for the Five Repeated Ordinal Measures for Neighborhood Satisfaction (825 Women)

Satisfaction	$\mu_{y_t^*}$	$\sigma_{y_t^* y_t^*}$	τ_{1_1}	τ_{2_1}	τ_{3_1}	τ_{4_1}
1992	1.321	1.067	0	1	1.655	2.442
1994	1.321	1.090	0	1	1.655	2.442
1996	1.366	1.147	0	1	1.655	2.442
1998	1.496	1.078	0	1	1.655	2.442
2000	1.550	1.115	0	1	1.655	2.442

Although this approach works well for ordinal variables, this cannot be used with dichotomous variables. Dichotomous variables have a single threshold, so there is not an option to set two thresholds to 0 and 1. Setting the single threshold to 0 is a possibility, but we must also either set the mean or the variance to a constant if we want to identify the auxiliary model. One set of restrictions that would enable the estimation of the change in the mean over time is to set the thresholds to 0 and the variances to 1 for all time periods. Estimation of the mean at each time would be possible. However, if it seems likely that the variances of y_{it}^* are changing over time, these restrictions become questionable. But unfortunately, without additional information, identification requires that we fix two out of the three of the mean, variance, or threshold of y_{it}^*.

In sum, one consequence of using ordinal repeated measures is that $y_{it} \neq \alpha_i + \lambda_t \beta_i + \epsilon_{it}$. That is, we can no longer assume that the ordinal variable has a linear relation to the random intercepts and random slopes. In this subsection we have shown how to construct a threshold model to link the ordinal variable, y_{it}, to its continuous underlying counterpart, y_{it}^*. We also showed that by setting the first two thresholds of a variable to 0 and 1 for ordinal variables with three or more categories, we can estimate the other thresholds and the means and standard deviations of y_{it}^*. With dichotomous variables, the choice is to constrain the threshold to zero and to restrict either the mean or standard deviation of y_{it}^*.

However, we have not discussed how to address the second assumption violation that concerns the moment structure hypotheses. We now turn to this topic.

Violations of Moment Structure Hypotheses Earlier we found that

$$\boldsymbol{\Sigma} \neq \boldsymbol{\Sigma}(\boldsymbol{\theta}) \tag{8.13}$$

$$\boldsymbol{\mu} \neq \boldsymbol{\mu}(\boldsymbol{\theta}) \tag{8.14}$$

when the observed variables are ordinal. This is true even when these moment structure hypotheses hold for the underlying continuous variables [i.e., $\boldsymbol{\Sigma}^* = \boldsymbol{\Sigma}(\boldsymbol{\theta})$ and $\boldsymbol{\mu}^* = \boldsymbol{\mu}(\boldsymbol{\theta})$]. We can address this problem by using the ordinal variables and the distributional assumptions that we make for y_{it}^* to develop consistent estimators of $\boldsymbol{\Sigma}^*$ and $\boldsymbol{\mu}^*$. We represent these consistent estimators as $\widehat{\boldsymbol{\Sigma}}^*$ and $\widehat{\boldsymbol{\mu}}^*$. The results from the preceding section provide $\widehat{\mu}^*$ and the main diagonal of $\widehat{\Sigma}^*$. The off-diagonal elements of $\widehat{\Sigma}^*$ (the covariances of the y_{it}^*s) remain to be determined.

We can refer to $\widehat{\Sigma}^*$ as the *polychoric covariance matrix*. Olsson (1979) proposed a very practical method by which we can estimate the polychoric *correlation* matrix. Given the variances that we estimated in the preceding section, we can convert the polychoric correlation matrix into a polychoric covariance matrix. One simplification suggested by Olsson (1979) is to estimate each polychoric correlation separately rather than trying to estimate all possible correlations simultaneously.

A starting point is the ML estimation of the polychoric correlation between a pair of ordinal variables. Combining these variables results in a $C_t \times C_u$ table where C_t and C_u are the number of categories.[8] The log likelihood of these variables would be (Olsson, 1979)

$$\ln L = K + \sum_{c_t=1}^{C_t} \sum_{c_u=1}^{C_u} N_{c_t c_u} \ln(\pi_{c_t c_u}) \tag{8.15}$$

where K is an irrelevant constant, $N_{c_t c_u}$ is the number of cases that fall into the $c_t c_u$ cell of the table, and

$$\pi_{c_t c_u} = P[y_t = c_t, y_u = c_u] \tag{8.16}$$

The probability of the y_t variable taking the value of c_t and of y_u taking the value of c_u will depend on the bivariate distribution of the variables underlying y_t and y_u and the values of their thresholds. When estimating the polychoric *correlation*, we standardize the underlying variables by setting their means to zero and variances to 1 and freely estimating all their thresholds. That is, we use the traditional identifying restriction of standardized variables. As we show below, we will be able to recover the polychoric covariances with a simple transformation.

Earlier, we assumed that the y_{it}^* values came from univariate normal distributions. Now we assume that each pair of these variables derives from a bivariate normal distribution with a correlation between them of ρ_{tu}. The ρ_{tu} is the polychoric correlation of y_t^* and y_u^* and it would enter the likelihood function as part of the probability on the right-hand side of Eq. (8.16).

Olsson (1979) proposed a two-step approach to calculating the polychoric correlations. First, the threshold values are computed from the univariate distributions under the traditional scaling with standardized underlying variables. Second, these threshold values are substituted into the log likelihood function of Eq. (8.15), and the value of ρ_{tu} that maximizes the log likelihood conditional on these values of the thresholds is calculated. The resulting ML estimator, $\widehat{\rho}_{tu}$, is a consistent estimator of ρ_{tu} (see Olsson, 1979). The same procedure is followed for each pair of ordinal variables in the analysis. We can then assemble all of the polychoric correlations into a matrix, say $\widehat{R}^*$, that will be a consistent estimator of the population polychoric correlation matrix.

[8] Generally, there would be the same number of categories for repeated measures, but we keep it more general here by permitting the number of categories to differ.

The final step to obtaining the polychoric *covariance* matrix of $\widehat{\boldsymbol{\Sigma}}^*$ is to calculate

$$\widehat{\boldsymbol{\Sigma}}^* = D\widehat{R}^* D \tag{8.17}$$

where

$$D = \begin{bmatrix} \sqrt{\widehat{\sigma}_{y_1^* y_1^*}} & 0 & \cdots & 0 \\ 0 & \sqrt{\widehat{\sigma}_{y_2^* y_2^*}} & \cdots & 0 \\ \vdots & \vdots & \ddots & \vdots \\ 0 & 0 & \cdots & \sqrt{\widehat{\sigma}_{y_T^* y_T^*}} \end{bmatrix} \tag{8.18}$$

is a diagonal matrix of the estimated standard deviations of the underlying repeated measures. We can use the $\widehat{\sigma}_{y_t^* y_t^*}$ estimate that comes from the univariate analyses described in the preceding subsection. The resulting $\widehat{\boldsymbol{\Sigma}}^*$ is a consistent estimator of $\boldsymbol{\Sigma}^*$.

Below is the $\widehat{\boldsymbol{\Sigma}}^*$ for the five time points of the neighborhood satisfaction data running from 1992 to 2000:

$$\widehat{\boldsymbol{\Sigma}}^* = \begin{bmatrix} 1.139 & & & & \\ 0.574 & 1.188 & & & \\ 0.495 & 0.714 & 1.316 & & \\ 0.381 & 0.518 & 0.624 & 1.162 & \\ 0.415 & 0.419 & 0.535 & 0.676 & 1.244 \end{bmatrix} \tag{8.19}$$

Theoretically it would be possible to use a ML estimator to estimate $\boldsymbol{\Sigma}^*$, the means, and the thresholds of all y_i^*'s simultaneously (e.g., Lee et al., 1990). However, the computational burden is quite heavy, so contemporary users rely on the multiple-step approach we outlined above. In addition, there is some evidence that the gains of simultaneous estimation are not that great compared to estimation in stages. For example, Olsson (1979) and Jöreskog (2001a,b) find that jointly estimating the thresholds and polychoric correlation for a pair of variables leads to results that are quite similar to those of first estimating the thresholds and then estimating the correlation conditional on these threshold values.[9] However, as computational advances are made, simultaneous estimation might become more feasible.

One drawback of the pairwise approach to estimating the polychoric correlations (covariances) is that there is no guarantee that the resulting matrix will be positive definite. Nonpositive definite matrices can cause estimation difficulties for some fitting functions. Wothke (1993) discusses the nature of this problem and some ways to address nonpositive definite matrices in SEM. This problem appears less frequently in practice when the sample size is large, but can be a problem in more modest sample sizes (e.g., $N < 200$). Another challenge is how best to incorporate missing data into the two-step procedures. The use of full information

[9]Sparseness in bivariate tables can create estimation difficulties for this bivariate approach, and possibly for the simultaneous approach.

ML estimates the polychoric moment structures and the associated latent curve model simultaneously, and thus naturally allows for missing data (e.g., Neale et al., 2002). However, the empirical evaluation of multiple integrals often makes this a highly computationally burdensome endeavor. Further work is needed to determine optimal approaches for fitting these models in the presence of missing data.

In sum, we have given an overview of how to develop a consistent estimator of $\boldsymbol{\Sigma}^*$ and $\boldsymbol{\mu}^*$. We thus now have methods for correcting the violation of assumptions that occurs if we were to use the covariance matrix and means of the ordinal variables. Any consistent estimator for SEMs could be applied to $\widehat{\boldsymbol{\Sigma}}^*$ and $\widehat{\boldsymbol{\mu}}^*$. For example, we could apply the ML fitting function to these and we would obtain consistent estimators of the latent curve model parameters. However, the resulting asymptotic standard errors and the chi-square test statistic for the model are unlikely to be accurate. The problem is that the distributional assumptions that underlie these significance tests are typically violated. This takes us to the third consequence of using ordinal variables, the failure of the distributional assumptions and the need to correct for it.

Distributional Assumptions and Estimators The classic estimator in SEM for fitting polychoric correlation matrices is the *weighted least squares* (WLS) *estimator*:[10]

$$F_{\text{WLS}} = [\widehat{\boldsymbol{\rho}}^* - \boldsymbol{\rho}^*(\boldsymbol{\theta})]'\mathbf{W}^{-1}[\widehat{\boldsymbol{\rho}}^* - \boldsymbol{\rho}^*(\boldsymbol{\theta})] \tag{8.20}$$

where $\widehat{\boldsymbol{\rho}}^*$ is a vector that contains all the estimated polychoric covariances (or correlations) and the means of the $\mathbf{y}^*$ variables[11], $\boldsymbol{\rho}^*(\boldsymbol{\theta})$ represents the model-implied values of the polychoric covariances and the means, $\boldsymbol{\theta}$ contains all of the model parameters, and $\mathbf{W}$ is a weight matrix. Values of $\widehat{\boldsymbol{\theta}}$ are chosen that minimize this weighted sum of squared deviations of the polychoric covariances and means from their model-implied counterparts.

According to asymptotic theory, the optimal weight $\mathbf{W}$ occurs when $\mathbf{W} = \boldsymbol{\Sigma}_{\widehat{\rho}^*\widehat{\rho}^*}$, where $\boldsymbol{\Sigma}_{\widehat{\rho}^*\widehat{\rho}^*}$ is the asymptotic covariance matrix of $\widehat{\boldsymbol{\rho}}^*$. This leads to an estimator with several desirable properties, including consistency and asymptotic efficiency among all other possible $\mathbf{W}$'s. Furthermore, the asymptotic covariance matrix of $\widehat{\boldsymbol{\theta}}$ is

$$\text{ACOV}(\widehat{\boldsymbol{\theta}}) = N^{-1}\left[\frac{\partial\boldsymbol{\rho}^*(\boldsymbol{\theta})}{\partial\boldsymbol{\theta}}\right]'\boldsymbol{\Sigma}^{-1}_{\widehat{\rho}^*\widehat{\rho}^*}\left[\frac{\partial\boldsymbol{\rho}^*(\boldsymbol{\theta})}{\partial\boldsymbol{\theta}}\right] \tag{8.21}$$

Estimates of the asymptotic covariance matrix elements result when sample estimates of the unknowns replace their population counterparts. Furthermore, $(N-1)$ F_{WLS} forms a test statistic with an asymptotic chi-square distribution that allows for a test of the moment structure hypotheses. Jöreskog (1994) and Muthén and Satorra (1995) provide consistent estimators of $\boldsymbol{\Sigma}_{\widehat{\rho}^*\widehat{\rho}^*}$.

[10] Sometimes this estimator is called the *arbitrary distribution function* (ADF).

[11] In some cases where the thresholds are not taken as given, the thresholds also would be included in $\boldsymbol{\rho}^*$.

In practice, researchers have found that very large sample sizes are required for the preceding asymptotic properties to hold, particularly when the model is large (i.e., has many parameters). The specific sample size will depend on the number of indicators per factor, the R^2 values of the indicators, and other factors that vary by model. In addition, in large models the computational burden of finding the inverse of $\widehat{\mathbf{\Sigma}}_{\widehat{\rho}^*\widehat{\rho}^*}$ can be prohibitive. This has led researchers to seek alternative estimators. Jöreskog and Sörbom (1981) suggested diagonal WLS (DWLS) with the same fitting function F_{WLS} but with $\mathbf{W} = \text{diag}\left(\mathbf{\Sigma}_{\widehat{\rho}^*\widehat{\rho}^*}\right)$. With this diagonal matrix, the inverse is much easier to find. The main problem with this alternative is that the optimal weight matrix is not used, and as a result, formula (8.21) and the test statistic might not be accurate. Browne (1984) provided results that permit derivation of the appropriate asymptotic covariance matrix of the parameter estimates when $\mathbf{W}$ differs from the optimal weight matrix. Muthén et al. (1997) make use of these results to provide the asymptotic covariance matrix of $\widehat{\boldsymbol{\theta}}$ when derived from F_{WLS} with $\mathbf{W} = \text{diag}\left(\mathbf{\Sigma}_{\widehat{\rho}^*\widehat{\rho}^*}\right)$. Furthermore, these authors use Satorra's (1992) results to provide an adjusted chi-square test statistic for testing the moment structure hypotheses. Jöreskog et al. (1999) provide a corrected version of the ML estimator to apply to the polychoric matrix. Both sets of authors suggest that these estimators have better small-sample properties than those of F_{WLS} with $\mathbf{W} = \mathbf{\Sigma}_{\widehat{\rho}^*\widehat{\rho}^*}$. Flora (2002) and Flora and Curran (2004) present evidence in support of this, although more research is needed to make firm conclusions as to their relative properties. Furthermore, the problem of sparseness in cross-tabulations of categorical variables has not been studied adequately. Zero or near-zero cells can hamper estimation of the polychoric correlations (covariances).

In sum, we have several estimators that we can apply to the polychoric covariance matrix and mean vector that take account of the third and final problem of the distributional assumption violations that occurs with categorical variables.

8.1.3 Empirical Example

We estimate an unconditional latent curve model for the neighborhood satisfaction variables for the five repeated ordinal measures from 1992 to 2000. We analyze the polychoric covariance matrix and means reported in the preceding sections and use the corrected ML estimator from PRELIS and LISREL (Jöreskog et al., 1999) to estimate the model. The $T_{\text{ML-P}}$ test statistic is from the ML fitting function applied to the polychoric covariance matrix and means, but with a correction to the usual chi-square statistic that takes account of the distribution of the data (Jöreskog et al., 1999). We do not have substantive reasons to expect particular functional forms in neighborhood satisfaction over this period. Therefore, we explore several different functional forms for possible changes. Table 8.3 provides the estimates for linear, freed loadings, and quadratic models for these data, and Table 8.4 gives the estimates of the means and variances of the intercepts and slopes for each of these models. (In the quadratic model, the linear term is represented by β_1 and the squared term is β_2.) The asymptotic standard errors are in parentheses and these standard errors take account that the polychoric matrix is analyzed (Jöreskog

Table 8.3 Estimation of Linear, Freed-Loading, and Quadratic Models for Neighborhood Satisfaction Repeated Measures, 1992–2000 ($N = 825$)

	Satisfaction1	Satisfaction2	Satisfaction3	Satisfaction4	Satisfaction5
			Linear Model		
α	1.00	1.00	1.00	1.00	1.00
β	0.00	1.00	2.00	3.00	4.00
VAR(ϵ_t)	0.57	0.58	0.69	0.53	0.54
	(0.068)	(0.054)	(0.060)	(0.054)	(0.071)
R^2	0.53	0.49	0.44	0.54	0.58
			Freed-Loading Model		
α	1.00	1.00	1.00	1.00	1.00
β	0	−0.27	0.094	0.68	1.00
		(0.14)	(0.11)	(0.11)	
VAR(ϵ_t)	0.71	0.42	0.65	0.53	0.48
	(0.062)	(0.063)	(0.062)	(0.055)	(0.082)
R^2	0.46	0.63	0.48	0.54	0.62
			Quadratic Model		
α	1.00	1.00	1.00	1.00	1.00
β_1	0.00	1.00	2.00	3.00	4.00
β_2	0	1.00	4.00	9.00	16.00
VAR(ϵ_t)	0.44	0.51	0.56	0.54	0.23
	(0.095)	(0.055)	(0.064)	(0.058)	(0.100)
R^2	0.62	0.56	0.56	0.57	0.81

Table 8.4 Unconditional Model Means and Variances of the Random Intercepts and Slopes ($N = 825$ women)

	Linear Model		Freed-Loading Model		Quadratic Model		
	α	β	α	β	α	β_1	β_2
Mean	1.29	0.06	1.36	0.19	1.31	0.01	0.01
	(0.03)	(0.01)	(0.03)	(0.04)	(0.04)	(0.03)	(0.01)
Variance	0.65	0.04	0.61	0.42	0.71	0.35	0.02
	(0.05)	(0.01)	(0.03)	(0.11)	(0.08)	(0.08)	(0.00)

et al., 1999, pp. 192–195). It might seem counterintuitive that we can examine functional form with these data, but if we have the means and standard deviations of the repeated measures, we can examine these issues as we did in Chapter 4. The coefficients for α and β are fixed in the linear and quadratic models, but are estimated for the middle three coefficients of β in the freed-loading model.

Table 8.5 lists statistics on the overall fit of these competing models. The test statistic, $T_{\text{ML-P}}$, is the chi-square test statistic that uses the ML fitting function

Table 8.5 Overall Fit Measures for Linear, Freed-Loading, and Quadratic Models for Neighborhood Satisfaction Unconditional Model, 1992–2000 ($N = 825$)

	T_{ML-P}	df	Prob.	RMSEA	IFI	TLI
Linear model	66.58	10	0.00	0.083	0.967	0.967
Freed-loading model	37.68	7	0.00	0.073	0.982	0.974
Quadratic model	11.75	6	0.068	0.034	0.997	0.994

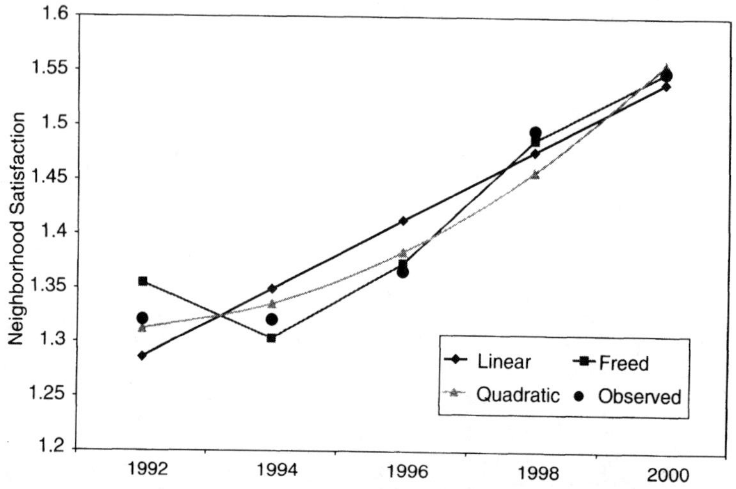

FIGURE 8.2 Plot of observed means and implied means from linear, freed-loading, and quadratic models for neighborhood satisfaction unconditional model, 1992–2000.

and takes into account that the polychoric matrix is analyzed (see Jöreskog et al., 1999).[12] In large samples, this test statistic should follow a chi-square distribution. Table 8.5 also reports the RMSEA, TLI, and IFI. The T_{ML-P} for each hypothesized model is part of all three of these fit indices. The baseline model test statistic that enters the calculation of the TLI and IFI is calculated using the T_{ML-P} statistic that results with a model that freely estimates the variances and means of the underlying indicators but constrains their covariances to zero. The behavior of T_{ML-P} and the fit indices we construct using it are not as well studied in the context of models with ordinal and dichotomous observed variables. Hence, we must be cautious in their interpretation.

Keeping this qualification in mind, Table 8.5 suggests that the quadratic model has the best fit of the three. Figure 8.2 plots the implied means from each of the models against the observed means. All of these results suggest that the quadratic

[12]We used PRELIS 2.53 and LISREL 8 to estimate these models. The test statistic, T_{ML-P}, is referred to as c_3 or as the Satorra–Bentler scaled chi-square statistic (Jöreskog et al., 1999, pp. 194–196).

model has the best fit to the data.[13] Examining the quadratic model and the results in Table 8.4 indicates that the mean of neither β_1 nor β_2 is significantly different from zero. This means that the average slope and squared slope is zero. But the statistically significant variances of both terms suggests that neighborhoods differ from each other in their slopes and squared slopes even though their means of these quantities are zero.

8.1.4 Conditional Model

Thus far we have concentrated on unconditional latent curve models with ordinal repeated measures. In this subsection we describe briefly modifications that make it possible to estimate a repeated ordinal measure model with covariate predictors. The covariates are observed variables that affect the random intercepts or slopes.

The level 1 and level 2 models and assumptions are the same as those described in Chapter 5:

$$y_{it}^* = \alpha_i + \lambda_t \beta_i + \epsilon_{it} \tag{8.22}$$

$$\alpha_i = \mu_\alpha + \gamma_{\alpha x_1} x_{1i} + \gamma_{\alpha x_2} x_{2i} + \zeta_{\alpha i} \tag{8.23}$$

$$\beta_i = \mu_\beta + \gamma_{\beta x_1} x_{1i} + \gamma_{\beta x_2} x_{2i} + \zeta_{\beta i} \tag{8.24}$$

with the exception that the level 1 model has the underlying y_{it}^* in place of y_{it}. The approach taken in the conditional model is similar to that of the unconditional model except that calculations are made conditional on the covariates. In brief, the approach taken involves the following steps: (1) do univariate probit regressions of each ordinal variable on the covariates and estimate the thresholds and probit regression coefficients; (2) conditional on the thresholds and coefficients from step (1), estimate all bivariate correlations among the disturbances; (3) based on step (2), calculate the polychoric covariance matrix y_{it}^*'s and x's to form $\widehat{\boldsymbol{\rho}}^*$; (4) estimate $\widehat{\boldsymbol{\Sigma}}_{\widehat{\rho}^*\widehat{\rho}^*}$; and (5) apply one of the estimators (Jöreskog et al., 1999; Muthén et al., 1997) described in Section 8.1.3. The probit regressions of step (1) are ordinal or dichotomous probit regressions where an ordinal or dichotomous variable is regressed on covariates. These regressions are similar to logistic regressions, with the exception that a different distributional assumption is made for the error term. Specifically, if ϵ_{it} from Eq. (8.22) comes from a logistic distribution, this is a logistic regression. Alternatively, here we assume that ϵ_{it} comes from a normal distribution and this leads to a probit regression. See Long (1997) for details.

[13]What would happen if the satisfaction variables were treated as if they were continuous rather than ordinal? Given that the ordinal measures have five categories, we would expect somewhat similar results when treated as continuous. We estimated the same models reported here, ignoring their ordinal nature. To make the units similar, we transformed the ordinal variables to have means and variances that matched those in Table 8.2 but did not set the covariances. The results with the ordinal variables were quite similar to those reported above, where we took account of their ordinal nature. However, this is no guarantee that similar robustness would be found with other examples or with ordinal variables with fewer categories or different distributions. Thus, it is prudent to take account of ordinal or dichotomous observed variables to check the robustness of results.

Table 8.6 Conditional Model for Estimates of the Latent Curve Parameters (α, β_1, β_2) of Satisfaction with Neighborhood Regressed on Predictors "Married" and "Living in Urban Area" ($N = 817$)

	α	β_1	β_2
Married	0.27 (0.049)	−0.082 (0.045)	0.013 (0.011)
Urban area	−0.17 (0.054)	0.052 (0.048)	−0.014 (0.012)
R^2	0.14	0.03	0.02

Although much of this method is similar to that of the unconditional model, there are differences. For one thing, the conditional model assumes that the disturbances of the level 1 and level 2 model come from bivariate normal distributions rather than assuming that the y_{it}^*'s come from bivariate normal distributions as we did for the unconditional model. The former is somewhat less restrictive than the latter. Another difference is that we estimate the bivariate polychoric correlation between the disturbances of the probit regressions rather than the zero-order polychoric correlation between the y_{it}^*s. A slight modification to this procedure is to fix the first two thresholds to 0 and 1 in the univariate probit regressions so that the variances of the disturbances are estimated rather than set to a variance of 1. In this case we would estimate the polychoric covariance of the disturbances rather than the polychoric correlation in the bivariate probit regressions.

We illustrate this procedure by extending our satisfaction with the neighborhood example to include two covariates: whether the woman is married and whether she lives in an urban area.[14] Table 8.6 reports the results for this conditional model when estimates of the trajectory parameters from the quadratic model are regressed on these two dummy variables.[15] Both variables have a statistically significant effect on the random intercepts, although they are opposite in sign. Being married has a significantly positive effect on the initial level of satisfaction, while being in an urban area has a significantly negative effect; each of these effects is net of the other. These two variables explain about 14% of the variation in the random intercept. In contrast, married and urban are largely insignificant in their impact on the linear and squared slopes, with the possible exception of the marginally significant negative effect of married on the linear term (β_1). The low R^2 values for β_1 and β_2 are consistent with these weak effects.

8.1.5 Summary

In this section we have discussed the analysis of repeated measures that are ordinal or dichotomous. We have reviewed the assumptions that would be violated by

[14]Although marital status and living in an urban area can also change over time, for expository purposes we treat them here as time-invariant predictors.

[15]Eight respondents were missing on marital status or urban, and this explains why there are 817 cases here compared to 825 in the analysis of just the satisfaction measures.

treating these as if they were continuous, and described methods to take account of their categorical nature. Key to the approach is to estimate the covariance matrix and vector of means of the continuous variables that underlie most such ordinal or dichotomous variables. As we showed, once we had the estimates of the means and covariances, we could estimate unconditional and conditional models for the repeated measures even to the point of exploring nonlinearity of the trajectories.

Although the material we presented will cover many situations encountered in practice, there are areas of latent curve modeling that require further development in SEMs. One such area is the treatment of other types and combinations of variables. If you have a mixture of continuous and ordinal variables, covariances of the continuous variables, polychoric covariances of the ordinal variables, and polyserial covariances of the continuous with ordinal variables could be analyzed. The polyserial association measures are devised to estimate the relation between a continuous variable and the underlying variable for an ordinal variable (e.g., Olsson et al. 1982) A less studied area in SEM is that of repeated nominal variables with three or more categories. The treatment of count variables that are repeated measures in SEM requires further development, but see Skrondal and Rabe-Heskein (2004) for promising new directions.[16] In contrast, multilevel modeling is further developed in its ability to handle count variables (e.g., Vonesh and Chinchilli, 1997).

Another issue is the treatment of missing data with ordinal and dichotomous measures. Some approaches look toward the direct ML approach that we discussed in regard to continuous variables to handle ordinal or dichotomous variables. For instance, Neale et al. (2002) have this option in the software package Mx (Neale et al., 2002) An alternative to ML for ordinal data with missing values is the method of multiple imputation (MI; Rubin, 1987; Schafer, 1997; Schafer and Graham, 2002). Here imputed data are categorized and then analyzed. Our expectation is that these and other areas of treating repeated measures will undergo technical developments that will solve these issues.

8.2 REPEATED LATENT VARIABLES WITH MULTIPLE INDICATORS

Nearly all of the literature on repeated measures focuses on repeated single indicators or observed variables. Measurement error is permitted through the disturbance term in the repeated measure. Placing the repeated measures in a SEM framework enables us to take advantage of major strengths of SEMs. These include the ability to incorporate multiple indicators of each latent variable, a method to determine the invariance of model parameters, and models that permit correlated disturbance variables for the same measures over time. There is a great deal of SEM literature that considers multiple indicators, invariance, and correlated disturbances, and there is a vast literature on growth curve models. Yet the literature that examines repeated latent variables each measured with multiple indicators combined with latent curve models is sparse. McArdle (1988), Duncan and Duncan (1996), and

[16]When there are few values of the count variables, one possibility is to treat them as ordinal variables.

Sayer and Cumsille (2001) are exceptions. In this chapter we examine issues that arise when we synthesize these two well-established research areas by combining repeated latent variables with multiple indicators and latent curve models.

In this section we focus on examining the change in *latent* variables rather than modeling change in *observed* variables. We start by contrasting the single repeated measure that we have relied on in the rest of the book with a repeated latent variable with multiple indicators. Then we discuss how to build a multiple-indicator repeated latent variable model. We present several ways in which the disturbances of the indicators might be structured. We then discuss the issue of invariance of parameters in multiple-indicator models so that we can explore the degree to which the indicator and latent variable relations stay the same over time. Dichotomous and ordinal multiple indicators alone or mixed with continuous indicators share these complications and give rise to new issues of invariance that we also examine. We introduce these issues with unconditional models, but also will explore them in the context of conditional LCMs.

8.2.1 Single Repeated Measure versus Multiple Indicators

In this section we look briefly at some of the limitations of the usual repeated measure approach that the multiple-indicator approach can overcome. Consider the usual single-indicator level 1 model as

$$y_{it} = \alpha_i + \beta_i \lambda_t + \epsilon_{it} \tag{8.25}$$

where we make the same assumptions described earlier. Suppose that

$$\epsilon_{it} = \zeta_{it} + \upsilon_{it}$$

where ζ_{it} is a disturbance capturing the departure of the repeated variables from the trajectory and υ_{it} is a disturbance made up of specific error and measurement error. *Specific error*, a term that comes from the factor analysis literature, refers to error that is due to the particular measure used (e.g., Harman, 1967). If we take the variance of ϵ_{it} and assume that ζ_{it} and υ_{it} are uncorrelated, the VAR(ϵ_{it}) in the single repeated measure is a combination of the disturbance variance due to the trajectory disturbance (ζ_{it}) and the disturbance of the indicator (υ_{it}). A single-indicator repeated measure analysis generally will not permit separate estimates of the two components. An advantage of the multiple-indicator approach is that it can separate these. In fact, we can estimate and compare the reliabilities of each of the multiple indicators since the multiple-indicator approach gives us an estimate of R^2 for each indicator.[17]

Another advantage of the multiple indicators is that we can have models that permit the disturbances for the same indicator over time to correlate. Single

[17] Strictly speaking, the unique component variance consists of the variance due to the specific variance and the pure error variance. The specific variance is considered a component of reliability. This implies that the reliability estimate that comes from the R^2 of an indicator is a conservative estimate of the reliability of the indicator. For further discussion, see Bollen (1989b, Chap. 6).

indicators might still have disturbances that correlate over time, but typically this is not considered. To do so would require careful consideration of the identification of such a model. However, below we present several models that take account of possible correlation of the disturbances in multiple-indicator models.

Sometimes multiple indicators of a repeated latent variable are incorporated into the usual single repeated measures model through the use of a *scale index*, or *composite*. That is, the multiple indicators are combined to form a single variable at each time. An advantage of composite measures is that they permit the researcher to apply standard repeated measures approaches to the composites rather than incorporating a multiple indicator measurement model. Another advantage is that fewer parameters are estimated using composites than using multiple indicators. In small samples this can be helpful in that convergence problems and improper solutions will be less frequent. But using composites is not without costs. One issue is that aggregating the indicators at each time point assumes that the relations of the latent variable to the multiple indicators are the same at each time point. In other words, invariant measures are assumed. If they are not invariant, the latent curve model using composites will confound measurement differences with real change in the latent variables.

Another issue is the possible dependence among the disturbances of the indicators that make up the composite. It is quite possible that there will be correlated disturbance variables for repeated single or multiple indicators. Correlated disturbances of individual indicators are not removed by aggregating the individual indicators. So it might be necessary to estimate LCMs with the composites where the disturbances are permitted to correlate over time. There are fewer options to capturing this dependence with composites than with multiple indicators, as we will see in the sections that follow. Once a multiple-indicator analysis establishes the measurement properties, a composite approach might be suitable. But in most areas of research the knowledge is not complete enough to make this a desirable alternative.

Another possibility is that there are autoregressive relations among the repeated latent variables or even among the repeated multiple indicators. The autoregressive latent trajectory (ALT) model considers a single indicator with an autoregressive relation among the repeated measures (Curran and Bollen, 2001; Bollen and Curran, 2004), and we discussed this model in Chapter 7. A simple extension of this model permits the same autoregressive relation where the lagged latent variable influences the current value. If the indicator at time $t - 1$ affects itself at time t, an autoregressive relation could be built among these as well. In this latter case, the cause might be memory effects in responding to the same question at multiple points in time.

8.2.2 Multiple-Indicator Model

In a multiple-indicator model, the repeated "measure" is a latent variable. The level 1 model for the latent variable is

$$\eta_{it} = \alpha_i + \beta_i \lambda_t + \zeta_{it} \tag{8.26}$$

where η_{it} is the repeated latent variable for case i and time t; ζ_{it} is the disturbance for case i and time t with a mean of zero and uncorrelated with α_i, β_i, and λ_t; and the other terms are as defined before. The equation includes no *fixed* intercept or constant. Thus, when estimating this model, the fixed regression constant should be set to zero. The only intercept is a random one (i.e., α_i).

In the unconditional model the level 2 equations are

$$\alpha_i = \mu_\alpha + \zeta_{\alpha i} \tag{8.27}$$

$$\beta_i = \mu_\beta + \zeta_{\beta i} \tag{8.28}$$

where the disturbances ($\zeta_{\alpha i}$, $\zeta_{\beta i}$) have means of zero and are uncorrelated with ζ_{it} and λ_t. Thus far this unconditional model is very similar to the single-indicator model that we have dealt with for most of the book, with the exception that we are now modeling a repeated *latent* variable.

The difference becomes greater when we add a measurement model to the latent curve model so that we have multiple indicators for each time:

$$y_{jit} = \nu_{jt} + \Lambda_{jt}\eta_{it} + \upsilon_{jit} \tag{8.29}$$

where $j = 1, 2, ldots, J$ and J is the total number of indicators for the ith case in the tth time period, ν_{jt} is the intercept for the jth indicator in the tth time period, Λ_{jt} is the factor loading for the jth indicator at the tth time period, and υ_{jit} is the disturbance for the ith case in the tth time period and the jth indicator. We assume that υ_{jit} has a mean of zero and is uncorrelated with η_{it}, α_i, β_i, and λ_t. Note that λ_t and Λ_{jt} refer to different parameters. The λ_t is a scalar that is a function of time, such as $\lambda_t = t - 1$ in the case of the linear LCM. The Λ_{jt} is a different scalar that refers to a coefficient of the effect of the latent variable η_{it} on the indicator, y_{jit}. To simplify our discussion we assume that we have the same set of indicators for each wave. However, we could have different numbers of indicators for different waves as long as we had at least one indicator in common. The common indicator would anchor the latent variable and enable over-time comparisons.

Another complication that can occur is when the primary construct represented by the latent variables changes in its meaning or relevance over time. For instance, a researcher might measure aggression by biting or taking toys from others among a group of toddlers. Although these are suitable measures for aggression in toddlers, the same behaviors are less relevant for teens. Other measures of aggression would be needed for the teens, and the researcher would need to question whether the concept of aggression is equivalent among these age groups even when its indicators change. Thus, thought must be given to the appropriateness of a concept and its indicators for the full range of subjects.

Equation (8.29) provides an equation for a confirmatory factor analysis (Bollen, 1989b, Chap. 7) where there are multiple measures for the latent variable in each time period and the collection of Λ_{jt} factor loadings gives the coefficient effects of the latent variables on the multiple indicators. Although this part of the model is a confirmatory factor analysis (CFA), there are aspects of it in the LCM context that set it apart from the usual CFA. One is that rather than having different factors and different indicators as is typical in CFA, we typically have the same indicators and the same latent variable at different points in time. This leads to the possibilities that some or all of the parameter values at one point in time are equal to those at another point in time. We refer to this as the issue of invariance, and we discuss this in more detail later. A second issue is that since we have the same indicators at different times, it is likely that the disturbance (υ_{jit}) for the same indicator at two points in time is associated. It is to this issue that we now turn.

Dependence Among Disturbances of Same Indicator If we use the same measure over time, it is common in SEM to determine whether the disturbance of a measure at one time is correlated with the disturbance of the same measure at another time. In a multiple-indicator LCM, we can do the same. One possibility is to have covariances of consecutive disturbances as free parameters [$\text{COV}(\upsilon_{ji,t}, \upsilon_{ji,t+1}) \neq 0$]. Figure 8.3 illustrates a three-wave latent curve model with multiple indicators and correlated disturbances. The model has three indicators for each time.

The curved two-headed arrows linking the indicators' disturbances represent the distinct possibility that the disturbance at one time is associated with the disturbance of the same variable at the next time. In the path diagram we represent this association as lasting only one time period. It is possible that there is sufficient stability in the disturbances to last more than one time period so that the disturbances should be allowed to correlate for two or more periods. In this situation, there would be a curved two-headed arrow connecting the first wave disturbance

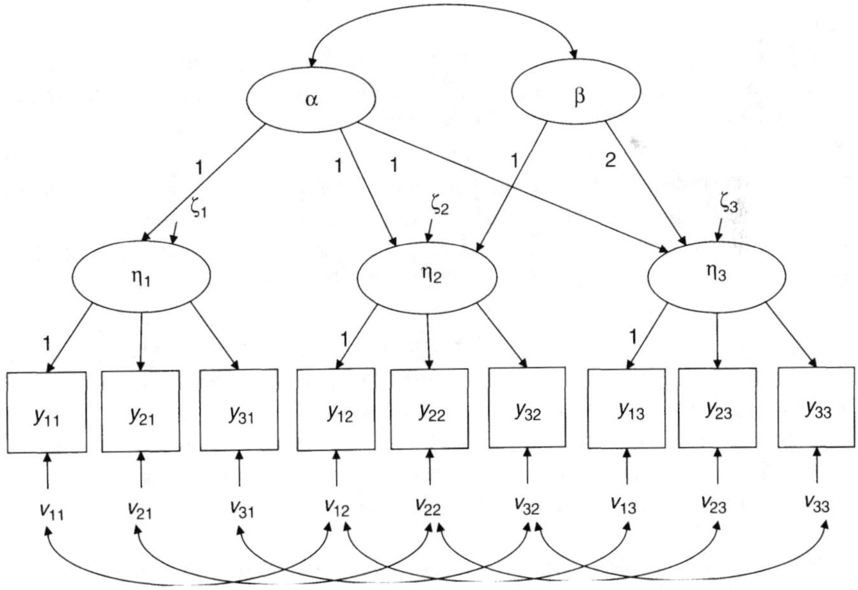

FIGURE 8.3 Multiple-indicator latent curve model with correlated disturbances.

with its counterpart at the second and third waves of data. If it were reasonable to assume that the covariance of the disturbances for the same indicator were the same for equal lags, an equality constraint could be set such that the time 1 and time 2 covariance equals the time 2 and time 3 covariance for the same disturbance variable at different times. This equality constraint could be tested. If it holds, it would reduce the number of distinct estimated parameters.

Second, a first-order autoregressive linkage between the disturbances also is possible:

$$\upsilon_{ji,t+1} = \rho_{j,t+1,t}\upsilon_{ji,t} + \omega_{ji,t+1} \tag{8.30}$$

where $\rho_{j,t+1,t}$ is the autoregressive coefficient that gives the impact of the prior disturbance j on the current one and $\omega_{ji,t+1}$ is a disturbance term with a mean of zero and uncorrelated with $\upsilon_{ji,t}$ and with the other disturbances and exogenous variables in the model. Assuming this structure to the disturbance contrasts with the correlated disturbance model in that it hypothesizes a structural relation between prior and current values rather than assuming just an association. The autoregressive structure permits indirect effects of lagged values of the disturbance on later values.[18]

A third way to capture the dependence of disturbances is to use "method" latent variables to capture the time-invariant component of the disturbance for the same indicator. For example, we might have a design in which the mother, father, and child all report on the same construct (e.g., family stress). In this situation, each repeated measure would be influenced by both the substantive latent variable (e.g., family stress) and a method factor (e.g., reporter) that corresponds to each measure. Figure 8.4 is a path diagram of such a model with three repeated latent variables, each measured with three indicators.

This structure implies that there is a time-invariant component (e.g., a stable disturbance component) that is present in the same indicator over time that creates an association not captured by the substantive latent variable that also influences the indicator. If a researcher hypothesizes that this method factor has the same effect on the indicator at each point in time, the factor loadings would be set to 1. This would establish equality, scale the factor, and help to identify the model The method factors for different indicators are uncorrelated in Figure 8.4. This represents the hypothesis that the method factors for each indicator are distinct and unrelated to each other.

Typically, there is insufficient information to dictate which of these three ways of capturing the dependence among the disturbances of the indicators is best for a particular application. However, we have enough information to know that if we ignore the dependence when it is present, we are likely to have a distorted view of the LCM that holds for the substantive variables. Therefore, it is prudent to consider one or more of these patterns of dependence when estimating these models.

[18]Although not a method for modeling associations between disturbances, an ALT model, such as discussed in Chapter 7, would be a way to model dependence among the repeated indicators through incorporating autoregressive relations among lagged indicators.

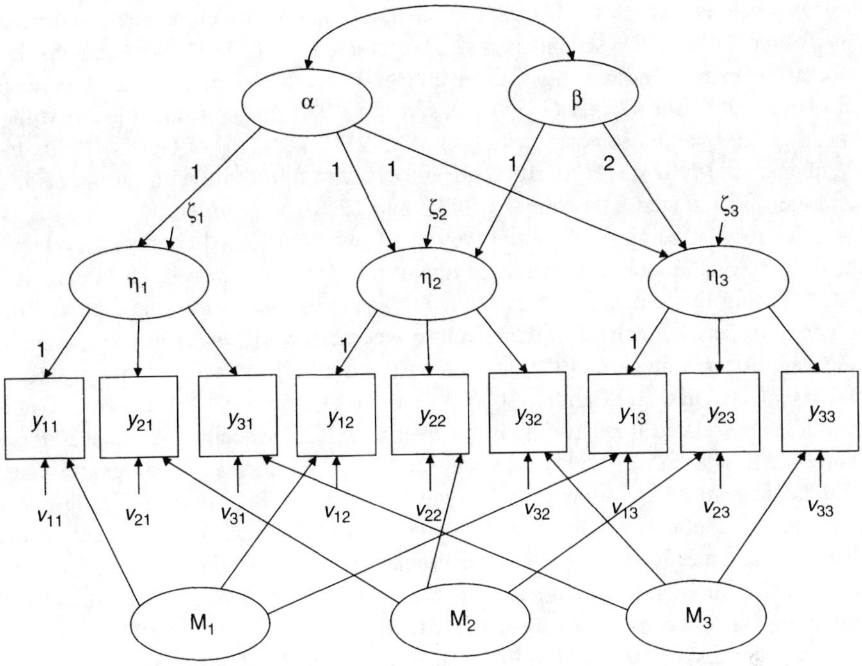

FIGURE 8.4 Multiple-indicator latent curve model with method factors.

Identification Identification of these models is complicated by the addition of the measurement model for the multiple indicators of the repeated latent variables. We omit proofs but provide some guidelines on identification here. A simple way to approach identification is to use the Two-Step Rule (Bollen, 1989b, pp. 328–331). The first step is to treat the model as a measurement model ignoring the latent curve parameters. For instance, in Figure 8.3 we would concentrate on the three latent variables, their indicators, and their correlated disturbances. We would exclude α_i and β_i and treat the latent variables as correlated factors. The resulting model would be a confirmatory factor analysis with correlated disturbance terms. We can then use any of the rules of identification for CFA to determine whether this model would be identified (Bollen, 1989b, pp. 238–256; Davis, 1993). The second step to identification is to treat the three correlated factors as if they were observed variables and to determine whether the latent curve parameters structural model would be identified. Based on our earlier results on the identification of unconditional latent curve models (see Chapter 2), the model is identified with at least three repeated variables. Hence, this model would be identified. Similarly, we could use the Two-Step Rule to address the identification of the other models in this section.

Empirical Example This example is a LCM for the latent variable of the degree of liberal democracy in national political systems over three waves of data: 1980, 1983, and 1986. In each year there are three measures of liberal democracy that we will use as indicators. The first is an index of liberal democracy constructed by Bollen (1998), the second is an index created by Gurr (1990), and the last liberal democracy index is by Vanhanen (1990). All three indices were taken from Bollen (1998). The Bollen (1998) measure for 1980 ranges from 0 to 10 with a mean of 3.997 and a standard deviation of 3.894. The ranges of Gurr's (1990) and Vanhanen's (1990) scales for 1980 are 0 to 10 and 0 to 40.1, with means of 3.65 and 8.23, and standard deviations of 4.03 and 12.10, respectively.

Prior to examining the latent curve model, we constructed confirmatory factor analysis (CFA) models with the three liberal democracy latent variables intercorrelated. Fitting this model involves two complications. One is missing data and the other is excess kurtosis. To address these we use the ML estimates where missing data are permitted, and the asymptotic standard errors take account of excess kurtosis (see Yuan and Bentler, 2000).[19] The first CFA model had the disturbance variables correlated for the same index over time. The second CFA replaced these correlated disturbances with a separate method factor for each democracy index. The latent democracy variable's scale was set by restricting the factor loading to 1 and the intercept to 0 for Bollen's index in each wave of data. Although each model had an adequate fit, the correlated disturbance model was slightly better. Specifically, using the corrected chi-square (Yuan and Bentler, 2000, p. 177), the model allowing disturbances to covary had a chi square of 34.34 on 15 df ($p = 0.003$) with a RMSEA $= 0.088$, IFI $= 0.976$, and TLI $= 0.942$. The model with method factors had a chi-square of 43.92 on 18 df ($p < 0.001$) with a RMSEA $= 0.093$, IFI $= 0.968$, and TLI $= 0.935$. The model without any disturbance covariances had a chi-square of 215.92 on 24 df, RMSEA $= 0.22$, IFI $= 0.763$, and TLI $= 0.639$.

Given these results, we formulated a LCM that corresponds to Figure 8.3, with the exception that there are three pairs of correlated disturbances for each index rather than just the two shown in the Figure. Table 8.7 reports the test statistic and the p-values for the uncorrected and the corrected chi-square test statistic that takes account of excess kurtosis. The corrected chi-square is much less than half of the uncorrected chi-square test statistic, although its *p-value* is 0.003. Considering the TLI, IFI, and RMSEA, the results suggest an adequate overall model fit.[20]

Table 8.8 presents the estimates of the intercepts and factor loadings from this LCM. We see that all three indices have significant factor loadings and high R^2 values. The Bollen liberal democracy index that scales the latent variable (i.e., $\nu_j = 0$ and $\Lambda_{j1} = 1$) has the highest R^2 values, ranging from 0.92 to 0.96. The Gurr democracy index also has high R^2 values (0.86 to 0.89) and the Vanhanen democracy index has lower R^2 values (0.77 to 0.80). The Gurr democracy index has intercepts that are within sampling fluctuations of zero and slopes close to 1, making it similar to the Bollen democracy index with respect to these parameters.

[19]The estimates come from Mplus 3.0 with the MLR estimator (Muthén and Muthén, 2004, pp. 365–369).

[20]By the usual standards of the RMSEA, the fit is borderline. However, given the components of fit and the other fit indices, we feel that this model is a reasonable approximation to the data. Also note that there was a small negative error variance for the first latent liberal democracy variable. We reestimated the model after setting this to a small positive value and find only a small decrease in model fit.

Table 8.7 Overall Fit Measures from Latent Curve Model for Liberal Democracy, Multiple-Indicator Model from 1980, 1983, and 1986 ($N = 166$)

T_{ML}	df	Prob.	Corrected T_{ML}	df	Prob.	RMSEA[a]	IFI[a]	TLI[a]
93.86	16	<0.01	35.88	16	0.003	0.09	0.98	0.94

[a]Fit indices calculated using the Yuan–Bentler "corrected" T_{ML}, df.

Table 8.8 Maximum Likelihood Estimates of Intercepts (ν_{jt}), Slopes (Λ_{jt}), and R^2 values of Democracy Indices on Liberal Democracy Latent Variables for 1980, 1983, and 1986 ($N = 166$)

	1980			1983			1986		
	ν_j	Λ_{j1}	R^2	ν_j	Λ_{j2}	R^2	ν_j	Λ_{j3}	R^2
Bollen index of democracy	0	1	0.96	0	1	0.92	0	1	0.93
Gurr index of democracy	−0.05 (0.23)[a]	0.96 (0.05)	0.89	−0.18 (0.27)	0.96 (0.06)	0.87	−0.18 (0.27)	0.95 (0.06)	0.86
Vanhanen index of democracy	−2.85 (0.50)	2.82 (0.17)	0.80	−3.12 (0.64)	2.98 (0.19)	0.79	−2.81 (0.64)	3.01 (0.20)	0.77

[a]Estimated "robust" asymptotic standard errors in parentheses.

Table 8.9 Mean, Variance, and Covariance of Random Intercepts (α_i) and Random Slopes (β_i) ($N = 166$)

	α		β
Mean	4.01 (0.30)[a]		0.18 (0.08)
Variance	15.15 (1.01)		0.56 (0.27)
Covariance		−0.90 (0.38)	

[a]Estimated "robust" asymptotic standard errors in parentheses.

In contrast, the Vanhanen democracy index has a negative intercept and a larger positive slope than the other two indices.

Table 8.9 provides estimates of the mean, variance, and covariance of the random intercepts and random slopes for the liberal democracy latent variable model. This shows that the mean initial level of liberal democracy was 3.95, which is statistically significantly different from zero. The variance of the random intercepts also differs significantly from zero, suggesting variability in the initial levels of liberal democracy. The estimated mean slope is 0.21, suggesting that on average there is a positive slope over this period. This, too, is statistically significant. The positive mean slope is consistent with the worldwide democratization trend that occurred during the 1980s. The estimated variance in the slope is 0.39, which is marginally statistically significant. This is modest evidence of variability around the estimated mean slope of 0.21. The statistically significant negative covariance of the intercepts and slopes suggests that high initial levels of political democracy tend to accompany lower slopes, and vice versa.

Also noteworthy is that the trajectories described by the random intercepts and slopes explain nearly all of the variances in the repeated latent democracy variable (R^2 values of 1.00, 0.97, 0.97 for 1980, 1983, 1986, respectively).[21] In this case the latent curve provides an excellent description of the trajectory of the latent democracy variables.

8.2.3 Invariance: Continuous Multiple Indicators

The issue of invariance emerged in Chapter 6, where we were comparing different groups and whether their LCMs were similar. Here we turn to invariance again, but our question of similarity is invariance over time rather than over group. Invariance is important to scientific study, since we want to know whether changes that we observe are due to real changes in the phenomena being studied or are changes due to a changing relation between the latent variables and their indicators. Thus, we want to ensure that we have sufficient invariance to study change.

The preceding discussion of repeated latent variables with multiple indicators applies whether the indicators are continuous, dichotomous/ordinal, or a mixture. There are issues of measurement comparability and identification that are usefully discussed by separately treating continuous and noncontinuous multiple indicators. In this section we give attention to continuous measures in unconditional LCMs. We focus on invariance or whether the relation between the indicator and the latent variable remains the same for each wave of data. Invariance is not an either/or situation. Rather, it is a matter of degree of invariance (e.g., Meredith, 1993). That is, some parameters might be the same over time, whereas others might change. Our discussion of invariance is divided into the topics of model form, scaling, factor loadings, intercepts, disturbance variances of indicators, and disturbance variances of the latent variable.

Model Form The first stage of invariance is to assume that the model form is the same over time. That is, the latent variables should have the same dimensions, the latent curve should be of the same structure, and other parameters should be common, if not the same value, at each time. In other words, at each time the matrices and the positions of the fixed and free parameters should be the same if the model forms are the same.

[21] In fact, in 1980 the liberal democracy latent variable has a very small negative variance ("improper solution") estimate for the disturbance. This appears more due to sampling variability than to other potential causes of negative error variances (Chen et al., 2001).

Scaling As with any other latent variable, we must assign a scale to the repeated latent variables. The two most common ways are by using a scaling indicator or by standardizing the latent variable to a mean of zero and a variance of 1. The latter option is particularly undesirable in the case of repeated measures since if the latent variable is forced to have a mean of zero and variance of 1 at each time point, we are in essence removing any possible trend in the means of the repeated latent variable and we are forcing the variance to be constant over time. We recommend a scaling indicator approach from Bollen (1989b, pp. 306–309), where the most reliable and valid indicator of the latent variable is chosen, its factor loading is set to 1, and its intercept is set to zero. This scaling provides the latent variable with a scale that is in a similar metric as the scaling indicator and gives the latent variable a mean of the scaling indicator. To optimize comparability, the same scaling indicator should be used for each wave of data. The implicit assumption in this strategy is that the relation between the latent variable and the scaling indicator does not change. We say more about sensitivity checks that are available to probe this assumption next.

Factor Loadings Consider the measurement equation (8.29),

$$y_{jit} = \nu_{jt} + \Lambda_{jt}\eta_{it} + \upsilon_{jit} \tag{8.31}$$

An important first step in assessing invariance is to examine whether the factor loadings for the same indicator remain the same at each point in time. That is, does $\Lambda_{j1} = \Lambda_{j2} = \cdots = \Lambda_{jT}$ for all j? Failure to reject this hypothesis is evidence that the latent variable has the same impact on the indicators over time.

Suppose that we chose our scaling indicator poorly such that its loading on the latent variable is not invariant over time. In this case we might reject the equality constraints on the factor loadings solely because of the problem with the scaling indicator. We should perform sensitivity checks on this possibility by choosing another scaling indicator and repeating the invariance tests for the factor loadings with the exception that we not constrain the factor loadings for the original scaling variable. If we no longer reject these invariance hypotheses, this is evidence that the scaling indicator was the source of the invariance problem.

Intercepts If we want to extend this not just to the slope but also to the intercepts of the measurement model, we can examine the joint hypothesis of $\Lambda_{j1} = \Lambda_{j2} = \cdots = \Lambda_{jT}$ and $\nu_{j1} = \nu_{j2} = \cdots = \nu_{jT}$ for all j. By adding the test of intercepts, we are checking whether the value of the indicator over time is the same when the latent variable takes a value of zero. This test is important since it reveals whether shifts in the mean of an indicator reflect real changes in the latent variable or just changes in the intercept of the indicator. Having both the factor loadings and intercepts of the multiple indicators unchanging over time signifies a high degree of invariance. Strictly speaking, the means of the latent variable are comparable as long as the scaling indicator has an intercept equal to zero at each time. However, it would be unusual (although possible) to have the intercepts not invariant for all indicators except the scaling indicator. The researcher would need a high degree of confidence in the scaling indicator to make comparisons if intercepts for all other indicators except the scaling indicator are varying over time.

Disturbance Variances of Indicators As a next step in examining invariance, we can look at the variance and covariance parameters in the model. One possibility is to test whether the disturbance variances of the same indicator stay the same over time: $\sigma^2_{\upsilon_j 1} = \sigma^2_{\upsilon_j 2} = \cdots = \sigma^2_{\upsilon_j T}$ for all j. In the common situation of testing the invariance in a hierarchical fashion, we could add these equality restrictions on top of the restrictions on the intercepts and slopes that we described in the last subsection (i.e., $\Lambda_{j1} = \Lambda_{j2} = \cdots = \Lambda_{jT}$ and $\nu_{j1} = \nu_{j2} = \cdots = \nu_{jt}$ for all j). Equal disturbance variances over time suggest a constancy in the variances of the disturbances for the same indicator over time. Violation of this hypothesis might be due to the respondent becoming more familiar with the item and thus lessening the error in response.

Disturbance Variance of Latent Variable So far our tests of invariance have been limited to the measurement model parameters. We also can test aspects of the LCM for the repeated latent variables. Consider Eq. (8.32):

$$\eta_{it} = \alpha_i + \beta_i \lambda_t + \zeta_{it} \tag{8.32}$$

The means of the intercepts and slopes are constant over time, but the variance of the disturbance (ζ_{it}) might change over time. To test this we can introduce the restriction that $\psi_{11} = \psi_{22} = \cdots = \psi_{TT}$. Failure to reject this hypothesis would be consistent with the variance of the omitted influences on η_{it} staying constant over time and would be a further indication of invariance. Indeed, if a model maintains all these restrictions on coefficients, intercepts and variances simultaneously, it would be a highly invariant structure.

Hierarchy of Invariance In Chapter 6 we discussed the many possible hierarchies in which the degree of invariance could be tested across groups. The situation is analogous here except that our emphasis is on invariance over time. Theory or substantive knowledge might dictate that the equality constraints on parameters be introduced in a particular order. But the typical situation does not provide such clear guidance. However, lacking compelling alternatives, a reasonable ordering for testing invariance starts with model form, factor loadings, intercepts, disturbance variances of indicators, and disturbance variances of the latent variables.

A related question is: How much invariance is required? The answer to this question depends on the purpose to which the analysis is put and the assumptions about the quality of the model that a researcher is willing to make. Strictly speaking, all we need for many purposes is to have the factor loadings and the zero intercepts for the scaling indicators to hold over time. If it does, we will maintain consistent estimators of the means of the intercepts and slopes of the LCM and consistent estimators of the other parameters of the model as long as the rest of the model is specified correctly. However, if we find that the intercepts and factor loadings of the other indicators are noticeably different over time, this should raise doubt about the imposed equality of the intercepts and factor loadings of the scaling indicators over time. Or large changes in the disturbance variances for the same indicator over time should also raise suspicions about the quality of the model. In contrast, invariance of factor loadings and intercepts and evidence of only minor differences in the other model parameters increase confidence in the model specification and the ability to accurately assess the trajectory of the latent variable. It is thus important that these issues be closely considered within the context of the particular theoretical research question and empirical data set.

Empirical Example We return to the latent curve model for the repeated latent democracy variable from 1980 to 1986 to illustrate the examination of invariance. Table 8.10 contains the estimated corrected test statistics (Yuan and Bentler, 2000) from the ML estimates of the models that progress from the unconstrained model that we described in an earlier section to the increasingly restrictive forms of the model that test the invariance of the parameters over time. The results of these tests suggest that it is safe to assume that the model form and the factor loadings for the same indicator are the same over the seven-year period. The test of invariance of the intercepts is marginally significant, suggesting that at least one of the intercepts is not equal over time. However, referring to Table 8.8, we see that the unconstrained intercepts for the same democracy index do not exhibit large differences. If we consider these differences substantively unimportant, we could tentatively add the equality constraint for the intercepts to that of the factor loadings. If we then constrain either the disturbance variances for the same indicator or for the latent variable to be equal over time, we do not find statistically significant differences. Thus, most of the evidence supports the invariance of model parameters over time. Only the intercepts give a slight indication that they might vary.[22]

8.2.4 Dichotomous and Ordinal Multiple Indicators

The prior subsections assumed that the multiple indicators were continuous variables. In this section we discuss the issue of invariance when one or more of the indicators are ordinal or dichotomous. When we had continuous indicators, our starting point for examining invariance was the factor loadings and intercepts. With ordinal variables, we recommend that researchers start by examining the invariance of the thresholds across time. Suppose that y^*_{jit} is the continuous variable that

[22]Given the order in which we presented the material of this chapter, we determined the LCM for the democracy example without any invariance constraints and then turned to the invariance restrictions for this model. An alternative way of proceeding is to examine invariance first while treating the model as a CFA without any latent curve structure and then estimating the LCM. There is a possibility that some results might vary, depending on the order of analysis; we are unaware of any analytic or empirical research that demonstrates this.

Table 8.10 Tests of Invariance for Multiple-Indicator Latent Curve Model for Liberal Democracy ($N = 166$)

Models	Corrected T_{ML}	df	Prob.	χ^2 Diff.	df	Prob.
No equality constraints	35.88[a]	16	0.00			
$\Lambda_{j1} = \Lambda_{j2} = \cdots = \Lambda_{jT}$	44.65	20	0.00	8.08	4	0.089
$+\nu_{j1} = \nu_{j2} = \cdots = \nu_{jT}$	56.89	24	0.00	18.78	4	0.001
$+\sigma^2_{\upsilon_{j1}} = \sigma^2_{\upsilon_{j2}} = \cdots = \sigma^2_{\upsilon_{jT}}$	67.93	30	0.00	9.78	6	0.134
$+\psi_{11} = \psi_{22} = \cdots = \psi_{TT}$	71.60	32	0.00	4.16	2	0.125

[a]Corrected Yuan–Bentler T_{ML}.

underlies the ordinal indicator y_{jit} such that

$$y_{jit} = c_{jt} \qquad \text{when} \quad \tau_{c_j-1,t} < y^*_{jit} \leq \tau_{c_j,t} \tag{8.33}$$

where $c_{jt} = 1, 2, 3, ldots, C_{jt}$ the total number of ordered categories, $\tau_{c_j-1,t}$, $\tau_{c_j,t}$ are the lower and upper thresholds for category c_{jt} with $\tau_{0_j,t} = -\infty$ and $\tau_{C_j,t} = +\infty$, and the thresholds are ordered from lowest to highest. Equation (8.33) connects the jth ordinal variable to its underlying continuous counterpart such that as the continuous variable lies between two thresholds, the ordinal variable will register in the category that falls between those thresholds.

A starting point for an invariance hypothesis for ordinal variables is to test $\tau_{c_{j1}} = \tau_{c_{j2}} = \cdots = \tau_{c_{jT}}$ for each j and for all c. If the first two thresholds are set to 0 and 1, as described previously, the first two thresholds are automatically set equal. However, with ordinal variables with four or more categories, we can test the equality of the remaining thresholds. If we fail to reject the null hypothesis that the thresholds for the same variable over time are equal, we can explore other invariance hypotheses. Indeed, we can follow the same steps suggested for continuous multiple indicators in examining scaling and the invariances of factor loadings, intercepts, disturbance variances for indicators, and disturbance variances of the latent variables.

What if one or more of the thresholds for an indicator are not equal over time? A strict interpretation of invariance would conclude that the measures are not invariant and would stop the search for the degree of invariance. Furthermore, the LCM would not be estimated. A more exploratory approach would free those thresholds that appear to be unequal and would continue the examination of invariance by moving along the recommended hierarchy of invariance. Or if it is just one or two of the multiple indicators that has changing thresholds, those faulty indicators might be removed from the analysis, and invariance could be examined for the remaining indicators.

8.2.5 Conditional Models

The preceding discussion has concentrated on invariance in unconditional models with multiple indicators. Invariance in conditional models extends the range of parameters for which invariance is checked, but it does not create any special problems. Generally, invariance of the coefficients of the covariates that influence the latent curve parameters is checked for invariance prior to checking the invariance of the variances or covariances of the covariates. Furthermore, in a conditional model, a researcher might place the test of the covariates coefficients higher in the hierarchy than the tests of the disturbance variances for the indicators or latent variables. The latter point raises a broader issue. We have given a *suggested* hierarchy in the checking of invariance. Substantive concerns or specific hypotheses might lead researchers to a different ordering in which the invariance is checked. We would encourage researchers to feel free to modify the order of invariance that we have described.

8.2.6 Summary

In this section we discussed how we can build LCMs for repeated latent variables. We assumed that a researcher has multiple indicators for each latent variable. Much of what we discussed was the same as the previous material when the focus was repeated observed variables. The key differences were the use of repeated latent variables rather than repeated observed variables and the added complexity that accompanies multiple-indicator latent variable models. SEMs are ideally suited for handling these complexities, and the end result is a LCM in which we can separate the error in indicators from the disturbance error that leads to departures from the latent curve trajectory for the latent variable. We also showed how to examine issues of invariance in the measurement model and the trajectory over time. Finally, we briefly described the extension of these models when some of the indicators are ordinal or dichotomous.

8.3 LATENT COVARIATES

In the preceding section we demonstrated LCMs for repeated latent variables. The addition of observed covariates to predict the random trajectory parameters was straightforward. Similarly, we can take advantage of the ability of SEMs to handle latent exogenous variables to address the situation where there are measurement errors in the time-invariant covariates and multiple indicators to measure them. The level 1 model in this situation would be the same as before,

$$\eta_{it} = \alpha_i + \beta_i \lambda_t + \zeta_{it} \tag{8.34}$$

Although we have written this as a repeated latent variable model, the following discussion applies equally to the latent curve models fitted to manifest repeated measures described in prior chapters. The measurement equations for the repeated latent variable are

$$y_{jit} = \nu_{jt} + \Lambda_{jt}\eta_{it} + \upsilon_{jit} \tag{8.35}$$

where all terms and assumptions are the same as before.

The difference comes in the definition of the level 2 model, that is,

$$\alpha_i = \mu_\alpha + \gamma_{\alpha\xi_1}\xi_{i1} + \gamma_{\alpha\xi_2}\xi_{i2} + \cdots + \gamma_{\alpha\xi_Q}\xi_{iQ} + \zeta_{\alpha i} \quad (8.36)$$

$$\beta_i = \mu_\beta + \gamma_{\beta\xi_1}\xi_{i1} + \gamma_{\beta\xi_2}\xi_{i2} + \cdots + \gamma_{\beta\xi_Q}\xi_{iQ} + \zeta_{\beta i} \quad (8.37)$$

where $\xi_{i1}, \xi_{i2}, \ldots, \xi_{iQ}$ are the Q latent time-invariant covariates that are uncorrelated with the disturbances of all equations and $\gamma_{\alpha\xi_q}$ and $\gamma_{\beta\xi_q}$ are regression coefficients giving ξ_{iq}'s effects on α_i and β_i, respectively. To complete the model with the latent covariates, we need to construct a measurement model just like was done for the repeated latent variables in Eq. (8.35). The measurement model for the latent time-invariant covariates is

$$\mathbf{x}_i = \boldsymbol{\nu}_{xi} + \boldsymbol{\Lambda}_x\boldsymbol{\xi}_i + \boldsymbol{\delta}_i \quad (8.38)$$

where $\boldsymbol{\nu}_{xi}$ is a vector of intercepts for the multiple indicators in $\mathbf{x}_i$, $\boldsymbol{\Lambda}_x$ is a matrix of factor loadings that gives the coefficients of $\boldsymbol{\xi}_i$ on $\mathbf{x}_i$, $\boldsymbol{\xi}_i$ is the vector of all the latent covariates (the ξ_{iq}'s) and $\boldsymbol{\delta}_i$ is the vector of unique components ("disturbances") that are part of $\mathbf{x}_i$ not explained by $\boldsymbol{\xi}_i$. The $\boldsymbol{\delta}_i$ has a mean of zero and is uncorrelated with $\boldsymbol{\xi}_i$ and all other disturbances. We index these variables by the subscript i to indicate that they can differ over cases but are not changing over time. If a time-invariant covariate has no or negligible error, its unique component and intercept are set to zero ($\nu_{ji} = 0$; $\delta_{ji} = 0$) and its factor loading is set to 1. Thus, a researcher can have a mixture of observed and latent covariates.

For example, suppose that we are interested in the trajectory of life satisfaction following retirement. We have a latent covariate of status of position held during the last year of work and two observed covariates of age and gender. Figure 8.5 is the path diagram of a LCM with the latent covariate and two observed variable covariates influencing the random intercept and random slope of a repeated life satisfaction latent variable. To avoid crowding the diagram, the measurement model

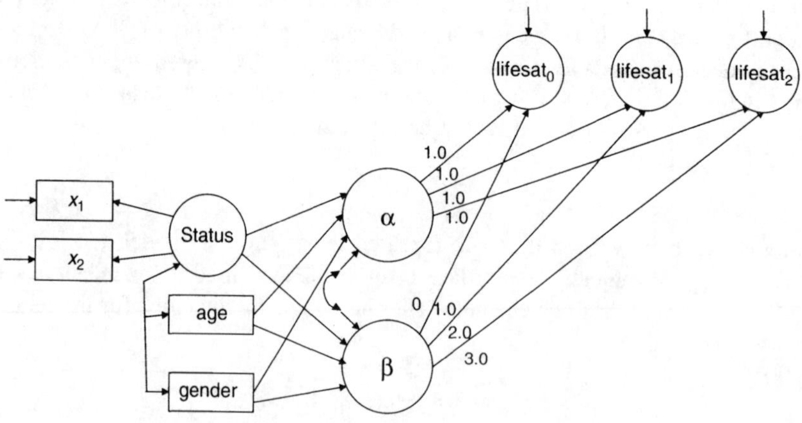

FIGURE 8.5 Path diagram of LCM with two observed and one latent covariate.

of the repeated latent variable is not shown. The model structure described above can incorporate this situation such that we would estimate the impact of the latent position status variable on the parameters of the life satisfaction trajectory while taking account of measurement error.

In Chapter 7 we considered time-varying covariates that had direct influences on the repeated measures net of the effects of the latent curve model. It treated only *observed* time-varying covariates. Using a SEM approach, there is no reason not to allow *latent* time-varying covariates. To do so, requires a modification of the level 1 equation to

$$\eta_{it} = \alpha_i + \beta_i \lambda_t + \gamma_t \xi_{it} + \epsilon_{it} \tag{8.39}$$

where ξ_{it} is a latent time-varying covariate, γ_t is its coefficient, and η_{it} is the repeated latent variable. The repeated latent variable is influenced by the random intercept, random slope, and the latent time-varying covariate. In Chapter 7 we used w_{it} to represent the time-varying covariates that entered the level 1 equation. Now that we are allowing measurement error and multiple indicators of the time-varying covariates, we use w_{kit} where the k indexes the multiple indicators. Equation (8.38) can still serve as the measurement model for the latent time-varying covariate if we expand what is included in its vectors. Now the elements of $\boldsymbol{\xi}_i$ include both the time-invariant and time-varying latent variables, and the elements of $\mathbf{x}_i$ include both the multiple indicators of the time-invariant latent variables and the time-varying multiple indicators, w_{kit}. Figure 8.6 is a path diagram example of a repeated latent

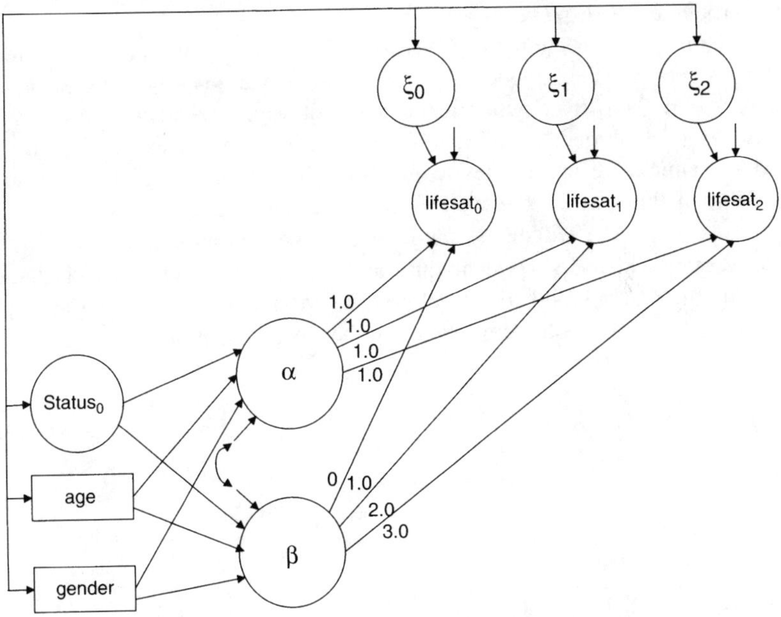

FIGURE 8.6 Path diagram of LCM with time-invariant and time-varying latent covariate, each measured with two indicators.

variable with a single time-invariant latent variable measured with two indicators (x_1 and x_2) and a time-varying latent variable measured with two indicators at each wave. In this model, none of the measurement models are shown to simplify the diagram.

In sum, in this section we have shown that *latent* time-invariant or *latent* time-varying covariates are permissible in the SEM approach to latent curve models. This added flexibility permits researchers to construct more realistic models where either the repeated variable or the covariates are permitted to have measurement error and multiple indicators.

8.4 CONCLUSIONS

In this chapter we have discussed two extensions of the LCM that are useful in practice. These were the use of dichotomous or ordinal repeated measures and the modeling of repeated latent variables with multiple indicators. The incorporation of the LCM into the SEM framework opens up many more extensions than we have discussed. For example, the SEM literature has a considerable amount of discussion of methods of assessing model fit (e.g., Bollen, 1989b, pp. 256–281; Bollen and Long, 1993; Hu and Bentler, 1999) that carries over to LCMs. There are SEM methods for calculating statistical power of the chi square test that apply here as well (e.g., Satorra and Saris, 1985; Matsueda and Bielby, 1986; MacCallum et al., 1996; Muthén and Curran, 1997). Outliers and influential cases detection techniques developed for SEM apply to LCMs (e.g., Bollen and Arminger, 1991; Cadigan, 1995; Arbuckle, 1997, pp. 238–40).

Although this connection of LCMs to SEM opens many doors, there remain challenges. For instance, modeling repeated count variables or repeated nominal variables with more than two categories is not well studied with current SEM approaches (see Skrondal and Rabe-Hesketh, 2004, for an exception). In these situations, alternative approaches to growth curve modeling might prove useful until SEM methods tackle these problems. Although our focus has been on SEM approaches to LCMs, we should mention that there is an exciting trend toward the convergence of the SEM techniques and multilevel techniques such that each approach stimulates the others' development. We anticipate that both approaches will benefit from the continuing interaction between practitioners of each.

References

Aber, M. S., & McArdle, J. J. (1991). Latent growth curve approaches to modeling the development of competence. In M. Chandler & M. Chapman (Eds.), *Criteria for Competence* (pp. 231–258). Hillsdale, NJ: Lawrence Earlbaum Associates.

Adam, J. (1978). Sequential strategies and the separation of age, cohort, and time-of-measurement contributions to developmental data. *Psychological Bulletin, 85*, 1309–1316.

Agresti, A. (1990). *Categorical Data Analysis.* New York: Wiley.

Aiken, L. S., & West, S. G. (1991). *Multiple Regression: Testing and Interpreting Interactions.* Newbury Park, CA: Sage.

Akaike, H. (1973). Information theory and an extension of the maximum likelihood principle. In B. N. Petrov & F. Csaki (Eds.), *Proceedings of the 2nd International Symposium on Information Theory* (pp. 267–281). Budapest: Akademiai Kiado.

Akaike, H. (1987). Factor analysis and AIC. *Psychometrika, 52*, 317–332.

Allison, P. D. (2002). *Missing Data.* Thousand Oaks, CA: Sage.

Amemiya, T., & Anderson, T. W. (1990). Asymptotic chi-square tests for a large class of factor analysis models. *Annals of Statistics, 18*, 1453–1463.

Anderson, J., & Gerbing, D. W. (1984). The effects of sampling error on convergence, improper solutions and goodness-of-fit indices for maximum likelihood confirmatory factor analysis. *Psychometrika, 49*, 155–173.

Anderson, T. W. (1960). Some stochastic process models for intelligence test scores. In K. J. Arrow, S. Karlin, & P. Suppes (Eds.), *Mathematical Methods in the Social Sciences* (pp. 205–220). Stanford, CA: Stanford University Press.

Anderson, T. W., & Rubin, H. (1956). Statistical inference in factor analysis. *Proceedings of the 3rd Berkeley Symposium for Mathematical Statistics Problems, 5*, 111–150.

Arbuckle, J. L. (1996). Full information estimation in the presence of incomplete data. In G. A. Marcoulides & R. E. Schumacker (Eds.), *Advanced Structural Equation Modeling* (pp. 243–278). Hillsdale, NJ: Lawrence Erlbaum Associates.

Arbuckle, J. L. (1997). *AMOS 3.6 User's Guide*, 3.6 ed. Chicago: SmallWaters Corp.

Arminger, G., & Schoenberg, R. (1989). Pseudo maximum likelihood estimation and a test for misspecification in mean and covariance structure models. *Psychometrika, 54*, 409–425.

Arminger, G., & Stein, P. (1997). Finite mixtures of covariance structure models with regressors. *Sociological Methods and Research, 26*, 148–182.

Arminger, G., Stein, P., & Wittenberg, J. (1999). Mixtures of conditional mean- and covariance-structure models. *Psychometrika, 64*(4), 475–494.

Baker, G. A. (1954). Factor analysis of relative growth. *Growth, 18*, 137–143.

Baltes, P. B. (1968). Longitudinal and cross-sectional sequences in the study of age and generation effects. *Human Development, 11*, 145–171.

Bartlett, M. S. (1937). The statistical conception of mental factors. *British Journal of Psychology, 28*, 97–104.

Bauer, D. J. (2003). Estimating multilevel linear models as structural equation models. *Journal of Educational and Behavioral Statistics, 28*, 135–167.

Bauer, D. J., & Curran, P. J. (2003). Distributional assumptions of growth mixture models: implications for overextraction of latent trajectory classes. *Psychological Methods, 8*, 338–363.

Bauer, D. J., & Curran, P. J. (2004). The integration of continuous and discrete latent variable models: potential problems and promising opportunities. *Psychological Methods, 9*(1), 3–29.

Bentler, P. M. (1990). Comparative fit indexes in structural models. *Psychometrika, 107*(2), 238–246.

Biesanz, J. C., & Bollen, K. A. (2003). A common framework for factor score estimates. Manuscript under review.

Biesanz, J. C., Deeb-Sossa, N., Papadakis, A. M., Bollen, K. A., & Curran, P. J. (2004). The role of coding time in estimating and interpreting growth curve models. *Psychological Methods, 9*(1), 30–52.

Bishop, Y. M. M., Fienberg, S. E., & Holland., P. W. (1975). *Discrete Multivariate Analysis: Theory and Practice*. Cambridge, MA: MIT Press.

Bollen, K. A. (1987). Total, direct, and indirect effects in structural equation models. In C. C. Clogg (Ed.), *Sociological Methodology, 1987* (pp. 37–69). Washington, DC: American Sociological Association.

Bollen, K. A. (1989a). A new incremental fit index for general structural equation models. *Sociological Methods and Research, 17*, 303–316.

Bollen, K. A. (1989b). *Structural Equations with Latent Variables*. New York: Wiley.

Bollen, K. A. (1990). Overall fit in covariance structure models: two types of sample size effects. *Psychological Bulletin, 107*(2), 256–259.

Bollen, K. A. (1998). *Cross National Indicators of Liberal Democracy, 1950–1990* Ann Arbor, Michigan: Interuniversity Consortium for Political and Social Research Study 2532.

Bollen, K. A. (2002). Latent variables in psychology and the social sciences. *Annual Review of Psychology, 53*, 605–634.

Bollen, K. A. & Curran, P. J. (2004). Autoregressive latent trajectory (ALT) models: a synthesis of two traditions. *Sociological Methods and Research, 32*, 336–383.

Bollen, K. A., & Arminger, G. (1991). Observational residuals in factor analysis and structural equation models. *Sociological Methodology, 21*, 235–262.

Bollen, K. A., & Long, J. S. (Eds.) (1993a). *Testing Structural Equation Models*. Newbury Park, CA: Sage.

Bollen K. A. & Long J. S. (1993b). Introduction. In K.A. Bollen and J.S. Long (Eds.), *Testing Structural Equation Models*. Newbury Park, CA: Sage.

Bollen, K. A., & Stine, R. (1990). Direct and indirect effects: classical and bootstrap estimates of variability. *Sociological Methodology, 20*, 115–140.

Bollen, K. A., & Stine, R. A. (1992). Bootstrapping goodness-of-fit measures in structural equation models. *Sociological Methods and Research, 21*, 205–229.

Boomsma, A. (1982). The robustness of LISREL against small sample sizes in factor analysis models. In K. G. Joreskog & H. Wold (Eds.), *Systems Under Indirect Observation, Part I* (pp. 149–173). Amsterdam: North-Holland.

Browne, M. W. (1984). Asymptotically distribution-free methods for the analysis of covariance structures. *British Journal of Mathematical and Statistical Psychology, 37*(1), 62–83.

Browne, M. W. (1987). Robustness of statistical inference in factor analysis and related models. *Biometrika, 74*, 375–384.

Browne, M. W., & Du Toit, S. H. C. (1991). Models for learning data. In L. M. Collins and J. L. Horn (Eds.), *Best Methods for the Analysis of Change* (pp. 47–68). Washington, DC: American Psychological Association.

Browne, M. W. (1993). Structured latent curve models. In C. M. Cuadras & C. R. Rao (Eds.), *Multivariate Analysis: Future Directions*, Vol. 2 (pp. 171–197). New York: Elsevier Science.

Browne, M. W., & Cudeck, R. (1993). Alternative ways of assessing model fit. In K. A. Bollen & J. S. Long (Eds.), *Testing Structural Equation Models*. Newbury Park, CA: Sage.

Browne, M. W., & du Toit, S. H. C. (1991). Models for learning data. In L. M. Collins & J. L. Horn (Eds.), *Best Methods for the Analysis of Change* (pp. 47–68). Washington, DC: American Psychological Association.

Bryk, A. S., & Raudenbush, S. W. (1992). *Hierarchical Linear Models: Applications and Data Analysis Methods*. Newbury Park, CA: Sage.

Cadigan, N. G. (1995). Local influence in structural equations models. *Structural Equation Modeling, 2*(1), 13–30.

Carrig, M., Wirth, R. J., & Curran, P. J. (2004). A SAS macro for estimating and visualizing individual growth curves. *Structural Equation Modeling, 11*, 132–149.

Chen, F., Bollen, K. A., Paxton, P., Curran, P. J., & Kirby, J. B. (2001). Improper solutions in structural equation models. *Sociological Methods and Research, 29*(4), 468–508.

Cohen, J. (1978). Partialled products are interactions; partialled powers are curve components. *Psychological Bulletin, 85*, 858–866.

Cohen, J., & Cohen, P. (1983). *Applied Multiple Regression/Correlation Analysis for the Behavioral Sciences*, 2nd ed. Hillsdale, NJ: Lawrence Erlbaum Associates.

Costner, H. L., & Schoenberg, R. (1973). Diagnosing indicator ills in multiple indicator models. In A. S. Goldberger & O. D. Duncan (Eds.), *Structural Equation Models in the Social Sciences* (pp. 167–200). New York: Seminar Press.

Curran, P. J., Stice, E. & Chassin, L. (1997). The relation between adolescent and peer alcohol use: A longitudinal random coefficients model. *Journal of Consulting and Clinical Psychology, 65*, 130–140.

Curran, P. J. (1997). Comparing three modern approaches to longitudinal data analysis: an examination of a single developmental sample. Presented at the symposium conducted at the meeting of the Society for Research on Child Development (Patrick J. Curran, Chair), April, Washington, DC.

Curran, P. J., & Bollen, K. A. (2001). The best of both worlds: combining autoregressive and latent curve models. In L. M. Collins & A. G. Sayer (Eds.), *New Methods for the Analysis of Change* (pp. 107–135). Washington, DC: American Psychological Association.

Curran, P. J., & Willoughby, M. T. (2003). Reconciling theoretical and statistical models of developmental processes. *Development and Psychopathology, 15*, 581–612.

Curran, P. J., Muthén, B. O., & Harford, T. C. (1998). The influence of changes in marital status on developmental trajectories of alcohol use in young adults. *Journal of Studies on Alcohol, 59*, 647–658.

Curran, P. J., Bauer, D. J., & Willoughby, M. T. (2004). Testing and probing main effects and interactions in latent curve analysis. *Psychological Methods, 9*, 220–237.

Curran, P. J., Bauer, D. J., & Willoughby, M. T. (in press). Testing and probing interactions in hierarchical linear growth models. In C. S. Bergeman & S. M. Boker (Eds.), *Methodological Issues in Aging Research*, Vol. 1. Mahwah, NJ: Lawrence Erlbaum Associates.

Davis, W. R. (1993). The FC1 rule of identification for confirmatory factor analysis: a general sufficient condition. *Sociological Methods and Research, 21*(4), 403–437.

du Toit, S. H. C., & Cudeck, R. (2001). The analysis of nonlinear random coefficient regression models with LISREL using constraints. In R. Cudeck, S. du Toit, & D. Sörbom (Eds.), *Structural Equation Modeling: Present and Future* (pp. 259–278). Lincolnwood, IL: Scientific Software.

Duncan, S. C., & Duncan, T. E. (1996). A multivariate latent growth curve analysis of adolescent substance use. *Structural Equation Modeling, 3*(4), 323–347.

Enders, C. K. (2001). The impact of nonnormality on full information maximum-likelihood estimation for structural equation models with missing data. *Psychological Methods, 6*(4), 352–370.

Flora, D. B. (2002). Evaluation of categorical variable methodology for confirmatory factor analysis with liker t-type data. Unpublished dissertation. University of North Carolina at Chapel Hill.

Flora, D. B., & Curran, P. J. (2004). An empirical evaluation of alternative methods of estimation for confirmatory factor analysis with ordinal data. *Psychological Methods, 9*, 466–491.

Gerbing, D. W., & Anderson, J. C. (1993). Monte Carlo evaluations of goodness-of-fit indices for structural equation models. In K. A. Bollen & J. S. Long (Eds.), *Testing Structural Equation Models* (pp. 40–65). Newbury Park, CA: Sage.

Glenn, N. D. (1976). Cohort analysts' futile quest: statistical attempts to separate age, period and cohort effects. *American Sociological Review, 41*(5), 900–904.

Gold, M. S., Bentler, P. M., & Kim, K. H. (2003). A comparison of maximum-likelihood and asymptotically distribution-free methods of treating incomplete nonnormal data. *Structural Equation Modeling, 10*(1), 47–79.

Gompertz, B. (1820). A sketch of an analysis and notation applicable to the estimation of the value of life contingencies. *Philosophical Transactions of the Royal Society, 110*, 214–294.

Gompertz, B. (1825). On the nature of the function expressive of the law of human mortality. *Philosophical Transactions of the Royal Society, 115*, 513–583.

Graham, J. W. (2003). Adding missing-data relevant variables to FIML-based structural equation models. *Structural Equation Modeling, 10*, 80–100.

Griliches, Z. (1957). Hybrid corn: an exploration in the economics of technical change. *Econometrica, 48*, 501–522.

Gurr, T. R. (1990). *Polity II: Political Structures and Regime Change, 1800–1986*, Ann Arbor, Michigan. Interuniversity Consortium for Political and Social Research Study 9263.

Haitovsky, Y. (1968). Missing data in regression analysis. *Journal of the Royal Statistical Society, Series B, 30*, 67–82.

Harman, H. H. (1967). *Modern Factor Analysis*, 2nd ed. Chicago: University of Chicago Press.

Heckman, J. T. (1979). Sample selection bias as a specification error. *Econometrica, 45*:153–161.

Heise, D. R. (1969). Separating reliability and stability in test–retest correlation. *American Sociological Review, 34*(1), 93–101.

Hipp, J. R., & Bauer, D. J. (in press). Local solutions in the estimation of growth mixture models. *Psychological Methods.*

Hipp, J. R., Bauer, D. J., Curran, P. J., & Bollen, K. A. (2004). Crimes of opportunity or crimes of emotion: testing two explanations of seasonal change in crime. *Social Forces, 82*(4), 1333–1372.

Hu, L.-T., & Bentler, P. M. (1999). Cutoff criteria for fit indexes in covariance structure analysis: conventional criteria versus new alternatives. *Structural Equation Modeling, 6*(1), 1–55.

Humphreys, L. G. (1960). Investigations of the simplex. *Psychometrika, 25*, 313–323.

Jedidi, K., Jagpal, H. S., & DeSarbo, W. S. (1994). Latent class structural equation models for marketing research. Unpublished paper.

Jedidi, K., Jagpal, H. S., & DeSarbo, W. S. (1997). Finite-mixture structural equation models for response-based segmentation and unobserved heterogeneity. *Marketing Science, 16*(1), 39–59.

Johnson, P. O., & Neyman, J. (1936). Tests of certain linear hypotheses and their applications to some educational problems. *Statistical Research Memoirs, 1*, 57–93.

Johnson, J. (1984). *Econometric Methods*. New York: McGraw-Hill.

Jones, B. L., Nagin, D. S., & Roeder, K. (2001). A SAS procedure based on mixture models for estimating developmental trajectories. *Sociological Methods and Research, 29*, 374–393.

Jones, M. P. (1996). Indicator and stratification methods for missing explanatory variables in multiple linear regression. *Journal of the American Statistical Association, 91*(433), 222–230.

Jöreskog, K. G. (1969). A general approach to confirmatory maximum likelihood factor analysis. *Psychometrika, 34*(2), 183–202.

Jöreskog, K. G. (1970). A general method for analysis of covariance structures. *Biometrika, 57*, 239–251.

Jöreskog, K. G. (1979). Statistical models and methods for analysis of longitudinal data. In K. G. Jöreskog & D. Sörbom (Eds.), *Advances in Factor Analysis and Structural Equation Models*. Cambridge, MA: Abt Books.

Jöreskog, K. G. (1994). On the estimation of polychoric correlations and their asymptotic covariance-matrix. *Psychometrika, 59*(3), 381–389.

Jöreskog, K. G. (2001a). *Analysis of Ordinal Variables*, Note 2, *Cross-Sectional Data*. pp. 1–26. www.ssicentral.com/lisrel/corner.htm.

Jöreskog, K. G. (2001b). *Analysis of Ordinal Variables*, Note 3, *Longitudinal Data*. pp. 1–26. www.ssicentral.com/lisrel/corner.htm.

Jöreskog, K. G., & Sörbom, D. (1981). *LISREL V: Analysis of Linear Structural Relationships by Maximum Likelihood*. Chicago: National Educational Resources.

Jöreskog, K. G., Sörbom, D., du Toit, S., & du Toit, M. (1999). *LISREL 8: New Statistical Features*. Chicago: Scientific Software.

Kenny, D. A., & Zautra, A. (1995). The trait–state–error model for multiwave data. *Journal of Consulting and Clinical Psychology, 63*, 52–59.

Lawley, D. N. (1940). The estimation of factor loadings by the method of maximum likelihood. *Proceedings of the Royal Society of Edinburgh, 60*, 64–82.

Lazarsfeld, P. F., & Henry, N. W. (1968). *Latent Structure Analysis*. Boston: Houghton Mifflin.

Lee, S.-Y., Poon, W. Y., & Bentler, P. M. (1990). Full maximum likelihood analysis of structural equation models with polytomous variables. *Statistics and Probability Letters, 9*, 91–97.

Li, F., Duncan, T. E., Duncan, S. C., & Acock, A. (2001). Latent growth modeling of longitudinal data: a finite growth mixture modeling approach. *Structural Equation Modeling, 8*(4), 493–530.

Li, K.-H., Meng, X.-L., Raghunathan, T. E., & Rubin, D. B. (1991). Significance levels from repeated *P*-values with multiple-inputed data. *Statistica Sinica, 1*, 65–92.

Little, R. J. A. (1988). A test of missing completely at random for multivariate data with missing values. *Journal of the American Statistical Association, 83*, 1198–1202.

Little, R. (1992). Regression with missing X's: A review. *Journal of the American Statistical Association, 87*, 1227–1237.

Little, R. J. A., & Rubin, D. B. (2002). *Statistical Analysis with Missing Data*, 2nd ed. Hoboken, NJ: Wiley.

Long, J. S. (1997). *Regression Models for Categorical and Limited Dependent Variables*. Newbury Park, CA: Sage.

MacCallum, R. C., Browne, M. W., & Sugawara, H. M. (1996). Power analysis and determination of sample size for covariance structure modeling. *Psychological Methods, 1*(2), 130–149.

MacCallum, R. C., Kim, C., Malarkey, W. B., & Kiecolt-Glaser, J. K. (1997). Studying multivariate change using multilevel models and latent curve models. *Multivariate Behavioral Research, 32*(3), 215–253.

Mardia, K. V. (1970). Measures of multivariate skewness and kurtosis with applications. *Biometrika, 57*, 519–530.

Mardia, K. V. (1974). Applications of some measures of multivariate skewness and kurtosis in testing normality and robustness studies. *Sankhya, series B, 36*, 115–128.

Matsueda, R. L., & Bielby, W. T. (1986). Statistical power in covariance structure models. In N. B. Tuma (Ed.), *Sociological Methodology*, Vol. 16 (pp. 120–158). Washington, DC: American Sociological Association.

McArdle, J. J. (1988). Dynamic but structural equation modeling of repeated measures data. In J. R. Nesselroade & R. B. Cattell (Eds.), *The Handbook of Multivariate Experimental Psychology*, 2nd ed. (pp. 561–614). New York: Plenum Press.

McArdle, J. J. (1989). Structural modeling experiments using multiple growth functions. In R. Kanfer, P. Ackerman, & R. Cudeck (Eds.), *Abilities, Motivation, and Methodology: The Minnesota Symposium on Learning and Individual Differences* (pp. 71–117). HIllsdale, NJ: Lawrence Erlbaum Associates.

McArdle, J. J. (2001). A latent difference score approach to longitudinal dynamic structural analyses. In R. Cudeck, S. du Toit, & D. Sorbom (Eds.), *Structural Equation Modeling: Present and Future* (pp. 342–380). Lincolnwood, IL: Scientific Software.

McArdle, J. J., & Hamagami, F. (2001). Latent difference score structural models for linear dynamic analyses with incomplete longitudinal data. In L. M. Collins & A. G. Sayer (Eds.), *New Methods for the Analysis of Change* (pp. 139–175). Washington, DC: American Psychological Association.

McDonald, R. P., & Marsh, H. W. (1990). Choosing a multivariate model: noncentrality and goodness of fit. *Psychological Bulletin, 107*(2), 247–255.

McDonald, R. P., & Mulaik, S. A. (1979). Determinacy of common factors: a nontechnical review. *Psychological Bulletin, 86*(2), 297–306.

McLachlan, G., & Basford, K. E. (1988). *Mixture Models: Inference and Applications to Clustering*. New York: Marcel Dekker.

Mehta, P. D., & West, S. G. (2000). Putting the individual back into individual growth curves. *Psychological Methods, 5*(1), 23–43.

Mehta, P. D., Neale, M. C., & Flay, B. R. (2004). Squeezing interval change from ordinal panel data: latent growth curves with ordinal outcomes. *Psychological Methods, 9*(3), 301–333.

Meredith, W. (1993). Measurement invariance, factor analysis and factorial invariance. *Psychometrika, 58*, 525–543.

Meredith, W., & Tisak, J. (1984). On "Tuckerizing" curves. Presented at the annual meeting of the Psychometric Society, Santa Barbara, CA.

Meredith, W., & Tisak, J. (1990). Latent curve analysis. *Psychometrika, 55*(1), 107–122.

Muthén, B. O. (2001). Latent variable mixture modeling. In G. A. Marcoulides & R. E. Schumacker (Eds.), *New Developments and Techniques in Structural Equation Modeling* (pp. 1–33). Mahwah, NJ: Lawrence Erlbaum Associates.

Muthén, B. O., & Curran, P. J. (1997). General longitudinal modeling of individual differences in experimental designs: a latent variable framework for analysis and power estimation. *Psychological Methods, 2*(4), 371–402.

Muthén, B. O., & Jöreskog, K. G. (1983). Selectivity problems in quasi-experimental studies. *Evaluation Review, 7*(2), 139–174.

Muthén, L. K. & Muthén, B. O. (2000). *Mplus User's Guide*. Second Edition. Los Angeles, CA: Muthén & Muthén.

Muthén, B. O., & Muthén, L. K. (2001). *MPlus: Statistical Analysis with Latent Variables User's Guide*, 2nd ed. Los Angeles: Stat Model.

Muthén, B. O., & Satorra, A. (1995). Complex sample data in structural equation modeling. *Sociological Methodology, 25*, 267–316.

Muthén, B. O., & Shedden, K. (1999). Finite mixture modeling with mixture outcomes using the EM algorithm. *Biometrics, 55*, 463–469.

Muthén, B. O., du Toit, S. H. C., & Spisic, D. (1997). Robust inference using weighted least squares and quadratic estimating equations in latent variable modeling with categorical and continuous outcomes. Unpublished paper.

Nagin, D. S. (1999). Analyzing developmental trajectories: a semiparametric, group-based approach. *Psychological Methods, 4*(2), 139–157.

Nagin, D. S. (2005). *Group-Based Models of Development*. Cambridge, MA: Harvard University Press.

Nagin, D., & Tremblay, R. E. (2001). Parental and early childhood predictors of persistent physical aggression in boys from kindergarten to high school. *Archives of General Psychiatry, 58*, 389–394.

Neale, M. C., Boker, S. M., Xie, G., & Maes, H. H. (2002). *Mx: Statistical Modeling*, 6th ed. Richmond, VA: Virginia Commonwealth University, Department of Psychiatry.

Olsson, U. (1979). Maximum likelihood estimation of the polychoric correlation coefficient. *Psychometrika, 44*(4), 443–460.

Olsson, U., Drasgow, F., & Dorans, N. J. (1982). The polyserial correlation coefficient. *Psychometrika, 47*, 337–347.

Pearl, R. (1924). *Studies in Human Biology*. Baltimore, MD: Williams & Wilkins.

Pearl, R. (1925). *The Biology of Population Growth*. New York: Alfred A. Knopf.

Pothoff, R. F. (1964). On the Johnson-Neyman technique and some extensions thereof. *Psychometrika, 29*, 241–256.

Quetelet, L. A. J. (1835). Sur l'homme et le développement de ses facultés ou essai de physique sociale. In E. Boyd, B. S. Savara, & J. F. Schilke (Eds.) (1980), *Origins of the Study of Human Growth* (pp. 317–332). Portland, OR: University of Oregon Health Sciences Center Foundation.

Quiroga, A.M. (1992). Studies of the polychoric correlation and other correlation measures for ordinal variables. Unpublished doctoral dissertation, University of Uppsala.

Raftery, A. E. (1993). Bayesian model selection in structural equation models. In K. A. Bollen & J. S. Long (Eds.), *Testing Structural Equation Models* (pp. 163–180). Newbury Park, CA: Sage.

Raftery, A. E. (1995). Bayesian model selection in social research. *Sociological Methodology, 25*, 111–163.

Rao, C. R. (1958). Some statistical methods for comparison of growth curves. *Biometrika, 51*, 83–90.

Raudenbush, S. W. (2001). Toward a coherent framework for comparing trajectories of individual change. In L. M. Collins & A. G. Sayer (Eds.), *New Methods for the Analysis of Change* (pp. 35–64). Washington, DC: American Psychological Association.

Reed, L. J. (1921). On the correlation between any two functions and its application to the general case of spurious correlation. *Journal of the Washington Academy of Sciences, 11*, 449–455.

Reed, L. J., & Love, A. G. (1932). Biometric studies on U.S. army officers: somatological norms, correlations, and changes with age. *Human Biology, 4*, 509–524.

Reed, L. J., & Pearl, R. (1927). On the summation of logistic curves. *Journal of the Royal Statistical Society, 90*, 729–746.

Robertson, T. B. (1908). On the normal rate of growth of an individual, and its biochemical significance. *Archiv für Entwicklungsmechanik der Organismen, 25*, 581–614.

Rogosa, D. R. (1980). Comparing nonparallel regression lines. *Psychological Bulletin, 88*, 307–321.

Rogosa, D. R. (1981). On the relationship between the Johnson–Neyman region of significance and statistical tests of parallel within group regressions. *Educational and Psychological Measurement, 41*, 73–84.

Rogosa, D. R., & Saner, H. M. (1995). Longitudinal data analysis examples with random coefficent models. *Journal of Educational and Behavioral Statistics, 20*, 149–170.

Rubin, D. B. (1976). Inference and missing data. *Biometrika, 63*(3), 581–592.

Rubin, D. B. (1987). *Multiple Imputation for Nonresponse in Surveys*. New York: Wiley.

Saris, W. E., & Satorra, A. (1993). Power evaluations in structural equation models. In K. A. Bollen & J. S. Long (Eds.), *Testing Structural Equation Models* (pp. 181–204). Newbury Park, CA: Sage.

Saris, W. E. & Satorra, A. (1993). Power evaluations in structural equation models. In K.A. Bollen and J.S. Long (Eds.), *Testing Structural Equation Models*. Newbury Park, CA: Sage.

SAS Institute (2000). *SAS/STAT User's Guide*, 2nd ed. Vol. Version 8. Cary, NC: SAS.

Satorra, A. (1990). Robustness issues in structural equation modeling: a review of recent developments. *Quality and Quantity, 24*, 367–386.

Satorra, A. (1992). Asymptotic robust inferences in the analysis of mean and covariance structures. *Sociological Methodology, 22*, 249–278.

Satorra, A., & Bentler, P. M. (1988). Scaling corrections for chi-square statistics in covariance structure analysis. *Proceedings of the Business and Economic Statistics Section of the American Statistical Association*, pp. 308–313.

Satorra, A., & Bentler, P. M. (1994). Corrections to test statistics and standard errors in covariance structure analysis. In A. von Eye & C. C. Clogg (Eds.), *Latent Variables Analysis: Applications for Developmental Research* (pp. 399–419). Newbury Park, CA: Sage.

Satorra, A., & Saris, W. E. (1985). Power of the likelihood ratio test in covariance structure analysis. *Psychometrika, 50*, 83–90.

Sayer, A. G., & Cumsille, P. E. (2001). Second-order latent growth models. In L. M. Collins & A. G. Sayer (Eds.), *New Methods for the Analysis of Change* (pp. 179–200). Washington, DC: American Psychological Association.

Schafer, J. L. (1997). *Analysis of Incomplete Multivariate Data*. New York: Chapman & Hall.

Schafer, J. L., & Graham, J. W. (2002). Missing data: Our view of the state of the art. *Psychological Methods, 7*, 147–177.

Schwarz, G. (1978). Estimating the dimensions of a model. *Annals of Statistics, 6*, 461–464.

Shapiro, A. (1987). Robustness properties of the MDF analysis of moment structures. *South African Statistical Journal, 21*, 39–62.

Skrondal, A. and Rabe-Hesketh, S. (2004). *Generalized Latest Variable Modeling: Multilevel, Longitudinal and Structural Equation Models*. Boca Raton, FL: chapman & Hall/CRC.

Snijders, T. A. B., & Bosker, R. (1999). *Multilevel Analysis: An Introduction to Basic and Advanced Multilevel Modeling*. Newbury Park, CA: Sage.

Steiger, J. H. (1979). Factor indeterminacy in the 1930's and the 1970's: some interesting parallels. *Psychometrika, 44*, 157–167.

Steiger, J. H. (1989). *EzPATH: A Supplementary Module for SYSTAT and SYGRAPH* (computer program manual). Evanston, IL: Systat, Inc.

Steiger, J. H., & Lind, J. C. (1980). Statistically based tests for the number of common factors. Presented at the Psychometric Society, Iowa City, IA.

Steiger, J. H., Shapiro, A., & Browne, M. W. (1985). On the multivariate asymptotic distribution of sequential chi-square statistics. *Psychometrika, 50*, 253–264.

Thomson, G. H. (1939). *The Factorial Analysis of Human Ability*. Boston: Houghton Mifflin.

Tucker, L. R. (1958). Determination of parameters of a functional relation by factor analysis. *Psychometrika, 23*, 19–23.

Tucker, L. R., & Lewis, C. (1973). A reliability coefficient for maximum likelihood factor analysis. *Psychometrika, 38*, 1–10.

U.S. Department of Justice (1995). *Uniform Crime Reports*. Washington, DC: U.S. Government Printing Office.

U.S. Department of Justice (Ed.) (2000). *Uniform Crime Reporting Program Data:* [*United States*]; *1975–1997* [*Offenses Known and Clearances by Arrest, 1990*] (computer file). Ann Arbor, MI: Inter-university Consortium for Political and Social Research (producer and distributor).

Vanhanen, T. (1990). *The Process of Democratization: A Comparative Study of 147 States, 1980–88*. New York: Crane Russak.

Verhulst, P. F. (1845). Recherches mathématiques sur la loi d'accroissement de la population. *Nouveaux Mémoires de l'Académie Royale des Sciences et Belles-Lettres de Bruxelles, 18*, 1–38.

Verhulst, P. F. (1847). Deuxième mémoire sur la loi d'accroissement de la population. *Mémoires de l'Académie Royale des Sciences et Belles-Lettres de Bruxelles, 20*, 1–32.

Vonesh, E. F., & Chinchilli, V. M. (1997). *Linear and Nonlinear Models for the Analysis of Repeated Measurements*. New York: Marcel Dekker.

Werts, C. E., Jöreskog, K. G., & Linn, R. L. (1971). Comment on the estimation of measurement error in panel data. *American Sociological Review, 36*, 110–112.

White, H. (1980). A heteroscedasticity-consistent covariance matrix estimator and a direct test for heteroscedasticity. *Econometrica, 48*, 817–838.

Wiley, D. E., & Wiley, J. A. (1970). The estimation of measurement error in panel data. *American Sociological Review, 35*, 112–117.

Wilkinson, L., & The Talk Force on Statistical Inference. (1999). Statistical methods in psychology journals: guidelines and explanations. *American Psychologist, 54*, 594–604.

Willett, J. B., & Sayer, A. G. (1994). Using covariance structure analysis to detect correlates and predictors of individual change over time. *Psychological Bulletin, 116*(2), 363–381.

Winship, C., & Mare, R. D. (1992). Models for sample selection bias. *Annual Review of Sociology, 18*, 327–350.

Wishart, J. (1938). Growth-rate determinations in nutrition studies with the bacon pig, and their analysis. *Biometrika, 30*(1–2), 16–28.

Wothke, W. (1993). Nonpositive definite matrices in structural modeling. In K. A. Bollen & J. S. Long (Eds.), *Testing Structural Equation Models* (pp. 256–293). Newbury Park, CA: Sage.

Yuan, K.-H., & Bentler, P. M. (2000). Three likelihood-based methods for mean and covariance structure analysis with nonnormal missing data. *Sociological Methodology, 30*, 165–200.

Yung, Y.-F. (1994). Finite mixtures in confirmatory factor-analytic models. Unpublished Ph.D. dissertation, UCLA, Los Angeles.

Yung, Y.-F. (1997). Finite mixtures in confirmatory factor-analysis models. *Psychometrika, 62*(3), 297–330.

Zeger, S. L., & Harlow, S. D. (1987). Mathematical models from laws of growth to tolls for biological analysis: fifty years of growth. *Growth, 51*, 1–21.

Author Index

Subject Index

WILEY SERIES IN PROBABILITY AND STATISTICS

ESTABLISHED BY WALTER A. SHEWHART AND SAMUEL S. WILKS

Editors: *David J. Balding, Noel A. C. Cressie, Nicholas I. Fisher, Iain M. Johnstone, J. B. Kadane, Geert Molenberghs. Louise M. Ryan, David W. Scott, Adrian F. M. Smith, Jozef L. Teugels*

Editors Emeriti: *Vic Barnett, J. Stuart Hunter, David G. Kendall*

The ***Wiley Series in Probability and Statistics*** is well established and authoritative. It covers many topics of current research interest in both pure and applied statistics and probability theory. Written by leading statisticians and institutions, the titles span both state-of-the-art developments in the field and classical methods.

Reflecting the wide range of current research in statistics, the series encompasses applied, methodological and theoretical statistics, ranging from applications and new techniques made possible by advances in computerized practice to rigorous treatment of theoretical approaches.

This series provides essential and invaluable reading for all statisticians, whether in academia, industry, government, or research.

† ABRAHAM and LEDOLTER · Statistical Methods for Forecasting
AGRESTI · Analysis of Ordinal Categorical Data
AGRESTI · An Introduction to Categorical Data Analysis
AGRESTI · Categorical Data Analysis, *Second Edition*
ALTMAN, GILL, and McDONALD · Numerical Issues in Statistical Computing for the Social Scientist
AMARATUNGA and CABRERA · Exploration and Analysis of DNA Microarray and Protein Array Data
ANDĚL · Mathematics of Chance
ANDERSON · An Introduction to Multivariate Statistical Analysis, *Third Edition*
* ANDERSON · The Statistical Analysis of Time Series
ANDERSON, AUQUIER, HAUCK, OAKES, VANDAELE, and WEISBERG · Statistical Methods for Comparative Studies
ANDERSON and LOYNES · The Teaching of Practical Statistics
ARMITAGE and DAVID (editors) · Advances in Biometry
ARNOLD, BALAKRISHNAN, and NAGARAJA · Records
* ARTHANARI and DODGE · Mathematical Programming in Statistics
* BAILEY · The Elements of Stochastic Processes with Applications to the Natural Sciences
BALAKRISHNAN and KOUTRAS · Runs and Scans with Applications
BARNETT · Comparative Statistical Inference, *Third Edition*
BARNETT and LEWIS · Outliers in Statistical Data, *Third Edition*
BARTOSZYNSKI and NIEWIADOMSKA-BUGAJ · Probability and Statistical Inference
BASILEVSKY · Statistical Factor Analysis and Related Methods: Theory and Applications
BASU and RIGDON · Statistical Methods for the Reliability of Repairable Systems
BATES and WATTS · Nonlinear Regression Analysis and Its Applications
BECHHOFER, SANTNER, and GOLDSMAN · Design and Analysis of Experiments for Statistical Selection, Screening, and Multiple Comparisons
BELSLEY · Conditioning Diagnostics: Collinearity and Weak Data in Regression

*Now available in a lower priced paperback edition in the Wiley Classics Library.
†Now available in a lower priced paperback edition in the Wiley–Interscience Paperback Series.

† BELSLEY, KUH, and WELSCH · Regression Diagnostics: Identifying Influential Data and Sources of Collinearity
BENDAT and PIERSOL · Random Data: Analysis and Measurement Procedures, *Third Edition*
BERRY, CHALONER, and GEWEKE · Bayesian Analysis in Statistics and Econometrics: Essays in Honor of Arnold Zellner
BERNARDO and SMITH · Bayesian Theory
BHAT and MILLER · Elements of Applied Stochastic Processes, *Third Edition*
BHATTACHARYA and WAYMIRE · Stochastic Processes with Applications
† BIEMER, GROVES, LYBERG, MATHIOWETZ, and SUDMAN · Measurement Errors in Surveys
BILLINGSLEY · Convergence of Probability Measures, *Second Edition*
BILLINGSLEY · Probability and Measure, *Third Edition*
BIRKES and DODGE · Alternative Methods of Regression
BLISCHKE AND MURTHY (editors) · Case Studies in Reliability and Maintenance
BLISCHKE AND MURTHY · Reliability: Modeling, Prediction, and Optimization
BLOOMFIELD · Fourier Analysis of Time Series: An Introduction, *Second Edition*
BOLLEN · Structural Equations with Latent Variables
BOLLEN and CURRAN · Latent Curve Models: A Structural Equation Perspective
BOROVKOV · Ergodicity and Stability of Stochastic Processes
BOULEAU · Numerical Methods for Stochastic Processes
BOX · Bayesian Inference in Statistical Analysis
BOX · R. A. Fisher, the Life of a Scientist
BOX and DRAPER · Empirical Model-Building and Response Surfaces
* BOX and DRAPER · Evolutionary Operation: A Statistical Method for Process Improvement
BOX, HUNTER, and HUNTER · Statistics for Experimenters: Design, Innovation, and Discovery, *Second Editon*
BOX and LUCEÑO · Statistical Control by Monitoring and Feedback Adjustment
BRANDIMARTE · Numerical Methods in Finance: A MATLAB-Based Introduction
BROWN and HOLLANDER · Statistics: A Biomedical Introduction
BRUNNER, DOMHOF, and LANGER · Nonparametric Analysis of Longitudinal Data in Factorial Experiments
BUCKLEW · Large Deviation Techniques in Decision, Simulation, and Estimation
CAIROLI and DALANG · Sequential Stochastic Optimization
CASTILLO, HADI, BALAKRISHNAN, and SARABIA · Extreme Value and Related Models with Applications in Engineering and Science
CHAN · Time Series: Applications to Finance
CHARALAMBIDES · Combinatorial Methods in Discrete Distributions
CHATTERJEE and HADI · Sensitivity Analysis in Linear Regression
CHATTERJEE and PRICE · Regression Analysis by Example, *Third Edition*
CHERNICK · Bootstrap Methods: A Practitioner's Guide
CHERNICK and FRIIS · Introductory Biostatistics for the Health Sciences
CHILÈS and DELFINER · Geostatistics: Modeling Spatial Uncertainty
CHOW and LIU · Design and Analysis of Clinical Trials: Concepts and Methodologies, *Second Edition*
CLARKE and DISNEY · Probability and Random Processes: A First Course with Applications, *Second Edition*
* COCHRAN and COX · Experimental Designs, *Second Edition*
CONGDON · Applied Bayesian Modelling
CONGDON · Bayesian Statistical Modelling
CONOVER · Practical Nonparametric Statistics, *Third Edition*
COOK · Regression Graphics

*Now available in a lower priced paperback edition in the Wiley Classics Library.
†Now available in a lower priced paperback edition in the Wiley–Interscience Paperback Series.

COOK and WEISBERG · Applied Regression Including Computing and Graphics
COOK and WEISBERG · An Introduction to Regression Graphics
CORNELL · Experiments with Mixtures, Designs, Models, and the Analysis of Mixture Data, *Third Edition*
COVER and THOMAS · Elements of Information Theory
COX · A Handbook of Introductory Statistical Methods
* COX · Planning of Experiments
CRESSIE · Statistics for Spatial Data, *Revised Edition*
CSÖRGŐ and HORVÁTH · Limit Theorems in Change Point Analysis
DANIEL · Applications of Statistics to Industrial Experimentation
DANIEL · Biostatistics: A Foundation for Analysis in the Health Sciences, *Eighth Edition*
* DANIEL · Fitting Equations to Data: Computer Analysis of Multifactor Data, *Second Edition*
DASU and JOHNSON · Exploratory Data Mining and Data Cleaning
DAVID and NAGARAJA · Order Statistics, *Third Edition*
* DEGROOT, FIENBERG, and KADANE · Statistics and the Law
DEL CASTILLO · Statistical Process Adjustment for Quality Control
DEMARIS · Regression with Social Data: Modeling Continuous and Limited Response Variables
DEMIDENKO · Mixed Models: Theory and Applications
DENISON, HOLMES, MALLICK and SMITH · Bayesian Methods for Nonlinear Classification and Regression
DETTE and STUDDEN · The Theory of Canonical Moments with Applications in Statistics, Probability, and Analysis
DEY and MUKERJEE · Fractional Factorial Plans
DILLON and GOLDSTEIN · Multivariate Analysis: Methods and Applications
DODGE · Alternative Methods of Regression
* DODGE and ROMIG · Sampling Inspection Tables, *Second Edition*
* DOOB · Stochastic Processes
DOWDY, WEARDEN, and CHILKO · Statistics for Research, *Third Edition*
DRAPER and SMITH · Applied Regression Analysis, *Third Edition*
DRYDEN and MARDIA · Statistical Shape Analysis
DUDEWICZ and MISHRA · Modern Mathematical Statistics
DUNN and CLARK · Basic Statistics: A Primer for the Biomedical Sciences, *Third Edition*
DUPUIS and ELLIS · A Weak Convergence Approach to the Theory of Large Deviations
* ELANDT-JOHNSON and JOHNSON · Survival Models and Data Analysis
ENDERS · Applied Econometric Time Series
† ETHIER and KURTZ · Markov Processes: Characterization and Convergence
EVANS, HASTINGS, and PEACOCK · Statistical Distributions, *Third Edition*
FELLER · An Introduction to Probability Theory and Its Applications, Volume I, *Third Edition,* Revised; Volume II, *Second Edition*
FISHER and VAN BELLE · Biostatistics: A Methodology for the Health Sciences
FITZMAURICE, LAIRD, and WARE · Applied Longitudinal Analysis
* FLEISS · The Design and Analysis of Clinical Experiments
FLEISS · Statistical Methods for Rates and Proportions, *Third Edition*
† FLEMING and HARRINGTON · Counting Processes and Survival Analysis
FULLER · Introduction to Statistical Time Series, *Second Edition*
FULLER · Measurement Error Models
GALLANT · Nonlinear Statistical Models
GEISSER · Modes of Parametric Statistical Inference
GEWEKE · Contemporary Bayesian Econometrics and Statistics

*Now available in a lower priced paperback edition in the Wiley Classics Library.
†Now available in a lower priced paperback edition in the Wiley–Interscience Paperback Series.

GHOSH, MUKHOPADHYAY, and SEN · Sequential Estimation
GIESBRECHT and GUMPERTZ · Planning, Construction, and Statistical Analysis of Comparative Experiments
GIFI · Nonlinear Multivariate Analysis
GIVENS and HOETING · Computational Statistics
GLASSERMAN and YAO · Monotone Structure in Discrete-Event Systems
GNANADESIKAN · Methods for Statistical Data Analysis of Multivariate Observations, *Second Edition*
GOLDSTEIN and LEWIS · Assessment: Problems, Development, and Statistical Issues
GREENWOOD and NIKULIN · A Guide to Chi-Squared Testing
GROSS and HARRIS · Fundamentals of Queueing Theory, *Third Edition*
† GROVES · Survey Errors and Survey Costs
* HAHN and SHAPIRO · Statistical Models in Engineering
HAHN and MEEKER · Statistical Intervals: A Guide for Practitioners
HALD · A History of Probability and Statistics and their Applications Before 1750
HALD · A History of Mathematical Statistics from 1750 to 1930
† HAMPEL · Robust Statistics: The Approach Based on Influence Functions
HANNAN and DEISTLER · The Statistical Theory of Linear Systems
HEIBERGER · Computation for the Analysis of Designed Experiments
HEDAYAT and SINHA · Design and Inference in Finite Population Sampling
HELLER · MACSYMA for Statisticians
HINKELMANN and KEMPTHORNE · Design and Analysis of Experiments, Volume 1: Introduction to Experimental Design
HINKELMANN and KEMPTHORNE · Design and Analysis of Experiments, Volume 2: Advanced Experimental Design
HOAGLIN, MOSTELLER, and TUKEY · Exploratory Approach to Analysis of Variance
HOAGLIN, MOSTELLER, and TUKEY · Exploring Data Tables, Trends and Shapes
* HOAGLIN, MOSTELLER, and TUKEY · Understanding Robust and Exploratory Data Analysis
HOCHBERG and TAMHANE · Multiple Comparison Procedures
HOCKING · Methods and Applications of Linear Models: Regression and the Analysis of Variance, *Second Edition*
HOEL · Introduction to Mathematical Statistics, *Fifth Edition*
HOGG and KLUGMAN · Loss Distributions
HOLLANDER and WOLFE · Nonparametric Statistical Methods, *Second Edition*
HOSMER and LEMESHOW · Applied Logistic Regression, *Second Edition*
HOSMER and LEMESHOW · Applied Survival Analysis: Regression Modeling of Time to Event Data
† HUBER · Robust Statistics
HUBERTY · Applied Discriminant Analysis
HUNT and KENNEDY · Financial Derivatives in Theory and Practice
HUSKOVA, BERAN, and DUPAC · Collected Works of Jaroslav Hajek—with Commentary
HUZURBAZAR · Flowgraph Models for Multistate Time-to-Event Data
IMAN and CONOVER · A Modern Approach to Statistics
† JACKSON · A User's Guide to Principle Components
JOHN · Statistical Methods in Engineering and Quality Assurance
JOHNSON · Multivariate Statistical Simulation
JOHNSON and BALAKRISHNAN · Advances in the Theory and Practice of Statistics: A Volume in Honor of Samuel Kotz
JOHNSON and BHATTACHARYYA · Statistics: Principles and Methods, *Fifth Edition*

*Now available in a lower priced paperback edition in the Wiley Classics Library.
†Now available in a lower priced paperback edition in the Wiley–Interscience Paperback Series.

JOHNSON and KOTZ · Distributions in Statistics
JOHNSON and KOTZ (editors) · Leading Personalities in Statistical Sciences: From the Seventeenth Century to the Present
JOHNSON, KOTZ, and BALAKRISHNAN · Continuous Univariate Distributions, Volume 1, *Second Edition*
JOHNSON, KOTZ, and BALAKRISHNAN · Continuous Univariate Distributions, Volume 2, *Second Edition*
JOHNSON, KOTZ, and BALAKRISHNAN · Discrete Multivariate Distributions
JOHNSON, KOTZ, and KEMP · Univariate Discrete Distributions, *Third Edition*
JUDGE, GRIFFITHS, HILL, LÜTKEPOHL, and LEE · The Theory and Practice of Econometrics, *Second Edition*
JUREČKOVÁ and SEN · Robust Statistical Procedures: Aymptotics and Interrelations
JUREK and MASON · Operator-Limit Distributions in Probability Theory
KADANE · Bayesian Methods and Ethics in a Clinical Trial Design
KADANE AND SCHUM · A Probabilistic Analysis of the Sacco and Vanzetti Evidence
KALBFLEISCH and PRENTICE · The Statistical Analysis of Failure Time Data, *Second Edition*
KASS and VOS · Geometrical Foundations of Asymptotic Inference
† KAUFMAN and ROUSSEEUW · Finding Groups in Data: An Introduction to Cluster Analysis
KEDEM and FOKIANOS · Regression Models for Time Series Analysis
KENDALL, BARDEN, CARNE, and LE · Shape and Shape Theory
KHURI · Advanced Calculus with Applications in Statistics, *Second Edition*
KHURI, MATHEW, and SINHA · Statistical Tests for Mixed Linear Models
* KISH · Statistical Design for Research
KLEIBER and KOTZ · Statistical Size Distributions in Economics and Actuarial Sciences
KLUGMAN, PANJER, and WILLMOT · Loss Models: From Data to Decisions, *Second Edition*
KLUGMAN, PANJER, and WILLMOT · Solutions Manual to Accompany Loss Models: From Data to Decisions, *Second Edition*
KOTZ, BALAKRISHNAN, and JOHNSON · Continuous Multivariate Distributions, Volume 1, *Second Edition*
KOTZ and JOHNSON (editors) · Encyclopedia of Statistical Sciences: Volumes 1 to 9 with Index
KOTZ and JOHNSON (editors) · Encyclopedia of Statistical Sciences: Supplement Volume
KOTZ, READ, and BANKS (editors) · Encyclopedia of Statistical Sciences: Update Volume 1
KOTZ, READ, and BANKS (editors) · Encyclopedia of Statistical Sciences: Update Volume 2
KOVALENKO, KUZNETZOV, and PEGG · Mathematical Theory of Reliability of Time-Dependent Systems with Practical Applications
LACHIN · Biostatistical Methods: The Assessment of Relative Risks
LAD · Operational Subjective Statistical Methods: A Mathematical, Philosophical, and Historical Introduction
LAMPERTI · Probability: A Survey of the Mathematical Theory, *Second Edition*
LANGE, RYAN, BILLARD, BRILLINGER, CONQUEST, and GREENHOUSE · Case Studies in Biometry
LARSON · Introduction to Probability Theory and Statistical Inference, *Third Edition*
LAWLESS · Statistical Models and Methods for Lifetime Data, *Second Edition*
LAWSON · Statistical Methods in Spatial Epidemiology
LE · Applied Categorical Data Analysis
LE · Applied Survival Analysis

*Now available in a lower priced paperback edition in the Wiley Classics Library.
†Now available in a lower priced paperback edition in the Wiley–Interscience Paperback Series.

LEE and WANG · Statistical Methods for Survival Data Analysis, *Third Edition*
LePAGE and BILLARD · Exploring the Limits of Bootstrap
LEYLAND and GOLDSTEIN (editors) · Multilevel Modelling of Health Statistics
LIAO · Statistical Group Comparison
LINDVALL · Lectures on the Coupling Method
LINHART and ZUCCHINI · Model Selection
LITTLE and RUBIN · Statistical Analysis with Missing Data, *Second Edition*
LLOYD · The Statistical Analysis of Categorical Data
LOWEN and TEICH · Fractal-Based Point Processes
MAGNUS and NEUDECKER · Matrix Differential Calculus with Applications in Statistics and Econometrics, *Revised Edition*
MALLER and ZHOU · Survival Analysis with Long Term Survivors
MALLOWS · Design, Data, and Analysis by Some Friends of Cuthbert Daniel
MANN, SCHAFER, and SINGPURWALLA · Methods for Statistical Analysis of Reliability and Life Data
MANTON, WOODBURY, and TOLLEY · Statistical Applications Using Fuzzy Sets
MARCHETTE · Random Graphs for Statistical Pattern Recognition
MARDIA and JUPP · Directional Statistics
MASON, GUNST, and HESS · Statistical Design and Analysis of Experiments with Applications to Engineering and Science, *Second Edition*
McCULLOCH and SEARLE · Generalized, Linear, and Mixed Models
McFADDEN · Management of Data in Clinical Trials
* McLACHLAN · Discriminant Analysis and Statistical Pattern Recognition
McLACHLAN, DO, and AMBROISE · Analyzing Microarray Gene Expression Data
McLACHLAN and KRISHNAN · The EM Algorithm and Extensions
McLACHLAN and PEEL · Finite Mixture Models
McNEIL · Epidemiological Research Methods
MEEKER and ESCOBAR · Statistical Methods for Reliability Data
MEERSCHAERT and SCHEFFLER · Limit Distributions for Sums of Independent Random Vectors: Heavy Tails in Theory and Practice
MICKEY, DUNN, and CLARK · Applied Statistics: Analysis of Variance and Regression, *Third Edition*
* MILLER · Survival Analysis, *Second Edition*
MONTGOMERY, PECK, and VINING · Introduction to Linear Regression Analysis, *Third Edition*
MORGENTHALER and TUKEY · Configural Polysampling: A Route to Practical Robustness
MUIRHEAD · Aspects of Multivariate Statistical Theory
MULLER and STOYAN · Comparison Methods for Stochastic Models and Risks
MURRAY · X-STAT 2.0 Statistical Experimentation, Design Data Analysis, and Nonlinear Optimization
MURTHY, XIE, and JIANG · Weibull Models
MYERS and MONTGOMERY · Response Surface Methodology: Process and Product Optimization Using Designed Experiments, *Second Edition*
MYERS, MONTGOMERY, and VINING · Generalized Linear Models. With Applications in Engineering and the Sciences
† NELSON · Accelerated Testing, Statistical Models, Test Plans, and Data Analyses
† NELSON · Applied Life Data Analysis
NEWMAN · Biostatistical Methods in Epidemiology
OCHI · Applied Probability and Stochastic Processes in Engineering and Physical Sciences
OKABE, BOOTS, SUGIHARA, and CHIU · Spatial Tesselations: Concepts and Applications of Voronoi Diagrams, *Second Edition*
OLIVER and SMITH · Influence Diagrams, Belief Nets and Decision Analysis

*Now available in a lower priced paperback edition in the Wiley Classics Library.
†Now available in a lower priced paperback edition in the Wiley–Interscience Paperback Series.

PALTA · Quantitative Methods in Population Health: Extensions of Ordinary Regressions
PANKRATZ · Forecasting with Dynamic Regression Models
PANKRATZ · Forecasting with Univariate Box-Jenkins Models: Concepts and Cases
* PARZEN · Modern Probability Theory and Its Applications
PEÑA, TIAO, and TSAY · A Course in Time Series Analysis
PIANTADOSI · Clinical Trials: A Methodologic Perspective
PORT · Theoretical Probability for Applications
POURAHMADI · Foundations of Time Series Analysis and Prediction Theory
PRESS · Bayesian Statistics: Principles, Models, and Applications
PRESS · Subjective and Objective Bayesian Statistics, *Second Edition*
PRESS and TANUR · The Subjectivity of Scientists and the Bayesian Approach
PUKELSHEIM · Optimal Experimental Design
PURI, VILAPLANA, and WERTZ · New Perspectives in Theoretical and Applied Statistics
† PUTERMAN · Markov Decision Processes: Discrete Stochastic Dynamic Programming
QIU · Image Processing and Jump Regression Analysis
* RAO · Linear Statistical Inference and Its Applications, *Second Edition*
RAUSAND and HØYLAND · System Reliability Theory: Models, Statistical Methods, and Applications, *Second Edition*
RENCHER · Linear Models in Statistics
RENCHER · Methods of Multivariate Analysis, *Second Edition*
RENCHER · Multivariate Statistical Inference with Applications
* RIPLEY · Spatial Statistics
RIPLEY · Stochastic Simulation
ROBINSON · Practical Strategies for Experimenting
ROHATGI and SALEH · An Introduction to Probability and Statistics, *Second Edition*
ROLSKI, SCHMIDLI, SCHMIDT, and TEUGELS · Stochastic Processes for Insurance and Finance
ROSENBERGER and LACHIN · Randomization in Clinical Trials: Theory and Practice
ROSS · Introduction to Probability and Statistics for Engineers and Scientists
† ROUSSEEUW and LEROY · Robust Regression and Outlier Detection
* RUBIN · Multiple Imputation for Nonresponse in Surveys
RUBINSTEIN · Simulation and the Monte Carlo Method
RUBINSTEIN and MELAMED · Modern Simulation and Modeling
RYAN · Modern Regression Methods
RYAN · Statistical Methods for Quality Improvement, *Second Edition*
SALTELLI, CHAN, and SCOTT (editors) · Sensitivity Analysis
* SCHEFFE · The Analysis of Variance
SCHIMEK · Smoothing and Regression: Approaches, Computation, and Application
SCHOTT · Matrix Analysis for Statistics, *Second Edition*
SCHOUTENS · Levy Processes in Finance: Pricing Financial Derivatives
SCHUSS · Theory and Applications of Stochastic Differential Equations
SCOTT · Multivariate Density Estimation: Theory, Practice, and Visualization
* SEARLE · Linear Models
SEARLE · Linear Models for Unbalanced Data
SEARLE · Matrix Algebra Useful for Statistics
SEARLE, CASELLA, and McCULLOCH · Variance Components
SEARLE and WILLETT · Matrix Algebra for Applied Economics
SEBER and LEE · Linear Regression Analysis, *Second Edition*
† SEBER · Multivariate Observations
† SEBER and WILD · Nonlinear Regression
SENNOTT · Stochastic Dynamic Programming and the Control of Queueing Systems
* SERFLING · Approximation Theorems of Mathematical Statistics

*Now available in a lower priced paperback edition in the Wiley Classics Library.
†Now available in a lower priced paperback edition in the Wiley–Interscience Paperback Series.

SHAFER and VOVK · Probability and Finance: It's Only a Game!
SILVAPULLE and SEN · Constrained Statistical Inference: Inequality, Order, and Shape Restrictions
SMALL and McLEISH · Hilbert Space Methods in Probability and Statistical Inference
SRIVASTAVA · Methods of Multivariate Statistics
STAPLETON · Linear Statistical Models
STAUDTE and SHEATHER · Robust Estimation and Testing
STOYAN, KENDALL, and MECKE · Stochastic Geometry and Its Applications, *Second Edition*
STOYAN and STOYAN · Fractals, Random Shapes and Point Fields: Methods of Geometrical Statistics
STYAN · The Collected Papers of T. W. Anderson: 1943–1985
SUTTON, ABRAMS, JONES, SHELDON, and SONG · Methods for Meta-Analysis in Medical Research
TANAKA · Time Series Analysis: Nonstationary and Noninvertible Distribution Theory
THOMPSON · Empirical Model Building
THOMPSON · Sampling, *Second Edition*
THOMPSON · Simulation: A Modeler's Approach
THOMPSON and SEBER · Adaptive Sampling
THOMPSON, WILLIAMS, and FINDLAY · Models for Investors in Real World Markets
TIAO, BISGAARD, HILL, PEÑA, and STIGLER (editors) · Box on Quality and Discovery: with Design, Control, and Robustness
TIERNEY · LISP-STAT: An Object-Oriented Environment for Statistical Computing and Dynamic Graphics
TSAY · Analysis of Financial Time Series, *Second Edition*
UPTON and FINGLETON · Spatial Data Analysis by Example, Volume II: Categorical and Directional Data
VAN BELLE · Statistical Rules of Thumb
VAN BELLE, FISHER, HEAGERTY, and LUMLEY · Biostatistics: A Methodology for the Health Sciences, *Second Edition*
VESTRUP · The Theory of Measures and Integration
VIDAKOVIC · Statistical Modeling by Wavelets
VINOD and REAGLE · Preparing for the Worst: Incorporating Downside Risk in Stock Market Investments
WALLER and GOTWAY · Applied Spatial Statistics for Public Health Data
WEERAHANDI · Generalized Inference in Repeated Measures: Exact Methods in MANOVA and Mixed Models
WEISBERG · Applied Linear Regression, *Third Edition*
WELSH · Aspects of Statistical Inference
WESTFALL and YOUNG · Resampling-Based Multiple Testing: Examples and Methods for *p*-Value Adjustment
WHITTAKER · Graphical Models in Applied Multivariate Statistics
WINKER · Optimization Heuristics in Economics: Applications of Threshold Accepting
WONNACOTT and WONNACOTT · Econometrics, *Second Edition*
WOODING · Planning Pharmaceutical Clinical Trials: Basic Statistical Principles
WOODWORTH · Biostatistics: A Bayesian Introduction
WOOLSON and CLARKE · Statistical Methods for the Analysis of Biomedical Data, *Second Edition*
WU and HAMADA · Experiments: Planning, Analysis, and Parameter Design Optimization
YANG · The Construction Theory of Denumerable Markov Processes
* ZELLNER · An Introduction to Bayesian Inference in Econometrics
ZHOU, OBUCHOWSKI, and McCLISH · Statistical Methods in Diagnostic Medicine

*Now available in a lower priced paperback edition in the Wiley Classics Library.
†Now available in a lower priced paperback edition in the Wiley–Interscience Paperback Series.